Lecture Notes in Statistics **133**
Edited by P. Bickel, P. Diggle, S. Fienberg, K. Krickeberg,
I. Olkin, N. Wermuth, S. Zeger

Springer
New York
Berlin
Heidelberg
Barcelona
Budapest
Hong Kong
London
Milan
Paris
Singapore
Tokyo

Dipak Dey
Peter Müller
Debajyoti Sinha (Editors)

Practical Nonparametric and Semiparametric Bayesian Statistics

Springer

Dipak Dey
Department of Statistics
University of Connecticut
Storrs, CT 06269

Peter Müller
ISDS
Duke University
Durham, NC 27708

Debajyoti Sinha
Department of Mathematics
University of New Hampshire
Durham, NH 03824

Library of Congress Cataloging-in-Publication Data

Practical nonparametric and semiparametric Bayesian statistics / Dipak Dey, Peter Müller, Debajyoti Sinha (editors).
p. cm. -- (Lecture notes in statistics ; 133)
Includes bibliographical references.
ISBN 0-387-98517-4 (softcover : alk. paper)
1. Bayesian statistical decision theory. 2. Nonparametric statistics. I. Dey, Dipak. II. Müller, Peter, 1963 Aug. 9- III. Sinha, Debajyoti. IV. Series: Lecture notes in statistics (Springer-Verlag) ; v. 133.
QA279.5.P73 1998
519.5'42--dc21 98-16460

Printed on acid-free paper.

Camera ready copy provided by the editors.
Printed and bound by Braun-Brumfield, Ann Arbor, MI.
Printed in the United States of America.

9 8 7 6 5 4 3 2 1

ISBN 0-387-98517-4 Springer-Verlag New York Berlin Heidelberg SPIN 10676439

Preface

Interest in Bayesian modeling using rich classes of nonparametric and semi-parametric methods has grown rapidly during the recent years due to great advances in computation. Computers are much faster, and computer algorithms and procedures have been developed that make it practical to apply Bayesian methods to complex applied problems. This volume is devoted to a series of papers that summarize an enormous but scattered literature on nonparametrics and semiparametric Bayesian methods. For a practitioner and new researcher in this area, we hope this volume will fill the need for a complete text and reference book with the theory, available models, associated methodologies, and required algorithms. Our hope is that this book will also help researchers identify areas of important future research, promote regular use of such powerful data analytic methods and computational tools in standard statistical practice, and open new applications of semiparametric Bayesian methods.

The papers in this volume are divided into five parts: Dirichlet and related prior processes; Modeling random functions; Lévy and related processes; Prior elicitation and asymptotic properties; Case studies.

In Part I, Escobar and West introduce the Dirichlet process as a means to extend a standard hierarchical model by introducing uncertainty about conditional distributions in some of the levels of the hierarchical model. MacEachern discusses issues of efficient computational methods and illustrates computations in two fully explained examples. Newton, Quintana and Zhang explain an approximate method to calculate nonparametric Bayes estimates under Dirichlet process priors. Doss and Narasimhan introduce an interactive programming environment to compute estimators based on Dirichlet process priors. Using importance sampling, the program interactively computes inference for a range of prior assumptions. The last two chapters in Part I discuss the use of Dirichlet process priors to relax parametric assumptions in specific modeling contexts. Kleinman and Ibrahim use mixture of Dirichlet process models for random effects distributions in normal linear models and generalized linear models. Finally in this part, Vlachos and Gelfand consider sequential design strategies in models involving Dirichlet process mixtures for the likelihood specification.

Chapters in the second part of the volume describe approaches for modeling random functions as linear combinations of terms from a given basis. Vidakovic reviews nonparametric Bayesian methods defined in the domain of the wavelet decompositions of a random function. Smith and Kohn compare hierarchical Bayesian models for irregular regression functions based on four different bases, using the "focused sampler" to explore the space of all possible subsets of the basis terms. This part is concluded by Ríos Insua and Müller introducing a reversible jump algorithm to estimate feedforward neural network nonparametric regression models.

Part III concerns the modeling of the intensity function of a point process. For many practical situations, for example, in survival analysis, a practical and conventional approach is to use a prior process, such as a Lévy process, on the hazard or cumulative intensity function. The chapter by Sinha and Dey presents an overview of Bayesian modeling, commonly used prior processes and methodologies for semiparametric analysis of survival data. Laud, Damien and Smith describe the use and posterior computation of a particular type of Lévy process (beta process) for semiparametric Bayesian analysis of univariate survival data. Wolpert and Ickstadt present the theory and method of a new simulation technique for Lévy random fields, including multivariate fields. Walker and Damien discuss posterior properties and simulation techniques for Lévy, neutral to the right (NTR) and mixture of Dirichlet processes. The last two chapters compare the similarities and dissimilarities among currently available techniques for simulating from different independent-increment stochastic processes. In the last chapter of Part III, Heikkinen presents a new dynamic prior process and associated posterior computations for the intensity function of a one- or two-dimensional point process.

Part IV deals with two important but still unfolding issues of semiparametric Bayesian analysis— prior elicitation and asymptotics. The chapter by Ibrahim and Sinha presents a method of prior elicitation for univariate survival models using a stage-0 sample. The method is also applicable to variable selection problems. In the last chapter of Part IV, Wasserman presents a brief overview of frequentists asymptotic properties of some nonparametric Bayesian procedures.

Part V concludes the volume with a discussion of case studies using nonparametric Bayesian approaches. Ickstadt, Wolpert and Lu study travel demand in Portland, Oregon, using a Poisson/Gamma random field introduced in the earlier chapter by Wolpert and Ickstadt. Müller and Rosner consider semiparametric pharmacokinetic and of the pharmacodynamic models with a nonparametric model for subject specific random effects. A mixture of normals with a Dirichlet process prior on the mixing measure is used. Liu and Arjas present a degradation model for fatigue crack growth using a hierarchical nonparametric model. Hirano studies longitudinal data models with nonparametric random effects distributions to model labor earnings of individuals in the U.S. The model assumes Dirichlet process mixtures for the nonparametric random effects distributions.

The cooperation of all contributors in the timely preparation of their manuscripts is greatly recognized. We decided early on that it was important to referee and critically evaluate the papers which were submitted for inclusion in this volume. For this substantial task, we relied on the service of numerous referees to whom we are most indebted. Among those who we wish to acknowledge are Omar Aguilar, Sanjib Basu, Ming–Hui Chen, Dario Gasbarra, Sujit Ghosh, Michael Lavine, Jaeyong Lee, Lynn Kuo, Ernst Linder, Todd Ogden, Nalini Ravishanker, Fabrizio Ruggeri, and Ananda Sen.

We are also grateful to John Kimmel and the Springer-Verlag referees to consider our proposal in a very timely manner. Our special thanks go for their encouragements in this project to Jim Berger and our spouses, Rita, Gautami and Rumi.

Dipak Dey, Peter Müller and Debajyoti Sinha
Storrs, CT, Durham, NC and Durham, NH

Contents

Contributors

Elias Arjas, *Elja.Arjas@rni.helsinki.fi*, Rolf Nevanlinna Institute, University of Helsinki, Finland.

Paul Damien, *damien@damien.bus.umich.edu*, Department of Statistics and Management Science, University of Michigan, Ann Arbor, MI.

Dipak Dey, *dey@stat.uconn.edu*, Department of Statistics, University of Conneticut, Storrs, CT.

Michael D. Escobar, *escobar@utstat.toronto.edu*, Department of Statistics, University of Toronto, Ontario, Canada.

Hani Doss, *doss@stat.ohio-state.edu*, Department of Statistics, The Ohio State University, Columbus, OH.

Alan E. Gelfand, *alan@stat.uconn.edu*, Department of Statistics, University of Conneticut, Storrs, CT.

Juhha Heikkinen, *Juha.Heikkinen@metla.fi*, Finnish Forest Research Institute,

Keisuke Hirano, *khirano@fas.harvard.edu*, Department of Economics, Harvard University, Cambridge, MA.

Joeseph G. Ibrahim, *ibrahim@jimmy.harvard.edu*, Department of Biostatistics, Harvard School of Public Health, Boston, MA.

Katja Ickstadt, *katja@stat.duke.edu*, Department of Statistics, University of North Carolina at Chapel Hill, Chapel Hill, NC.

Ken P. Kleinman, *kkleinma@neri.org*, NERI, Watertown, MA.

Robert Kohn, *robertk@agsm.unsw.edu.au*, Australian Graduate School of Management, University of New South Wales, Sydney, N.S.W. Australia.

Purushottam W. Laud, *laud@biostat.mcw.edu*, Division of Biostatistics, Medical College of Wisconsin, Milwaukee, WI.

Liping Liu, *liping@statms.stat.pku.edu.cn*, Department of Probability and Statistics, Peking University, P. R. China.

Xuedong Lu, *lu@stat.duke.edu*, Department of Civil and Environmental Engineering, Duke University, Durham, NC.

Steve N. MacEachern, *snm@stat.mps.ohio-state.edu*, Department of Statistics, The Ohio State University, Columbus, OH.

Peter Müller, *pm@stat.duke.edu*, Institute of Statistics and Decision Sciees, Duke University, Durham, NC.

B. Narasimhan, Division of Science, Penn State Erie, The Behrend College, Erie, PE.

Michael A. Newton, *newton@stat.wisc.edu*, Department of Statistics and Department of Biostatistics and Medical Informatics, University of Wisconsin-Madison, Madison, WI.

Fernando A. Quintana, *quintana@mat.puc.cl*, Departamento de Probabilidad y Estadística, Pontificia Universidad Católica de Chile, Santiago, Chile.

David Ríos Insua, *d.rios@escet.urjc.es*, ESCET, Universidad Rey Juan Carlos, Madrid, Spain.

Gary Rosner, *grosner@cstatz.mc.duke.edu*, Division of Biometry, Department of Community and Family Medicine, Duke University Medical Center, Durham, NC.

Debajyoti Sinha, *sinha@purabi.unh.edu*, Department of Mathematics, University of New Hampshire, Durham, NH.

Adrian F.M. Smith, *a.smith@ic.ac.uk*, Department of Mathematics, Imperial College, London, U.K.

Michael Smith, *mikes@smith.ecom.monash.edu.au*, Dept. of Econometrics and Business Statistics, Monash University, Clayton, Australia.

Brani Vidakovic, *brani@stat.duke.edu*, Institute of Statistics and Decision Sciences, Duke University, Durham, NC.

Pantelis Vlachos, *vlachos@stat.cmu.edu*, Department of Statistics, Carnegie Mellon University, Pittsburgh, PE.

Stephen Walker, *s.walker01@ic.ac.uk*, Department of Mathematics, Imperial College, London, U.K.

Larry Wasserman, *larry@stat.cmu.edu*, Department of Statistics, Carnegie Mellon University, Pittsburgh, PA.

Mike West, *mw@stat.duke.edu*, Institute of Statistics and Decision Sciences, Duke University, Durham, NC.

Robert Wolpert, *rlw@stat.duke.edu*, Institute of Statistics and Decision Sciences, Duke University, Durham, NC.

Yunlei Zhang, *bzhang@corp.cuc.com*, Cendent Corp., CT.

1

Computing Nonparametric Hierarchical Models

Michael D. Escobar
Mike West

ABSTRACT Bayesian models involving Dirichlet process mixtures are at the heart of the modern nonparametric Bayesian movement. Much of the rapid development of these models in the last decade has been a direct result of advances in simulation-based computational methods. Some of the very early work in this area, circa 1988-1991, focused on the use of such nonparametric ideas and models in applications of otherwise standard hierarchical models. This chapter provides some historical review and perspective on these developments, with a prime focus on the use and integration of such nonparametric ideas in hierarchical models. We illustrate the ease with which the strict parametric assumptions common to most standard Bayesian hierarchical models can be relaxed to incorporate uncertainties about functional forms using Dirichlet process components, partly enabled by the approach to computation using MCMC methods. The resulting methodology is illustrated with two examples taken from an unpublished 1992 report on the topic.

1.1 Introduction

It is news to no-one that the practice of hierarchical modeling – the mainstay of the applied Bayesian statisticians toolbox – has simply exploded in the last decade due to the development of Markov chain Monte Carlo (MCMC) methods for simulation-based implementation and analysis. At the conceptual heart of hierarchical modeling lies the structurally simplifying nesting/linking of submodels that allows the statistician to break a complex data structure and problem into components that are then naturally assembled in the tree structure of an encompassing hierarchical model. The methods of nested MCMC simulation analysis now standard in the Bayesian world are simply tailor-made for this hierarchical structure (Gilks, Richardson, and Spiegelhalter 1996, for example).

In applications of hierarchical models, as with all parametric models, there are almost inevitable concerns about the sensitivity of resulting inferences to assumed forms of component distributions. Hierarchical models require specification of distributions for parameters and, often, hyperparameters, about which there is usually considerable uncertainty. Hence it is of interest to combine the developments of modern approaches to nonparametric, and semiparametric, modeling of distribution functions with hierarchical models. The advances in applied nonparametric modeling in recent years have also been driven, in the main, by the development of MCMC methods, and the synthesis of models is eased and indeed naturally facilitated by these approaches to implementation and computation. This chapter describes and illustrates this synthesis using Dirichlet process models for uncertain functional forms of distributions for parameters in hierarchical models.

This chapter first provides some discussion and background on Dirichlet processes and of the implications and interpretations of the priors for the parameters of Dirichlet processes in specific modeling frameworks. We then describe MCMC algorithms for posterior sampling in models with such components, and some strategies for choosing prior distributions for the parameters of Dirichlet processes. Specific extensions to hierarchical models are keyed and illustrated in two data analyses.

We conclude this introduction with a collection of references related to the developments in this area in recent years. Much of the work in nonparametric modeling traces its origins back to the original papers of Freedman (1963), Ferguson (1973, 1974), and Blackwell and MacQueen (1973), though applications were quite limited until the late 1980s. The introduction of MCMC methods in this area began with Escobar (1988), later published in Escobar (1992, 1994). Various applied developments in the early 1990s, and more recently, include works on density estimation and related matters (Escobar and West 1995, based on an original technical report from 1991; West 1990, 1992), mixture deconvolution (West and Cao 1993, West and Turner 1994, Turner and West 1993), applications in hierarchical and prior modeling (MacEachern 1988, Escobar 1995, West 1997), regression and multivariate density estimation (Erkanli, Stangl and Müller 1993, Müller, Erkanli and West 1996, West, Müller and Escobar 1994), design (Bush and MacEachern 1996), and time series (Müller, West and MacEachern 1997), among others. More recent work has extended the range of applications in regression contexts and hierarchical/prior modeling in significant applications (Kuo and Mallick 1998, Mukhopadhyay and Gelfand 1997, West 1997).

Underlying these more applied developments were innovations in computation and the development of new MCMC schemes, including key contributions by Doss (1994), Bush and MacEachern (1996), MacEachern (1994), MacEachern and Müller (1998), West, Müller and Escobar (1994), and alternative, non-MCMC methods such as Liu (1996). Much more detail and perspective on algorithms and computation is provided in MacEachern (1998, this volume).

1.2 Notation and Perspectives

Consider a collection of random quantities $Y_i, (i = 1, \ldots, I)$, where Y_i is modeled as drawn from a specific distribution $F(Y_i|\theta_i, \zeta_i)$, independently over i given parameters θ_i and ζ_i. In any specific problem, the partition of parameters into the two sets θ_i and ζ_i will be specified. Write $Y = (Y_1, \ldots, Y_I)$, $\theta = (\theta_1, \ldots, \theta_I)$ and $\zeta = (\zeta_1, \ldots, \zeta_I)$. Each Y_i might be thought of as representing observations for an "individual". Let the parameters θ_i be independent draws from some prior distribution $G(\cdot|\lambda)$, characterized by a parameter λ. Denote the prior for ζ by H_1 and the prior for λ by H_2. It is assumed that Y and ζ are conditionally independent of λ given θ. It is also assumed that given Y, ζ and λ, quantities θ and G are conditionally independent of everything else in the hierarchical model. Here the term hierarchical model is meant in the broadest sense: it simply means a model with many levels, structured in terms of sequences of nested conditional distributions. A key point of this paper is to show how to model uncertainties about the functional form of G using Dirichlet processes.

A Dirichlet process prior for G is probability measure on the space of all probability measures G and is determined by two parameters: a distribution function $G_0(\cdot)$ that defines the "location" of the Dirichlet process prior, and a positive scalar precision parameter α. The defining property of the Dirichlet process is that for any partition $B_1, \ldots, B_m$ on the space of support of G_0 the vector of probabilities $(G(B_1), \ldots, G(B_m))$ has a Dirichlet distribution with parameter vector $(\alpha G_0(B_1), \ldots, \alpha G_0(B_m))$. From this definition, we see that the precision parameter determines the concentration of the prior for G around the prior "guess" G_0, and therefore measures the "strength of belief" in G_0. By way of notation we write $G \sim \mathcal{D}(G|G_0, \alpha)$. Here, we use the Dirichlet process to add a further stage to the hierarchical model, and this formally allows for the modeling of deviations away from the specific distribution G_0. So G_0 may be viewed as a "baseline" prior such as might be used in a typical parametric analysis, e.g., a normal prior in a normal linear model, and the framework enables the analysis of sensitivity to the assumptions of the baseline parametric model. For large values of α, a sampled G is very likely to be close to G_0. For small values of α, a sampled G is likely to put most of its probability mass on just a few atoms. In the latter case, when the mass of G is concentrated on a few atoms, the Dirichlet process is often used to model a finite mixture model. This is shown in Escobar and West (1995) and West, Müller, and Escobar (1994), and is discussed in more detail in Section 3.2.

The above notation is meant to be very general in order to accommodate almost any hierarchical model. To illustrate this, consider the following typical linear model. Assume one observes $X_i = \beta_{1i}W_{1i} + \beta_{2i}W_{2i} + \epsilon_i$ where the W's are known covariates. For $j = 1, 2$, define $\beta_j = (\beta_{j1}, \ldots, \beta_{jI})$. Assume that, for each i, the β_{1i} and β_{2i} are independently normally distributed, and that β_{ji} has mean and variance μ_j and τ_j^2, respectively, for each $j = 1, 2$. Assume the ϵ_i to be zero-mean, independent normals with variance σ^2. Also, assume that (μ_j, τ_j^2) has the usual normal/inverse gamma prior distribution with parameter vector A_j, and let σ^2 have an inverse gamma prior with parameter vector B. In this example, the X_i's corresponds to the Y_i's; we might now ask what corresponds to the θ_i's? The answer depends on which distribution one is uncertain about. For example, presently the β_{1i}'s are assumed to come from a normal population. However, maybe the normal assumption seems too restrictive – it might be desirable to have the β_{1i}'s from a distribution which is "near" a normal distribution. So, in this case θ_i would be β_{1i}, λ would be A_1, and ζ would be (β_2, σ^2) with $\zeta_i = (\beta_{2i}, \sigma^2)$. If one wishes to model uncertainty about the joint distribution of β_1 and β_2, then θ_i would be (β_{1i}, β_{2i}), λ would be a vector concatenating the elements of A_1 and A_2, and ζ would simply be the scalar σ^2 with each $\zeta_i = \sigma^2$.

Now suppose that, instead of being concerned about the distribution of the β's, one wishes to model uncertainty about the normality assumption for the ϵ_i's. One way of doing this is to first change the model by assuming that the ϵ_i's are from a normal distribution with mean 0 and variance σ_i^2 and the σ_i^2's are a sample from an uncertain distribution $G(\cdot)$ that does not necessarily put all its weight on a single atom (Brunner 1995). This generates the class of scale mixture of normals (see Andrews and Mallows 1974, West 1987). Here then, θ_i would be σ_i^2, the Y_i's would still be the X_i, and ζ would be the matrix with the jth column being $(\beta_{ji}; i = 1, \ldots, I)$ with $\zeta_i = (\beta_{1i}, \beta_{2i})$. The details of this are contained in Section 3.2, after a more technical description of the Dirichlet process.

The notation is designed so that given ζ, Y, and λ, then θ and G are conditionally independent of all remaining variables. Once θ and G are isolated, then the methods below can be used to model the uncertainty in the functional form of G for many hierarchical models. Note, from the example, that the variables θ, Y, λ, and ζ do not have to be scalars, but are allowed to be multidimensional. For an example where θ_i is uncountably infinite dimensional, see Tomlinson (1998). Also note that there is a symmetry between the prior and the likelihood. Both the prior and the likelihood are basically distribution functions and a Dirichlet process can be used to model uncertainty about either. Finally, note that Y does not need to be observed. Since everything in this paper is eventually conditional upon Y, it is clear that the framework can be embedded in a super-model which is being analysed using MCMC. Examples of this appear in Roeder et al (1998), Mukhopadhyay and Gelfand (1997), and West (1997).

1.3 Posterior Sampling in Dirichlet Process Mixtures

A specific MCMC algorithm is discussed in three parts. First, the computational scheme for the Dirichlet process itself is discussed, involving simulations of the parameters θ_i. This lies at the heart of more standard nonparametric models in density estimation and other applications (Escobar and West 1995, West, Müller and Escobar 1994). Then we discuss computational issues arising when we include learning about the precision parameter α of the Dirichlet process. Here we provide some additional commentary on the role of this parameter in a Dirichlet process model more generally. Finally, we describe how the overall hierarchical model analysis is structured by combining these components into the larger model structure.

1.3.1 Posterior sampling under a Dirichlet process

In our analysis, the random distribution G is integrated out of consideration, the analysis focusing entirely on the θ parameters assumedly generated from G. There are two basic components of the algorithm and approach. The first step uses an algorithm based on Escobar (1988, 1992, 1994) which uses a Polya urn scheme representation; the second step relies on the cluster structure of the θ's and is based on the work of MacEachern (1994). The combination of these two steps is introduced in West, Müller, and Escobar (1994), and Bush and MacEachern (1996).

Consider the distribution of $(Y_i|\theta_i, \zeta, \lambda)$ with ζ and λ held fixed in this subsection. Recall that the θ_i's come from G, and that $G \sim \mathcal{D}(G|G_0, \alpha)$. Assume here that both α and G_0 are known. This structure results in a posterior distribution which is a mixture of Dirichlet processes (Antoniak 1974). Using the Polya urn representation of the Dirichlet process (Blackwell and MacQueen 1973), the joint posterior distribution of $[\theta|Y, \zeta, \lambda]$ has the form

$$[d\theta|Y, \zeta, \lambda] \propto \prod_{i=1}^{n} f(Y_i|\theta_i, \zeta) \frac{\alpha\, G_0(d\theta_i|\lambda) + \sum_{k<i} \delta(d\theta_i|\theta_k)}{\alpha + i - 1},$$

where $f(Y_i|\theta_i, \zeta)$ is the density or the probability function of $F(Y_i|\theta_i, \zeta)$ at θ_i and ζ, and where $\delta(d\theta_i|\theta_k)$ is the distribution which is a point mass on θ_k. At this point, the random distribution G has been integrated out. Integrating out G simplifies the algorithm since we now sample only the θ's.

The above equation clarifies the effect of the precision parameter α. In the limiting case $\alpha \to \infty$ we have

$$[d\theta|Y, \zeta, \lambda] = \prod_{i=1}^{n} G_b(d\theta_i|Y_i, \zeta, \lambda) \propto \prod_{i=1}^{n} f(Y_i|\theta_i, \zeta_i)\; G_0(d\theta_i|\lambda).$$

Here $G_b(d\theta_i|Y_i, \zeta, \lambda) \propto f(Y_i|\theta_i, \zeta_i)G_0(d\theta_i|\lambda)$ is the "baseline" posterior, that is, it is the posterior assuming θ_i to come from the baseline prior G_0. As α gets very small, the estimation of θ_i is a little more complicated to understand. The posterior for θ_i is based largely on other θ_k's which are near Y_i, so that inference for θ_i heavily depends on Y_i and nearest "neighboring" Y_k's.

Gibbs sampling exploits the simple structure of the conditional posteriors for the elements of θ, resulting in the following conditional distribution. For each $i = 1, \ldots, I$,

$$[d\theta_i|\{\theta_k, k \neq i\}, Y, \zeta, \lambda] \sim q_0 \; G_b(d\theta_i|Y_i, \zeta, \lambda) + \sum_{k \neq i} q_k \; \delta(d\theta_i|\theta_k) \qquad (1)$$

with the following ingredients:

- $G_b(\theta_i|Y_i, \zeta, \lambda)$ is the baseline posterior distribution noted above;
- $q_0 \propto \alpha \int f(Y_i|\theta_i, \zeta_i)dG_0(\theta_i|\lambda)$, just α times the density of the marginal distribution of Y_i under the baseline prior $G_0(\cdot|\lambda)$;
- $q_k \propto f(Y_i|\theta_k, \zeta_i)$, the density of the marginal distribution of Y_i conditional on $\theta_i = \theta_k$;
- the quantities q_i are standardized to unit sum, $1 = q_0 + \sum_{k \neq i} q_k$.

Escobar and West (1995) give the above formulæ in normal models, and they appear for the general case in Escobar (1992; section 5, 1994).

Iterative posterior simulation generates new values of the θ_i from conditional distributions that are modified forms of the above structure, as discussed further below. The computations are very straightforward if we are assured that G_b is of manageable form, that the integrations required to compute q_0 may be cheaply performed, and that the density f is easily evaluated. There has been some criticism of the above method because of the potential difficulty in calculating the marginal distribution needed to obtain q_0. In regards to that, please note the following:

1. When G_0 is a conjugate prior, then the marginal is known analytically. In many large hierarchical models, conjugate priors are often used in the middle stages of the model. Also, since the interest is in finding distributions "around" the centering distribution G_0, the fact that G_0 is the conjugate prior is often not that strong an assumption because G's sampled from $\mathcal{D}(G_0, \alpha)$ can be fairly flexible.

2. If we must compute the marginal distribution numerically, we can often do so extremely quickly and accurately. Since the integration is over an individual θ_i and not over all of θ, the dimension of the integral is the dimension of the individual θ_i. Also, the parameters ζ

and λ are fixed for this integration. Therefore, the marginal is typically a low dimensional integral, and there are many efficient methods to numerically integrate low dimensional integrals. For example, if $f(Y_i|\theta_i, \zeta_i)G_0(d\theta_i|\lambda)$ can be put into the form of a smooth function times a normal or gamma density, then one can use Gaussian quadrature rules. For an overview of these and other integration methods, see Naylor and Smith (1982) or Evans and Swartz (1995). For an example of quadrature rules applied to this problem, see Tomlinson (1998).

3. It may be that the marginal distributions cannot be calculated efficiently. In such a case, the methods in this subsection might not apply directly. There is some work on methods to perform MCMC without calculating the marginal distribution needed for q_0. With such an alternative algorithm, one would use this technique to sample from the posterior conditional distribution for θ, and then the rest of the methods in this paper can be used to include a Dirichlet process enhancement into the hierarchical model. See, for example, MacEachern and Müller (1998) for a proposed alternative method to sample from the conditional distribution of θ.

The required computations are reduced by the fact that the distinct θ_i's typically reduce to fewer than I due to the clustering of the θ_i's inherent in the Dirichlet process (Antoniak 1974). Using the superscript $*$ to denote distinct values, suppose that the conditioning quantities θ_k's concentrate on $I^* \leq I-1$ distinct values θ_k^*, with some n_k taking this common value. Then, the above formula can be rewritten as:

$$[d\theta_i|\{\theta_k, k \neq i\}, Y, \zeta, \lambda] \sim q_0\ G_b(d\theta_i|Y_i, \zeta, \lambda) + \sum_{k=1}^{I^*} n_k q_k^*\ \delta(d\theta_i|\theta_k^*),$$

with $q_k^* \propto f(Y_i|\theta_k^*, \zeta)$, and $1 = q_0 + \sum_k n_k q_k^*$. Besides simplifying notation, the cluster structure of the θ_i can also be used to improve the efficiency of the algorithm.

Simulation of θ_i from the above conditional is now evidently straightforward: with conditional probability proportional to $n_k q_k^*$, the sampled value is one of those defining the existing "clusters;" otherwise, the sampled θ_i is drawn anew from the conditional posterior $G_b(d\theta_i|Y_i, \zeta, \lambda)$. When using the above conditional distribution in an MCMC algorithm, there may occur problems if the sum of the q_k^*'s becomes very large relative to q_0 on any iteration. This occurs when the Markov chain has "stabilized" on a small number of "clusters," and it is then unlikely to generate a "new" value of θ_k^*. In order to prevent the algorithm from getting stuck on a small set of θ_j^*'s in this way, it is helpful to "remix" the θ_j^*'s after every step. This improvement is used in Bush and MacEachern (1996) and West, Müller, and Escobar (1994), and it is a combination of the above algorithm with the algorithm developed in MacEachern (1994). The latter paper describes a method which considers the cluster membership structure. In that algorithm it is necessary to sample the new cluster membership structure by computing complex expressions which heavily rely on conjugacy, and which therefore relatively limit its applicability in hierarchical models. However, with the original Escobar method described above, each step of that algorithm produces the cluster memberships as a simple by-product. So the computational difficulties with the MacEachern algorithm are solved when combined with the Escobar algorithm. Also, the combined algorithm mixes better than the Escobar algorithm alone because the θ_k^*'s are resampled at each step providing more movement in the MCMC sampler's which in turn improves convergence.

Some notation is now introduced to define the remixing algorithm. Conditioning on I^*, introduce indicators $\mathcal{S}_i = j$ if $\theta_i \equiv \theta_j^*$ so that, given $\mathcal{S}_i = j$ and θ^*, $y_i \sim F_i(.|\theta_j^*)$. The cluster structure, which is called a configuration by West (1990), is defined by the set of indices $\mathcal{S} = \{\mathcal{S}_1, \ldots, \mathcal{S}_n\}$. The set $\mathcal{S}$ determines a one-way classification of the data $Y = \{y_1, \ldots, y_n\}$ into I^* distinct groups or clusters; the $n_j = \#\{\mathcal{S}_i = j\}$ observations in group j share the common parameter value θ_j^*. Now, define J_j as the set of indices of observations in group j; i.e., $J_j = \{i : \mathcal{S}_i = j\}$. Let $Y_{(j)} = \{Y_i : \mathcal{S}_i = j\}$ be the corresponding group of observations. Once the set $\mathcal{S}$ is known, the posterior analysis of the θ_j^*'s devolves into a collection of I^* independent analyses. Specifically, the θ_j^*'s are conditionally independent with posterior densities

$$p(\theta_j^*|Y, \mathcal{S}, I^*, \zeta, \lambda) \equiv p(\theta_j^*|Y_{(j)}, \mathcal{S}, I^*, \zeta, \lambda) \propto \prod_{i \in J_j} f_i(y_i|\theta_j^*, \zeta_i) dG_0(\theta_j^*, \lambda),$$

for $j = 1, \ldots, I^*$. Note that this is just the posterior of θ_j^* given several Y_i's sampled from the $F(.|\theta_j^*, \zeta_i)$.

In summary, in order to sample from the conditional distribution of θ_i given all the other parameters in the model, one needs to know:

(i) how to sample from the posterior distribution of θ_i given all the other parameters if G_0 were the prior distribution for θ_i,

(ii) how to evaluate the marginal distribution for Y_i assuming that G_0 is the prior distribution of θ_i,

(iii) how to evaluate the density function f of F, and

(iv) how to sample from the posterior of θ_j^* when there are several observations sampled from $f(Y_i|\theta_j^*, \zeta_i)$.

1.3.2 Learning about the precision parameter

The precision parameter, α, of the Dirichlet process is extremely important for the model. When α is small, then G tends to concentrate on a few atoms of probability. When α is large, then G is a distribution with many support points and the nonparametric model is "closer" to G_0, the baseline model. These features are to be borne in mind when considering priors for α. In this subsection, we discuss various effects of the parameter α and then issues arising in learning about α within the MCMC analysis.

Consider the following simple example to illustrate these points. Suppose the baseline parametric model is $Y_i = \mu_i + \epsilon_i$ with $\mu_i \sim N(\xi, \tau^2)$, $\epsilon_i \sim N(0, \sigma^2)$, $\sigma^2 \sim H_4$, and with some appropriate hyperpriors for ξ and τ^2. Now, if we were uncertain about the normality assumption for μ_i, we could identify μ_i as the θ parameter and adopt the Dirichlet process framework with $G_0 = N(\xi, \tau^2)$. A high degree of belief in normality would imply a prior on α favoring large values. Now, if instead of modeling uncertainty about the functional form of the distribution of μ_i, we consider modeling uncertainty about the constancy of variance of the ϵ_i. This can be done by relaxing the assumptions so that $\epsilon_i \sim N(0, \sigma_i^2)$, identifying $\theta_i = \sigma_i^2$ and basing the Dirichlet process around $G_0 = H_4$. To assign some prior weight to the original baseline model, the prior for α should have mass near zero. For more discussion about the prior for α and the model interpretations, see Escobar (1995, 1998) and Escobar and West (1995).

Learning about α from the data may be addressed with a view to incorporating α into the Gibbs sampling analysis. Assume a prior density $p(\alpha)$ (which may depend on the sample size I); this implies a prior for I^*, the number of distinct components, namely $P(I^*|I) = E[P(I^*|\alpha, I)]$. Using results in Antoniak (1974), the distribution of number of components may be written as

$$P(I^*|\alpha, I) = c_I(I^*) I!\, \alpha^{I^*} \Gamma(\alpha)/\Gamma(\alpha + I), \qquad (I^* = 1, 2, \ldots, I),$$

and $c_I(I^*) = P(I^*|\alpha = 1, I)$, not involving α. If required, the factors $c_I(I^*)$ are easily computed using recurrence formulæ for Stirling numbers. This is important, for example, in considering the implications for priors over I^* of specific choices of priors for α (and *vice-versa*) in the initial prior elicitation process. As an aside, note that there is a great deal of flexibility in representing prior opinions about I^* through choices of prior for α. Also, note that posterior distribution of α given I and I^* is conditionally independent of all the other variables in the model.

If the prior for α is a gamma distribution, or a mixture of gamma distributions, we have access to a neat data augmentation device for sampling α (Escobar and West 1995, West 1992). For example, let α have a gamma prior with a shape parameter $a > 0$ and scale parameter $b > 0$, namely $\alpha \sim Ga(a, b)$. When both a and b are small, then this prior puts weight on both large and small values of α. A data augmentation technique (Tanner and Wong 1987) can be used to generate conditional samples. At each Gibbs iteration, the currently sampled values of I^* and α allow us to draw a new value of α by first sampling, conditionally on α and I^* fixed at their most recent values, a latent variable η from the simple beta distribution

$(\eta|\alpha, I^*) \sim Be(\alpha+1, I)$, a beta distribution with mean $(\alpha+1)/(\alpha+I+1)$. Then, a new α value is sampled from a mixture of two gamma distributions based on the same I^* and the new η; in detail,

$$(\alpha|\eta, I^*) \sim \pi_\eta Ga(a+I^*, b-log(\eta)) + (1-\pi_\eta)Ga(a+I^*-1, b-log(\eta))$$

where the weights π_η are defined in odds form by

$$\pi_\eta/(1-\pi_\eta) = (a+I^*-1)/(I(b-log(\eta))).$$

Note that these distributions are well defined for all gamma priors, all η in the unit interval and all $I^* > 1$.

The above method of sampling α is used in the remaining sections of this paper. Before ending this section, we note an alternative prior for α used in Escobar (1994, 1998). Here $p(\alpha)$ is discrete, supported on a grid of points, $A_l, l = 1, \ldots, L$, with prior weights defined as $P(\alpha = A_l)$. Using Bayes theorem and the formula for $P(I^*|\alpha, I)$ defined above, the conditional posterior distribution is

$$P(\alpha = A_l|I^*, I) \propto P(\alpha = A_l)I!\ A_l^{I^*}\Gamma(A_l)\Gamma(A_l+I).$$

With the grid of points, it is necessary that L be large enough so that the MCMC chain can possible travel between different support points A_l. One choice of grid values is to let $\log(A_l)$ be equally spaced, with equal weight, and ranging from about n^{-1} to about n^2. See Escobar (1998) for an implementation of this algorithm.

As a final note about estimating α, Liu (1996) shows how to obtain the maximum likelihood estimate of α. Although this would not be useful to insert in the middle of a MCMC, the reader might use it in an empirical Bayes' approximation.

1.3.3 Complete Gibbs sampling schemes

We first revisit the model definition and structure. The primary parameters θ_i follow a prior distribution G which itself follows a Dirichlet process prior. The baseline prior G_0 depends on additional hyperparameters λ, and the priors H_1, H_2, and H_3 are adopted for the parameters ζ, λ and α, respectively. Inserting the Dirichlet process prior extends the standard hierarchical model as displayed in the following diagram:

$$\begin{array}{rcl} \multicolumn{3}{c}{\text{Standard Model}} \\ \hline [Y_i|\theta_i,\zeta] & \sim & F(Y_i|\theta_i,\zeta) \\ [\theta_i|G,\lambda] & \sim & G(\theta_i|\lambda) \\ [\zeta] & \sim & H_1(\zeta) \\ [\lambda] & \sim & H_2(\lambda) \\ & & \\ & & \end{array} \quad \overset{\mathcal{D}}{\Longrightarrow} \quad \begin{array}{rcl} \multicolumn{3}{c}{\text{Dirichlet Enhanced}} \\ \hline [Y_i|\theta_i,\zeta] & \sim & F(Y_i|\theta_i,\zeta) \\ [\theta_i|G] & \sim & G(\theta_i) \\ [G(\cdot)|\lambda,\alpha] & \sim & \mathcal{D}(G|G_0(\cdot|\lambda),\alpha) \\ [\zeta] & \sim & H_1(\zeta) \\ [\lambda] & \sim & H_2(\lambda) \\ [\alpha] & \sim & H_3(\alpha) \end{array}$$

Since the conditional posterior distributions for θ and α are presented in the earlier sections of this paper and since the random distribution function G is integrated out in the calculations, what remains to be specified is the conditional posterior distribution for ζ and λ. Since ζ is conditionally independent of G given Y and θ, the Dirichlet process enhancement does not change the conditional posterior distribution for ζ. With regard to the random variable λ, Blackwell and MacQueen (1973) show that the distinct values of θ_j^* are independent and identically distributed samples from $G_0(\cdot|\lambda)$. Hence, under the Dirichlet process enhancement, the conditional posterior distribution for λ is simply that from the standard model with the vector θ replaced by the vector of distinct elements θ^*. The Gibbs sampling analysis may now be implemented using the following conditional distributions of the full joint posterior for $[\theta, \zeta, \lambda, \alpha|Y]$:

(a) $[\theta_i|Y, \{\theta_k, k \neq i\}, \zeta, \lambda, \alpha]$ for each $i = 1, \ldots, n$.

(b) $[\theta_j^*|Y, \mathcal{S}, I^*, \zeta, \lambda]$ for each $j = 1, \ldots, I^*$.

(c) $[\zeta|Y, \theta, \lambda, \alpha]$, which, on observing the conditional independence structure of the posterior, reduces to $[\zeta|Y, \theta]$.

(d) $[\lambda|Y, \theta, \zeta, \alpha]$ which, again using the conditional independence structure, becomes just $[\lambda|\theta^*]$ where θ^* is the set of distinct values of θ.

(e) $[\alpha|Y, \theta, \zeta, \lambda]$, reducing to $[\alpha|\theta]$ as described in the previous section. This in turn is structured by first sampling $[\eta|\alpha, I^*]$ and then $[\alpha|\eta, I^*]$, where I^* is just the number of distinct elements in θ.

Gibbs sampling proceeds by simply iterating through (a)–(e) in order, sampling at each stage based on current values of all the conditioning variates. The following are some important, general features of these distributions:

- The distributions for generating the θ_i's in (a) are just those outlined in Section 3.1 for the simple Dirichlet process prior model. Thus, algorithms written to sample posteriors in such models may be directly implemented in the more complex hierarchical modeling framework without modification.

- The conditional posterior distribution (b) of θ_j^* is obtained by combining the baseline prior $G_0(\cdot|\lambda)$ with the likelihood $\prod_{i \in J(j)} f(Y_i|\theta_j^*, \zeta_i)$. That is, it is the posterior based on a random sample of the Y_i's which are drawn from the same θ_j^*'s. This is discussed in Section 3.1.

- The posterior (c) for sampling the parameter ζ of the sampling model is simply that found in the traditional, baseline hierarchical model used in the usual Gibbs sampling for such models. The Dirichlet process enhancement changes nothing here. This is true because ζ is conditionally independent of G given Y and θ.

- For λ, the hyperparameter of the baseline prior G_0, the required posterior (d) is exactly that arising from the prior $H_2(\lambda)$ combined with the likelihood $\prod_{i=1}^{I^*} G_0(d\theta_i^*|\lambda)$. That is, it is the posterior based on a random sample of (the distinct) θ_j^*'s drawn from the baseline prior G_0. This is true because the distinct values, θ_j^*, are independent and identically distributed from $G_0(\cdot|\lambda)$.

- The distributions for sampling α in (e) are precisely as discussed in Section 3.2.

Note that the above features simplify the inclusion of a Dirichlet process into a hierarchical model. The fact that there are no changes to the posterior samplers for ζ and very little change to the sampler for λ means that the Dirichlet process prior can be included in the middle of most Bayesian hierarchical models when computation uses MCMC methods – or essentially all applied models. Choice of priors for the Dirichlet process is simplified since G_0 is already modeled by the baseline parametric model. Hence prior specifications reduce to priors for α. Some examples and illustrations are now discussed.

1.4 An Example with Poisson-Gamma Structure

The numbers of a certain type of eye tracking anomaly for 101 subjects with schizophrenia are reported in Table 1, and a histogram of the data appears in Figure 1. The number of anomalies is believed to be associated with dysfunctions in the prefrontal cortex of the brain. A fuller description of the data along with a frequentist's mixture analysis of this data is contained in Sweeney et al (1993) and Pauler et al (1996). Consider modeling the number of anomalies, Y_i, as Poisson with individual means θ_i. This determines $F(Y_i|\theta_i)$, there being no additional ζ parameter. Further, take a baseline gamma prior G_0, under which the θ_i are independent gamma with specified shape λ_1 and scale λ_2, and set $\lambda = \{\lambda_1, \lambda_2\}$. Then the posterior distribution for known λ has the θ_i independent with $[\theta_i|Y_i, \lambda] \sim Ga(\lambda_1 + Y_i, \lambda_2 + 1)$.

Table 1

Outcome:	0	1	2	3	4	5	6	7	8	9
Count:	46	14	9	4	2	3	3	3	1	2

Outcome:	10	11	12	14	15	17	22	24	34
Count:	2	2	2	1	2	2	1	1	1

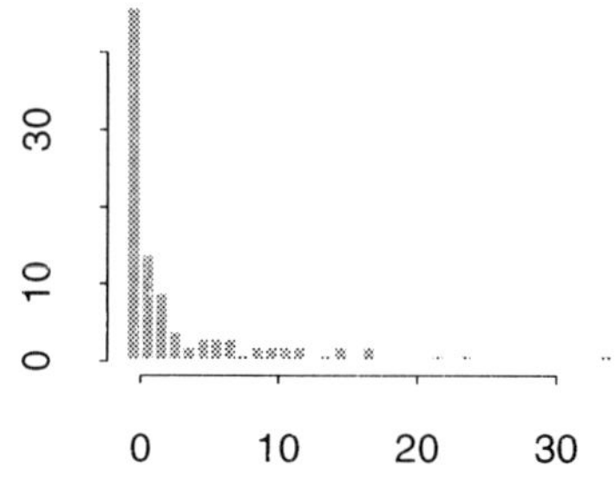

FIGURE 1. Histogram of the number of eye movement anomalies recorded for each of 101 subjects who are diagnosed with schizophrenia.

In the special case of a degenerate prior $[\lambda]$ and $\alpha \to \infty$, this model is complete.

Extending to a Dirichlet process analysis as outlined in Section 3 results in

$$\begin{aligned} [Y_i|\theta_i] &\sim F(Y_i|\theta_i), \\ [\theta_i] &\sim G(\theta_i), \\ [G|\alpha] &\sim \mathcal{D}(G_0, \alpha), \\ \alpha &\sim Ga(a, b). \end{aligned}$$

The required conditional posteriors (a)-(e) of Section 3.3 are easily derived. Note the special circumstances of no ζ parameter and assuming the prior for λ degenerate at the chosen values of λ_1 and λ_2; therefore steps (c) and (d) are vacuous. Step (e) is covered in Section 3.2, and the remaining steps (a) and (b) follow Section 3.1 to give

$$\begin{aligned} [\theta_i|Y, \{\theta_k, k \neq i\}, \alpha, \lambda] &\sim q_0\, Ga(\lambda_1 + Y_i, \lambda_2 + 1) + \sum_{k \neq i} q_k \delta(d\theta_i|\theta_k), \\ [\theta_j^*|Y, \mathcal{S}, \lambda] &\sim Ga(\lambda_1 + \sum_{i \in J_j} Y_i, \lambda_2 + n_j), \end{aligned}$$

with

$$q_0 \propto \alpha \binom{Y_i + \lambda_1 - 1}{Y_i} \left(\frac{\lambda_2}{1 + \lambda_2}\right)^{\lambda_1} \left(\frac{1}{1 + \lambda_2}\right)^{Y_i}$$

and, for each $k > 0$, $q_k \propto \theta_k^{Y_i} \exp(-\theta_k)/Y_i!$.

Two separate analyses are illustrated, based on baseline priors with λ set at either (0.4,0.1) or (1,1). The values (0.4,0.1) correspond to a method of moment estimate of (λ_1, λ_2) obtained from the complete data set. For the precision parameter, α, we consider two priors: $Ga(.01, .01)$ and $Ga(1, 1)$. The former prior gives reasonable mass to both high and low values of α. The $Ga(1, 1)$ prior favors relatively low values of α. The posteriors under these two priors are quite similar, it being the case that the choice of G_0 is more important for this (model:data) combination. For more on the effect of the prior values for the Dirichlet process on the model, see Escobar (1998).

In the analysis with $\lambda = (0.4, 0.1)$, Figures 2a, 2b, and 2c display the estimated marginal posteriors for θ_i when Y_i is equal to 4, 8, and 12. Under this prior, the inclusion of the Dirichlet process does not change the posterior distribution very much. The baseline model produces posterior distributions which are very close to the posteriors obtained under the two Dirichlet process models. The use of the Dirichlet process sometimes does lead to marginally increased spread in posteriors, reflecting additional uncertainty implied by the use of the Dirichlet structure. In part the concordance arises as the $Ga(0.4, 0.1)$ prior for the Poisson rates is not in conflict with the observed data.

Under the alternative $Ga(1, 1)$ prior, the data observed in the right tail of Figure 1 are "unusual." In this case the marginal distribution of Y_i under the standard model is negative binomial with parameters $p = .5$ and $r = 1$. That is, the marginal distribution of Y_i is equivalent to tossing a fair coin and counting the number of tails before the first appearance of a head. There is only a 0.1% chance of values greater than 9, yet 14% of the observations are greater than 9. When the Dirichlet process prior is used, the data downweights the $Ga(1, 1)$ component and results in posteriors very different than obtained under the standard/baseline model. Figures 2d, 2e, and 2f display the posterior margins for those θ_i for which Y_i is equal to 4, 8, and 12. Note the posterior distributions for $(\theta_i | Y = 12)$ in particular. Since the baseline prior puts very little probability on data values higher than 10, the baseline posterior distribution puts most of its mass below 10. However, the posteriors under each of the Dirichlet enhanced models are bimodal. The smaller modes appears as "reduced" versions of the baseline posterior, while the major modes favour values in the 10 to 20 range in each case. Hence the nonparametric analysis naturally adapts to and reflects the inherent conflict in the (model:data) match. For related discussion, see Escobar (1998).

Note finally that extensions to include learning about λ are straightforward. For an example of such a baseline parametric model, see George, Makov, and Smith (1994), who show how to construct conditional samplers to perform the analysis. Once this is done, the Dirichlet process enhancement may use the same algorithm to sample λ as the baseline model, but

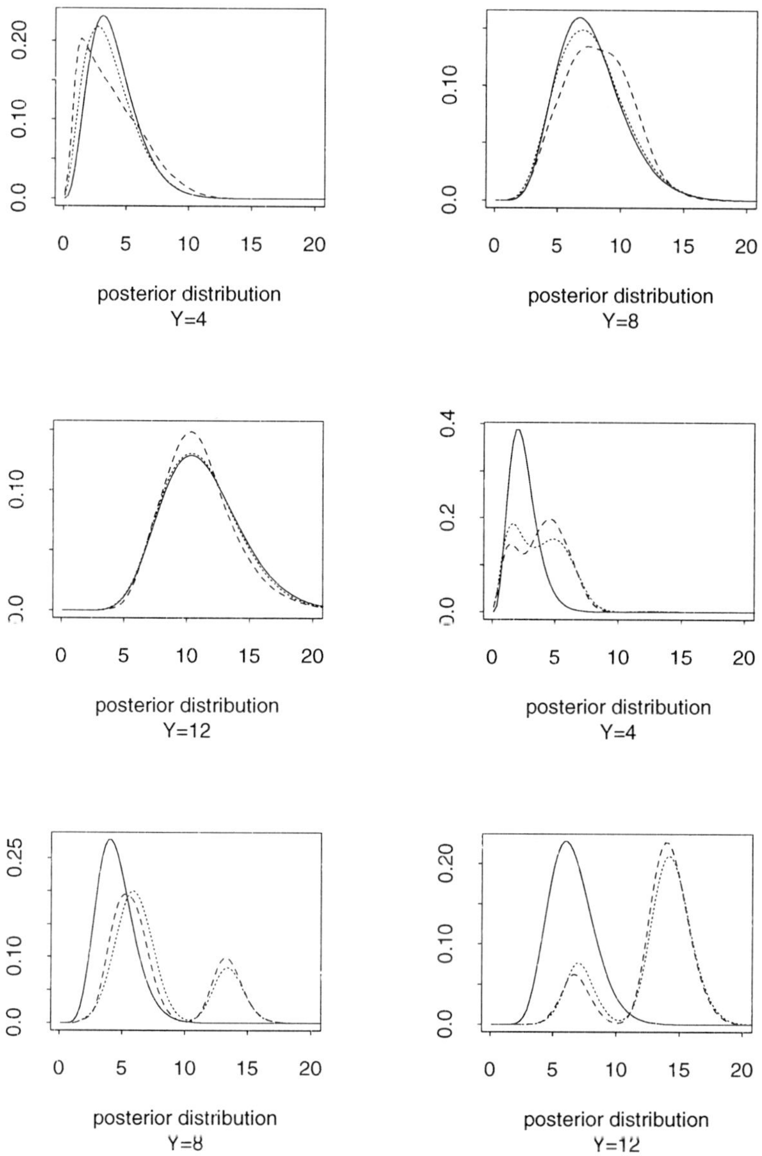

FIGURE 2. Fitted posterior distributions of θ under different conditions; (a, b, and c): the models where the baseline prior for θ is assumed to be $Ga(0.4, 0.1)$; (d, e, and f): the models where the baseline prior for θ is assumed to be $Ga(1, 1)$; (a and d): $Y_i = 4$; (b and e): $Y_i = 8$; (c and f): $Y_i = 12$. The solid lines "—": the baseline model; the dotted line "···": Dirichlet process with $\alpha \sim Ga(.01, .01)$; the dashed line "– – –": with $\alpha \sim Ga(1, 1)$.

the vector of θ_j^*'s would be substituted in for the vector of θ_i's. We do something along these lines in the following example.

1.5 An Example with Normal Structure

Gelfand et al (1990) present a data set of weights of rats measured at the same five time points. There are two groups of rats, a control group and a treatment group, with 30 rats in each group. Let Y_{ij} be the weight of the ith rat at the jth time point, with $i = 1, \ldots, I$ and $j = 1, \ldots, J_i$. First, each rat is modeled by a linear growth curve with normal errors,

$$[Y_{ij}|\theta_i, \sigma^2] \sim N(\theta_{i0} + \theta_{i1}X_j, \sigma^2),$$

where X_j is the age at the jth point. The line is defined by the parameter vector $\theta_i = (\theta_{i0}, \theta_{i1})$ and all rats have the same within-group variance σ^2. For G_0 we take the normal distribution

$$[\theta_i|\mu, \Sigma] \sim N(\mu, \Sigma).$$

Note here that $\lambda = \{\mu, \Sigma\}$ and $\zeta = \sigma^2$. Finally, uncertainty about the parameters σ^2, μ, and Σ is modeled with independent priors,

$$\begin{aligned} [\sigma^2] &\sim IG(0.5\nu_1, 0.5\nu_1\nu_2), \\ [\mu] &\sim N(\xi_1, \Xi_2), \\ [\Sigma^{-1}] &\sim W((\xi_3\Xi_4)^{-1}, \xi_3), \end{aligned}$$

where IG is the inverse gamma distribution and W is the Wishart distribution, and the ξ's, Ξ's, and ν's are prior parameters that are specified in any analysis. As shown in Gelfand et al (1990), the standard/baseline hierarchical model has conditional posterior distributions (where we drop the conditioning variates in each case, for clarity of display):

$$\begin{aligned} \sigma^2 &\sim IG(0.5(\nu_1 + n), 0.5S), \\ \theta_i &\sim N(D_i[\sigma^{-2}X_i^TY_i + \Sigma^{-1}\mu], D_i), \\ \mu &\sim N(V[I\Sigma^{-1}\bar{\theta} + \Xi_2^{-1}\xi_1], V), \\ \Sigma^{-1} &\sim W(U^{-1}, I + \xi_3), \end{aligned}$$

where $S = \sum_{ij}(Y_{ij} - \hat{Y}_{ij})^T(Y_{ij} - \hat{Y}_{ij}) + \nu_1\nu_2$, $\hat{Y}_{ij} = \theta_{i0} + \theta_{i1}X_j$, $n = \sum_{i=1}^I J_i$, $\bar{\theta} = I^{-1}\sum_{i=1}^I \theta_i$, $D_i = (\sigma^{-2}X_i^TX_i + \Sigma^{-1})^{-1}$, $V = (I\Sigma^{-1} + \Xi_2^{-1})^{-1}$, and $U = \sum_{i=1}^I(\theta_i - \mu)(\theta_i - \mu)^T + \xi_3\Xi_4$.

Now the above model is extended to incorporate a Dirichlet process prior for G with the baseline prior G_0 as the initial guess. This results in sampling from the following conditional distributions (again with conditioning variates suppressed, for clarity):

$$\begin{aligned}
\sigma^2 &\sim IG(0.5(\nu_1+n), 0.5S) \\
\theta_i &\sim q_0 N(D_i[\sigma^{-2}X_i^T Y_i + \Sigma^{-1}\mu], D_i) + \sum_{k\neq i} q_k \delta(\theta_i|\theta_k), \\
\theta_j^* &\sim N(D_j^*[\Sigma^{-1}\mu + \sigma^{-2}\sum_{i\in J(j)} X_i^T Y_i], D_j^*), \\
\mu &\sim N(V^*[I^*\Sigma^{-1}\bar{\theta}^* + \Xi_2^{-1}\xi_1], V^*), \\
\Sigma^{-1} &\sim W(U^{-1}, I^* + \xi_3),
\end{aligned}$$

where $\bar{\theta}^* = (I^*)^{-1}\sum_{i=1}^{I^*}\theta_i^*$, $D_j^* = (\Sigma^{-1} + \sigma^{-2}\sum_{i\in J(j)} X_i^T X_i)^{-1}$, $V^* = (I^*\Sigma^{-1} + \Xi_2^{-1})^{-1}$, $U = \sum_{j=1}^{I^*}(\theta_j^* - \mu)(\theta_j^* - \mu)^T + \xi_3\Xi_4$, $q_0 \propto \alpha f_i$ where f_i is the density at Y_i of the multivariate normal distribution $N(X_i\mu, X_i\Sigma X_i^T + \sigma^2\mathcal{I})$ (and where $\mathcal{I}$ is the identity matrix), and each q_k is proportional to the multivariate normal density of $Y_i \sim N(X_i\theta_k, \sigma^2\mathcal{I})$.

In extending from the traditional baseline analysis to the Dirichlet process analysis, note again that there is no change to the conditional distribution of σ^2. Note also that the conditional distributions of μ and Σ just have θ replaced by $\{\theta^*\}$. Also, a sample from the conditional of θ_i is either a sample from the conditional distribution of θ_i under the baseline analysis, or θ_i is set equal to a previously sampled value θ_k. The remixing step requires that we know the posterior distribution of θ_j^* when several subjects are sampled from that θ_j^*. Therefore, once a computer program is written to perform Gibbs sampling in the baseline analysis it is easily extended to perform analysis in the Dirichlet enhanced model.

In analysing this data, the same vague priors for σ, μ, and Σ are used as in Gelfand et al (1990). The parameter α is modeled with either a $Ga(1,1)$ prior or a $Ga(0.01, 0.01)$ prior. Analyses in Gelfand et al (1990) and again in Carlin and Polson (1991), report posterior inferences on μ. Here, by comparison, we report predictive inferences in terms of the predictive distribution for a future θ_i vector with $i = n+1$ (which, in the baseline analysis, is simply the posterior for μ convolved with a t-distribution). Results of the analysis of just the rats in the control group are reported in Figure 3. Figure 3a displays the plot of the maximum likelihood estimates of the intercept (θ_{0i}) and the slope (θ_{1i}). The other three plots display contours of the joint predictive distributions under different models. In the earlier studies with the standard hierarchical model, the posterior distribution for μ has appeared as unimodal, as in Figure 3b here. Carlin and Polson (1991) report that one observation has a large influence and maybe should be disregarded from any inferences from this data. These inferences contrast with the following observations in Dirichlet enhanced analyses.

Figures 3c and 3d display contour plots of the predictive distribution of θ for the two Dirichlet process analyses. Figure 3c is for the model with

$\alpha \sim Ga(.01, .01)$ and Figure 3d is for that with $\alpha \sim Ga(1, 1)$. The predictive distributions under both priors seem to contain ridges of probability, and the potential outlier discovered by Carlin and Polson appears to have been isolated with it's own "mini-mode" of probability. The $Ga(1, 1)$ prior seems to induce more severe clustering of the θ_i's, while the $Ga(.01, .01)$ induces more of a blend between the baseline predictive distribution and the predictive distribution obtained from a standard model with a straight $Ga(.01, .01)$ prior. It appears that the Dirichlet enhanced analysis leads to a natural and appropriate down weighting of the effect of outliers and influential points. Also, the contour plots of predictives under the enhanced analysis are naturally adjusted for population heterogeneity.

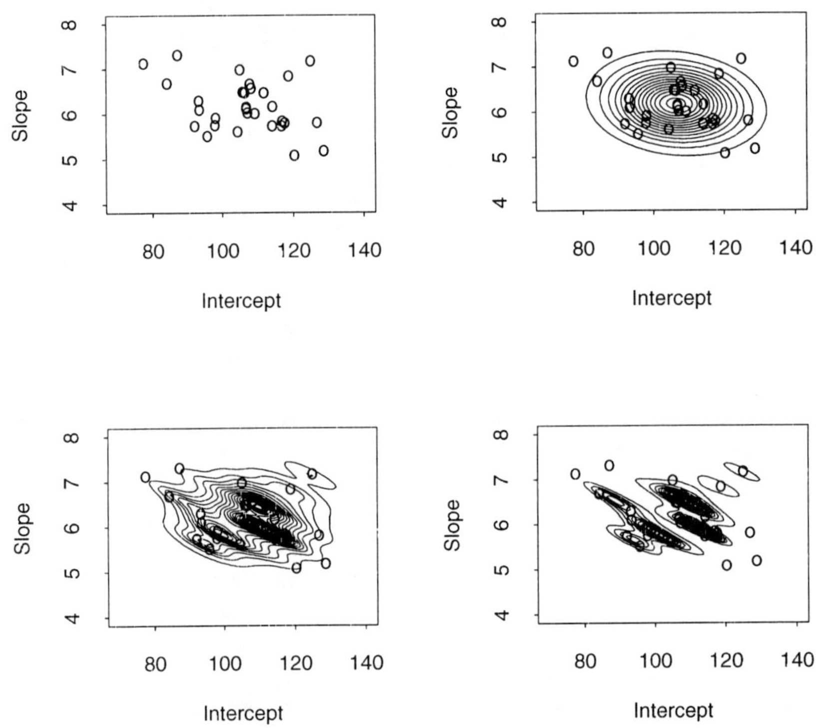

FIGURE 3. The dots are the MLE estimate of the intercept and slope of the linear growth curve of each subject consider alone. Figure (a): only the fitted MLE values are plotted. In the other three plots, the contour plots of the predicted distribution of the (intercept, slope) are shown for three different priors; (b) baseline model; (c) Dirichlet process with $\alpha \sim Ga(.01, .01)$; (d) Dirichlet process with $\alpha \sim Ga(1, 1)$.

Figure 4 displays marginal predictive distributions for the slopes of models for rats from both the control group and the treatment group. The lines are fit with the baseline prior model and with the Dirichlet process model with knowledge of the group membership for each rat. Basically, the control group and the treatment group are fit separately and the predictive distribution is simply the average of the predictive distributions for the two groups. Then, to see if the Dirichlet process model might be useful to control for unknown heterogeneity, the predictive distribution is calculated under the Dirichlet process model without knowledge of group membership. In Figure 4a, when $\alpha \sim Ga(.01, .01)$, the two predictive distributions are bimodal, and the predictive distribution fit without knowledge of group membership is somewhat flat between the modes of the other two predictive distributions. However, in Figure 4b, when $\alpha \sim Ga(1, 1)$, the predictive distribution obtained without knowledge of the group membership is very close to that in the analysis where group membership is known.

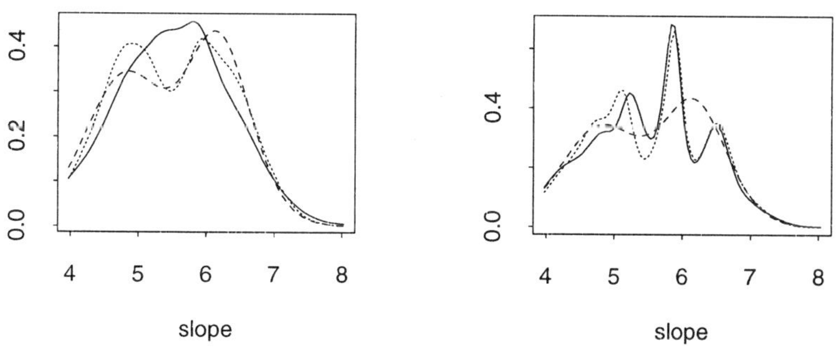

FIGURE 4. The predictive distribution for the slope parameter for the data which contain both treatment and control group. The dashed lines "— — —" are for the baseline model where the parameters are fit with knowledge of the grouping variable. The dotted lines "· · ·" are for the Dirichlet process model with knowledge of the grouping variable. The solid lines "—" are for the Dirichlet process model when there is no knowledge of the group membership of the subjects. Figure (a): Dirichlet process with $\alpha \sim Ga(.01, .01)$; (b) Dirichlet process with $\alpha \sim Ga(1, 1)$.

Acknowledgments: Michael D. Escobar was partially financed by the National Cancer Institute #RO1-CA54852-01, a National Research Service Award from NIMH Grant #MH15758, and by the Natural Sciences and Engineering Research Council of Canada. Mike West was partially financed by the National Science Foundation under grant DMS-9024793.

References

Andrews, D.F. and Mallows, C.L. (1974), "Scale mixtures of normal distributions," *Journal of the Royal Statistical Society, Series B*, 36, 99-102.

Antoniak, C.E. (1974), "Mixtures of Dirichlet processes with applications to nonparametric problems," *The Annals of Statistics*, 2, 1152-1174.

Blackwell, D., and MacQueen, J.B. (1973), "Ferguson Distribution via Polya Urn Schemes," *The Annals of Statistics*, 1, 353-355.

Brunner, L.J. (1995), " Bayesian linear regression with error terms that have symmetric unimodal densities," *Journal of Nonparametric Statistics*, 4, 335-348.

Bush, C.A. and MacEachern, S.N. (1996), "A semi-parametric Bayesian model for randomized block designs," *Biometrika*, 83, 275-285.

Carlin, B.P., and Polson, N.G. (1991), "An expected utility approach to influence diagnostics," *Journal Of the American Statistical Association*, 86, 1013-1021.

Doss, H. (1994), "Bayesian nonparametric estimation for incomplete data via successive substitution sampling," *The Annals of Statistics*, 22, 1763-1786.

Erkanli, A., Stangl, D., and Müller, P. (1993), "Analysis of ordinal data by the mixture of probit links," *Discussion Paper #93-A01*, ISDS, Duke University.

Escobar, M.D. (1988), Estimating the means of several normal populations by nonparametric estimation of the distribution of the means, Unpublished dissertation, Yale University.

Escobar, M.D. (1992), Invited comment of "Bayesian analysis of mixtures: some results on exact estimability and identification," by Florens, Mouchart, and Rolin. *Bayesian Statistics 4* (eds: J.M. Bernardo, J.O. Berger, A.P.Dawid and A.F.M.Smith). Oxford: University press, 142-144.

Escobar, M.D. (1994), "Estimating normal means with a Dirichlet process prior," *Journal of the American Statistical Association*, 89, 268-277.

Escobar, M.D., (1995), "Nonparametrics Bayesian Methods in Hierarchical Models," *The Journal of Statistical Inference and Planning*, 43, 97-106.

Escobar, M.D. (1998), "The effect of the prior on nonparametric Bayesian methods," (in preparation).

Escobar, M.D. and West, M. (1995), "Bayesian density estimation and inference using mixtures," *Journal of the American Statistical Association*, 90, 577-588.

Evans, M. and Swartz, T. (1995) "Methods for approximating integrals in statistics with special emphasis on Bayesian integration problems - with discussion," *Statistical Science*, 10, 254-272.

Ferguson, T.S. (1973), "A Bayesian analysis of some nonparametric problems," *The Annals of Statistics*, 1, 209-230.

Ferguson, T.S. (1974), "Prior distributions on spaces of probability measures," *The Annals of Statistics*, 2, 615-629.

Freedman, D. (1963), "On the asymptotic behavior of Bayes estimates in the discrete case," *Annals of Mathematical Statistics*, 34, 1386-1403.

Gelfand, A.E., Hills, S.E., Racine-Poon, A., and Smith, A.F.M. (1990), "Illustration of Bayesian inference in normal data models using Gibbs sampling.", *Journal Of the American Statistical Association*, 85, 972-985.

George, E.I., Makov, U.E., and Smith, A.F.M. (1994), "Fully Bayesian hierarchical analysis for exponential families via Monte Carlo computation," *Aspects of Uncertainty: A Tribute to D. V. Lindley* (eds: AFM Smith and PR Freeman), London: John Wiley and Sons, 181-199.

Gilks, W.R., Richardson, S., and Spiegelhalter, D.J., eds. (1996), *Markov Chain Monte Carlo in Practice*, London: Chapman and Hall.

Kuo, L. and Mallick, B. (1998), "Bayesian semiparametric inference for the accelerated failure," *Canadian Journal of Statistics*, to appear.

Liu, J.S. (1996), "Nonparametric hierarchical Bayes via sequential imputations," *The Annals of Statistics*, 24, 911-930.

MacEachern, S.N. (1988), Sequential Bayesian bioassay design. Unpublished dissertation. University of Minnesota.

MacEachern, S.N. (1994), "Estimating normal means with a conjugate style Dirichlet process prior," *Communications in Statistics: Simulation and Computation*, 23, 727-741.

MacEachern, S.N. and Müller, P. (1998), "Estimating mixture of Dirichlet process models," *Journal of Computational and Graphical Statistics*, to appear.

Mukhopadhyay, S. and Gelfand, A.E. (1997), "Dirichlet process mixed generalized linear models," *Journal Of the American Statistical Association*, 92, 633-639.

Müller, P., Erkanli, A., and West, M. (1996), "Bayesian curve fitting using multivariate normal mixtures," *Biometrika*, 83, 67-79.

Müller, P., West, M., and MacEachern, S.N. (1997), "Bayesian models for non-linear auto-regressions," *Journal of Time Series Analysis*, (in press).

Naylor, J.C. and Smith, A.F.M. (1982), "Applications of a method for the efficient computation of posterior distributions," *Applied Statistics*, 31, 214-235.

Pauler, D.K., Escobar, M.D., Sweeney, J.A., and Greenhouse, J. (1996), "Mixture Models for Eye-Tracking Data: A Case Study" *Statistics in Medicine*, 15, 1365-1376.

Roeder, K., Escobar, M., Kadane, J., and Balazs, I., (1998), "Measuring Heterogeneity in Forensic Databases", *Biometrika* to appear.

Sweeney, J.A., Clementz, B.A., Escobar, M.D., Li, S., Pauler, D.K., and Haas, G.L. (1993), "Mixture analysis of pursuit eye tracking dysfunction in schizophrenia," *Biological Psychiatry*, 34, 331-340.

Tanner, M.A. and Wong, W.H. (1987), "The calculation of posterior distributions by data augmentation (with discussion)," *Journal of the American Statistical Association*, 82, 528-550.

Tomlinson, G. (1998), Analysis of densities with Dirichlet process priors. Unpublished dissertation. University of Toronto.

Turner, D.A., and West, M. (1993), "Statistical analysis of mixtures applied to postsynpotential fluctuations," *Journal of Neuroscience Methods*, 47, 1-23.

West, M. (1987), "On scale mixtures of normal distributions," *Biometrika*, 74, 646-648.

West, M. (1990), "Bayesian kernel density estimation," *ISDS Discussion Paper #90-A02*, Duke University.

West, M. (1992), "Hyperparameter estimation in Dirichlet process mixture models," *ISDS Discussion Paper #92-A03*, Duke University.

West, M. (1997), "Hierarchical mixture models in neurological transmission analysis," *Journal Of the American Statistical Association*, 92, 587-608.

West, M., and Cao, G. (1993), "Assessing mechanisms of neural synaptic activity," In *Bayesian Statistics in Science and Technology: Case Studies*, (eds: C.A. Gatsonis, J.S. Hodges, R.E. Kass, and N.D. Singpurwalla), New York: Springer-Verlag.

West, M., Müller, P., and Escobar, M.D. (1994), "Hierarchial Priors and Mixture Models, with Applications in Regression and Density Estimation," *Aspects of Uncertainty: A Tribute to D. V. Lindley* (eds: AFM Smith and PR Freeman), London: John Wiley and Sons, 363-386.

West, M., and Turner, D.A. (1994), "Deconvolution of mixtures in analysis of neural synaptic transmission," *The Statistician*, 43, 31-43.

2

Computational Methods for Mixture of Dirichlet Process Models

Steven N. MacEachern

ABSTRACT This chapter lays out the basic computational strategies for models based on mixtures of Dirichlet processes. I describe the basic algorithm and give advice on how to improve this algorithm through a collapse of the state space of the Markov chain and through blocking of variates for generation. The computational methods are illustrated with a beta-binomial example and with the bioassay problem. Some advice is given for dealing with models that have little or no conjugacy present.

2.1 Introduction

The mixture of Dirichlet processes (MDP) model is one of extraordinary usefulness and generality. In its most common use, the model allows one to mimic a Bayesian, parametric hierarchical model a la Lindley and Smith (1972), but to relieve the strictures imposed by the parametric assumptions. A basic parametric model is

$$
\begin{aligned}
X_i|\theta_i, z_i &\sim F_{\theta_i, z_i}(.) \\
\theta_i|\nu &\sim G_{0,\nu} \\
\nu &\sim H_\nu(.)
\end{aligned}
$$

where $i = 1, \ldots, n$. The usual convention in writing a hierarchical model is followed, so that observations at a stage in the hierarchy are independent, and dependence upon all higher stages of the hierarchy is explicitly stated. Thus, for example, the $\theta_i|\nu$ are independent and identically distributed, from $G_{0,\nu}$.

As a specific example, the reader is encouraged to consider the simple normal theory hierarchical model. In this model, the data (the X_i), given the parameters and covariates, are normally distributed. The mid-level parameters (the θ_i), given the hyperparameter, are normally distributed. The hyperparameter (ν) is often assigned a conjugate prior distribution.

To formally match the notation, there are no covariates z_i, $\nu = (\mu, \tau^2)$, $F_{\theta_i,z_i} = N(\theta_i, \sigma^2)$, $G_{0,\nu} = N(\mu, \tau^2)$, and $H_\nu = N(\mu_0, \tau_0^2) \cdot IG(\alpha_0, \beta_0)$. In this description, $N(a, b)$ represents the normal distribution with mean a and variance b, and $IG(a, b)$ represents the inverse gamma distribution with shape a and scale b (hence mean $(b(a-1))^{-1}$ and variance $(b^2(a-1)^2(a-2))^{-1}$, provided these quantities exist). A model with unknown σ^2 would necessitate placing a distribution on σ^2, with a conjugate choice being an inverse gamma distribution.

In this simple hierarchical model, the θ_i are tied together through the portion of the model involving ν. Each of the θ_i contains information about ν which, in turn, provides information about $G_{0,\nu}$, and hence θ_j, $j \neq i$. Since θ_i contains information about θ_j, it will have an impact on inferences about θ_j. The model is appropriate when the θ_i are parameters for similar phenomena. See the example in Section 3.1 for a case where this modelling strategy, up to the form of the likelihood, is reasonably appropriate.

The presence of the z_i greatly broadens the scope of the hierarchical model. It allows one to include case-specific covariates, that, in more complex models allow for such features as X_i which arise from varying sample sizes, a regression structure which is separate from the portion of the model described above (and which would include a prior distribution over additional parameters), an indication of censoring, or qualitatively varying forms for F_{θ_i,z_i}. This last allows for the explicit combination of data that are observed at different resolutions or which arise from different sampling plans.

The MDP analog of the parametric model is

$$\begin{aligned} X_i|\theta_i, z_i &\sim F_{\theta_i,z_i}(.) \\ \theta_i|G &\sim G \\ G|M,\nu &\sim Dir(MG_{0,\nu}) \\ \nu &\sim H_\nu(.) \\ M &\sim H_M(.) \end{aligned}$$

where $i = 1, \ldots, n$. Most of the components of the MDP model match its parametric analog. One novel component is $G|(M, \nu) \sim Dir(MG_{0,\nu})$. The distribution on G is a Dirichlet process with base measure α. As is commonly done, the mass of the base measure, a positive quantity, is denoted by M. $G_{0,\nu}$ is a distribution function that is proportional to the measure α. A second novel component, $H_M(.)$, is the distribution over the mass of the base measure. The prior distribution has traditionally taken M and ν to be independent, as is assumed in this chapter, although this need not be the case.

The difference between the two models lies in the middle stage of the hierarchy, where, to create the MDP model, $(\theta_i|\nu) \sim G_{0,\nu}$ has been ripped out to be replaced by the nonparametric $(\theta_i|G) \sim G$; $(G|M, \nu) \sim Dir(MG_{0,\nu})$;

$M \sim H_M(.)$. It is this alteration that results in the additional flexibility of the MDP model, enabling the model to adapt to a wide variety of data.

The Dirichlet process provides a means of placing a distribution on the space of distribution functions. It satisfies the two goals outlined in Ferguson's (1973) seminal paper of having large support and being easy to work with. The support of the Dirichlet process is large, and, with appropriate choice of $G_{0,\nu}$ can be the set of all distributions on, say, the real line. With such large support, the distribution from which the θ_i are drawn is no longer restricted to lie in the set of distributions $G_{0,\nu}$, indexed by ν. This set, in a parametric hierarchical model, may be very small–say the set of normal distributions. The switch to the MDP model allows the data to adapt to a G that is skewed, has "shoulders", is multimodal, or departs from the parametric form $G_{0,\nu}$ in less evident ways.

The Dirichlet process is easy to work with when the hyperparameters, ν and M, are fixed and the θ_i are observed directly. In this instance, the distribution on G, after observing $\theta_1, \ldots, \theta_{n-1}$, would also be a Dirichlet process, with measure $MG_{0,\nu} + \sum_{i=1}^{n-1} \delta_{\theta_i}$, with δ_{θ_i} representing a point mass, of measure 1, at θ_i. The distribution for $\theta_n|(M, \nu, \theta_1, \ldots, \theta_{n-1})$ is just a rescaled version of this updated base measure: The distribution for $\theta_n|(M, \nu, \theta_1, \ldots, \theta_{n-1})$ can be written as a mixture, with $\theta_n \sim G_{0,\nu}$ with probability $M/(M+n-1)$ and θ_n set equal to θ_i, $i = 1, \ldots, n-1$ with probability $1/(M+n-1)$. Since the $\theta_i|G$ are independent and identically distributed, they are exchangeable, and so one can rewrite the conditional distribution for any $\theta_i|(M, \nu, \theta_1, \ldots, \theta_{i-1}, \theta_{i+1}, \ldots, \theta_n)$ as the mixture $\theta_i \sim G_{0,\nu}$ with probability $M/(M+n-1)$ and θ_i set equal to θ_j, for $j = 1, \ldots, i-1, i+1, \ldots, n$ with probability $1/(M+n-1)$. It is this representation of the conditional distribution of θ_n, which is most easily seen from the Polya urn scheme representation of the Dirichlet process (Blackwell and MacQueen, 1973), that lies both at the heart of the MDP modelling strategy and at the heart of the modern computational methods for fitting the MDP model.

To see the parallel between the MDP model and the parametric model that gives rise to it, consider the marginal distribution of a θ_i, say θ_1. As seen from the previous description, the distribution of $\theta_1|(M, \nu)$ is just $G_{0,\nu}$. With $n = 1$ the middle stage of the hierarchy collapses, and the two models are identical. The impact of the MDP model is seen when $n > 1$. In this instance, the θ_i divide into $k \leq n$ clusters, with values of θ_i within a cluster identical and with common cluster locations independent draws from $G_{0,\nu}$. It is this description of the clustering of the vector θ that leads to the view of the MDP model as a mixture model.

While the conditional distributions are tractable, the entirety of the MDP model is analytically intractable for all but the simplest cases or very small data sets, due to a combinatorial explosion in the number of possible cluster structures. As a result, one must turn to approximation in order to fit the model. With a modern computational environment, this most often means that a simulation based approximation is chosen. The early work on simulation methods was done by Kuo (1986) who developed an importance sampler for Dirichlet process based models, and Escobar (1994) who developed the first Markov chain Monte Carlo methods for fitting Dirichlet process based models. Further work has improved Escobar's initial algorithm and extended the methods to deal with the full MDP model (Escobar and West, 1995; MacEachern, 1994). Modifications have produced Markov chains that have been empirically shown to move through the parameter space more quickly and that naturally suggest more accurate estimators. These current algorithms are the focus of this chapter.

2.2 The Basic Algorithm

A basic computational algorithm that can be applied to a wide variety of problems extends Escobar's (1994) algorithm. The third step in this algorithm is due to Bush and MacEachern (1996). A nice description of this algorithm appears in West, Müller and Escobar (1994). This algorithm, still a Gibbs sampler (Gelfand and Smith, 1990) in the loose sense of the term, where parameters are generated from a sequence of conditional distributions, is easily programmed for the MDP model. It involves little effort beyond writing the Gibbs sampler for the corresponding parametric model.

In order to describe the algorithm, some additional notation is introduced. The vector θ^* consists of those elements of θ that lie in distinct clusters. It is of length $k \leq n$. The vector $s = (s_1, \ldots, s_n)$ indicates in which cluster each of the θ_i lies. If $s_i = j$, then θ_i lies in cluster j, and $\theta_i = \theta_j^*$. The number of s_i such that $s_i = j$ is denoted by n_j. In the algorithms below, a value will be generated for θ_i, conditional on θ_{-i}, the other parameters, and the data. When performing such a conditioning, the superscript $-$ denotes the fact that θ_i is left out of the conditioning. Thus n_j^- represents the number of θ's in cluster j, not counting θ_i, and k^- represents the number of clusters when θ_i is removed. If removal of θ_i shrinks the number of clusters, so that $k^- < k$, then relabel the kth cluster as cluster s_i: Set $s_j = s_i$ for all j with $s_j = k$, and set $\theta_{s_i}^* = \theta_k^*$. Since there are now only $k - 1$ clusters, remove θ_k^*.

The generations from the conditional distributions are

- Generate $\nu|\theta^*$. For this conditional distribution, ν has density proportional to $h(\nu) \prod_{i=1}^k g_{0,\nu}(\theta_i^*)$.

- Generate $(s_i, \theta_i)|(\nu, M, s_{-i}, \theta_{-i}, X)$. This distribution is derived by taking each θ_i, in turn, to be the last parameter arising from a Polya urn scheme. The parameter θ_i comes from cluster j with probability

$$q_j \propto (\frac{n_j^-}{M + n - 1})(f_{\theta_j^*, z_i}(x_i)),$$

for $j = 1, \ldots, k^-$. It comes from cluster $k^- + 1$ with probability

$$q_0 \propto (\frac{M}{M+n-1})(\int f_{\tilde{\theta},z_i}(x_i) g_{0,\nu}(\tilde{\theta}) d\tilde{\theta}).$$

In the sequel, the two terms in these probabilities are referred to as **1** and **2**, respectively. Term **1** comes from the prior probability of θ_i joining an established cluster or beginning a new cluster under the Polya urn scheme. Term **2** accounts for the appropriate marginal density of $X_i|\theta_j^*$.

In the event that $s_i = k^- + 1$, a value is needed for the newly created θ_k^*. This value is drawn from the distribution with density proportional to $g_{0,\nu}(\theta_k^*) f_{\theta_k^*,z_i}(x_i)$.

This step is implemented, for $i = 1, \ldots, n$. The values of parameters are updated, immediately upon their generation.

- Generate $\theta_i^*|(\nu, X)$. For this conditional distribution, θ_i^* has density proportional to $g_{0,\nu}(\theta_i^*) \prod_{j|s_j=i} f_{\theta_i^*,z_j}(x_j)$.

  This step is implemented for $i = 1, \ldots, k$.

If a distribution is placed on M, its value may be generated as an additional stage. Alternatively, the distribution of M may be removed through integration, as indicated in the next section on the collapse of the state space.

2.3 More Efficient Algorithms

There are a number of techniques that are known to improve the performance of Gibbs samplers in many problems. When regularity conditions are satisfied, these techniques increase the rate of convergence of the Markov chain that underlies the Gibbs sampler to its limiting distribution. The improvement in rate of convergence may be qualitative, with the rate moving from geometric to uniform, for example. Such improvement can be more useful than merely obtaining better constants for the same qualitative rate.

Two of the most popular (and easiest to implement) techniques are that of collapsing the state space and that of blocking several parameters together. The general heuristic that motivates both of these techniques arises from consideration of a single generation in the Gibbs sampler. To illustrate the heuristic, its application to a model with three parameters is examined. Call the parameters θ_1, θ_2, and θ_3. There are three Markov chains to compare. One of the chains is described by the generations $\theta_1|(\theta_2, \theta_3)$, $\theta_2|(\theta_1, \theta_3)$ and $\theta_3|(\theta_1, \theta_2)$, the second is described by the generations $(\theta_1, \theta_2)|\theta_3$, $(\theta_1, \theta_2)|\theta_3$ and $\theta_3|(\theta_1, \theta_2)$, and the third is described by the generations $(\theta_1, \theta_2)|\theta_3$, $(\theta_1, \theta_2)|\theta_3$ and $(\theta_1, \theta_3)|\theta_2$.

Compare the first two chains, generation by generation. For the first generation, chain 2 is preferable to chain 1. The values of both θ_1 and θ_2 are updated instead of merely updating θ_1, and this enables us to move closer to the joint posterior distribution of $(\theta_1, \theta_2, \theta_3)$. For the second generation, chain 2 again is preferable, since θ_1 and θ_2 are both updated instead of merely updating θ_2. The third generations are equivalent. Since, for each generation, chain 2 is at least as good as chain 1 and sometimes better, there is strong heuristic evidence that chain 2 is better.

Compare the second and third chains, generation by generation. The first two generations are identical. For the third generation, chain 3 appears preferable since both θ_1 and θ_3 are updated instead of merely updating θ_3. Following the earlier reasoning, chain 3 is preferable to chain 2.

The heuristic can be formalized by taking a fixed initial distribution for the state of the Markov chain and measuring the distance between the distribution on the state after a single generation and the limiting distribution of the chain. A chain appears preferable to another if it results in a smaller total variation distance after a single generation. Additional work yields theorems about which chain, as a complete collection of generations, is preferable, although the conditions are non-trivial.

The three chains described above are a standard Gibbs sampler, a Gibbs sampler that uses blocking, and a Gibbs sampler that collapses the state space. In practice, the implementation of chains 2 and 3 would be simplified: The redundant generation of $(\theta_1, \theta_2)|\theta_3$ would be eliminated from both chains. Additionally, θ_1 wouuld be entirely omitted from chain 3, leaving only the two generations $\theta_2|\theta_3$ and $\theta_3|\theta_2$. See Liu (1994) and Liu, Wong and Kong (1994) for theory on both collapse of the state space and blocking, and MacEachern (1994) for theory on the collapse of the state space. Subsections 3.1 and 3.2 illustrate the use of collapsing and blocking with the MDP model.

2.3.1 Collapsing

The collapse of the state space for a Gibbs sampler, illustrated with the chains above, exploits the idea that, since one is only simulating to avoid a difficult integration, one should perform as much integration as possible before simulating. As in the small example's chain 3, one can view any Gibbs sampler for which a parameter is completely marginalized (integrated out) as a heuristically improved Gibbs sampler where the marginalized parameter is generated, in a block, with each other parameter. This subsection illustrates how the collapse applies, in two different fashions, to the MDP model. A formula which is helpful in the following development is $P(\theta) = M^k \prod_{j=1}^{k}[(n_j - 1)! g_{0,\nu}(\theta_j^*)]/[\prod_{i=1}^{n}(M + i - 1)]$. The formula implicitly connects values of θ on the left side of the equation with values of θ^* on the right side of the equation. It may be derived from the Polya urn scheme representation of the Dirichlet process. When using the formula, one must, at times, be careful to work with the appropriate dominating measure.

Preintegration, or how to marginalize the mass parameter, M

The first part of the collapse is illustrated with a pre-integration over the distribution of the mass parameter, M. I first show how the integration works in a model with no data, and then apply the result to the model with data. Considering generation of $(M, \theta_i)|(\nu, s_{-i}, \theta_{-i})$, the Polya urn scheme leads to,

$$h_{M|(\nu,s_{-i},\theta_{-i})}(M) = \frac{h_M(M)M^{k^-} / \prod_{j=1}^{n-1}(M+j-1)}{\int h_M(M)M^{k^-} / \prod_{j=1}^{n-1}(M+j-1)dM}$$

and for $\theta_i|(M, \nu, s_{-i}, \theta_{-i})$ the mixture described in the introduction. Taking the product of these two distributions, one has

$$P(s_i = j|\nu, s_{-i}, \theta_{-i}) = n_j^- \int h_{M|(\nu,s_{-i},\theta_{-i})}(M) \frac{1}{M+n-1} dM \qquad \textbf{(1A)}$$

for $j = 1, \ldots, k^-$, and

$$P(s_i = k^- + 1|\nu, s_{-i}, \theta_{-i}) = \int h_{M|(\nu,s_{-i},\theta_{-i})}(M) \frac{M}{M+n-1} dM. \qquad \textbf{(1B)}$$

As before, if $s_i = k^- + 1$, the value of θ_i is drawn from $G_{0,\nu}$.

Evaluation of **1A** and **1B** can can be done quickly, to any degree of accuracy desired, before the simulation is begun, justifying the terminology pre-integration. When working with data, simply substitute **1A** for n_j^- and **1B** for M in step 2 of the basic algorithm.

How to marginalize the θ_i

The second technique relies on a conjugate structure for a portion of the parametric model. If the form $\int f_{\theta,z_i} g_{0,\nu}(\theta)d\theta$ is easily integrable, as is often the case when one chooses the basic algorithm, the mixing of the Markov chain can be improved with another collapse. This collapse, however, is dynamic, proceeding as the algorithm moves along. To accomplish this collapse, one integrates over the cluster locations. To do this, replace the marginal probabilities, $f_{\theta_j^*,z_i}(x_i)$, in step 2 of the basic algorithm with **2A**

$$\int f_{\theta_j^*,z_i}(x_i) g_{0,\nu,s_{-i},x}(\theta_j^*) d\theta_j^*, \qquad \textbf{(2A)}$$

where

$$g_{0,\nu,s_{-i},x}(\theta_j^*) = \frac{g_{0,\nu}(\theta_j^*) \prod_{l \neq i | s_l = j} f_{\theta_j^*,z_l}(x_l)}{\int g_{0,\nu}(\theta_j^*) \prod_{l \neq i | s_l = j} f_{\theta_j^*,z_l}(x_l) d\theta_j^*}$$

is the posterior density of θ_j^*, calculated as if it had prior density $g_{0,\nu}$ and updated with the data attached to cluster j. Make this replacement for calculation of $q_1, \ldots, q_{k^-}$. In the evaluation of q_0, **2A** matches the term $\int f_{\tilde{\theta},z_i}(x_i) g_{0,\nu}(\tilde{\theta}) d\tilde{\theta}$, and so no modification is needed. With conjugate forms, the integral **2A** reduces to calculation of a normalizing constant for a conjugate pair.

This collapse can be used in multivariate as well as univariate settings. The formulas for the collapse remain the same, with integrals taken over more than one dimension. A particularly common multivariate setting is the multivariate normal, where results on quadratic forms produce the normalizing constant. The normalizing constants can be programmed directly, with little effort.

Occasionally, even more integration can be performed, and the hyperparameter ν can be marginalized. Such an integration requires a great deal of conjugacy. An example is the simple hierarchical MDP model with normal likelihood, normal base measure, and normal distribution for ν–and with known variances. A second example is the conjugate normal hierarchical model where variances are unknown but the variances at different stages are tied together.

The baseball data

As a light-hearted example, I analyze the baseball batting average data of Efron and Morris (1975). The data set consists of the number of hits for players in the major leagues having exactly 45 at bats, as recorded in the April 29, 1970 edition of the *New York Times*. The goal of Efron and Morris' analysis is to predict the batting average, for each player, for the remainder of the season. They employ an empirical Bayesian analysis to produce predictions that are found to be much superior to predictions based on maximum likelihood. Here, an analysis based on the MDP model is used to produce predictions.

$$
\begin{aligned}
X_i|(\theta_i, z_i) &\sim Binomial(z_i, \theta_i) \\
\theta_i|G &\sim G \\
G|(M, \nu) &\sim Dir(MG_{0,\nu}), \text{ with } G_{0,\nu} \text{ the } beta(\alpha, \beta) \text{ distribution} \\
\alpha/(\alpha+\beta) &\sim beta(\nu_1, \nu_2) \\
\alpha+\beta &= 216.6 \\
M &\sim lognormal(2.81, 1.186^2)
\end{aligned}
$$

The model follows the simple hierarchical MDP model with F_{θ_i,z_i} the binomial(z_i, θ_i) distribution. In this instance, $z_i = 45$ for all i. The base measure, G_0, has a beta(α, β) shape. The parameters α and β will have a prior distribution chosen after a peek at the data. Given α and β, the prior mean for θ_i is $\alpha/(\alpha+\beta)$. The prior mean for X_i/z_i, integrating over

θ_i, is also $\alpha/(\alpha+\beta)$. Hence, I set the prior mean for $\alpha/(\alpha+\beta)$ equal to the overall batting average of the 18 players: $\sum_{i=1}^{18} X_i / \sum_{i=1}^{18} z_i = .265$. For convenience, I chose a beta (ν_1, ν_2) prior distribution for $\alpha/(\alpha+\beta)$. To find the two parameters of this prior distribution, I set $\nu_1/(\nu_1+\nu_2) = .265$, and selected $\nu_1+\nu_2$ to produce a sample standard deviation that matches the value $.265 * .735/810$ (where $810 = \sum_{i=1}^{18} z_i$). This results in $\nu_1+\nu_2 = 811$. Focusing on the marginal distribution for θ_i, I set the standard deviation to .030, creating a mean ± 2 standard deviation interval of width .120. To obtain such an interval, with $\alpha/(\alpha+\beta)$ fixed at .265, requires $\alpha+\beta = c = 216.6$. As a consequence of the prior distribution on the hyperparameter, the actual marginal prior distribution of θ_i will show some additional variability. The prior distribution on M was taken to be lognormal (μ, σ^2). The values of μ and σ were chosen to produce a mean of approximately 12 clusters, with a .1 chance of 6 or fewer clusters. The specific values chosen were $\mu = 2.81$ and $\sigma = 1.186$.

With the prior distribution specified, the analysis proceeds through simulation. Conditional generations for the Gibbs sampler are

- Generate $s_i|(s_{-i}, \alpha, X)$ from the distribution with

  $$q_j \propto \tilde{n}_j^- \frac{B(\alpha_j + X_i - 1, \beta_j + z_i - X_i - 1)}{B(\alpha_j, \beta_j)}, \; j = 1, \dots, k^-,$$

  and with

  $$q_0 \propto \tilde{M} \frac{B(\alpha + X_i - 1, \beta + z_i - X_i - 1)}{B(\alpha, \beta)},$$

  where $\alpha_j = \alpha + \sum_{l \neq i | s_l = j} X_l$ and $\beta_j = \beta + \sum_{l \neq i | s_l = j} (z_l - X_l)$ for $j = 1, \dots, k^-$, and where $\tilde{n}_j^-$ and $\tilde{M}$ are derived from the preintegration in **1A** and **1B**. Note that neither M nor θ_j^* appears in the formula for q_j. This generation relies on both collapses described earlier in this section.

  Perform this generation for $i = 1, \dots, n$, updating the vector s immediately upon generation of each s_i.

- Generate $\theta_j^*|(s, \alpha, X)$ from the beta $(\alpha + \sum_{l|s_l=j} X_l, \beta + \sum_{l|s_l=j} (z_l - X_l))$ distribution, for $j = 1, \dots, k^-$.

- Generate $\alpha|(\theta, s)$ from its posterior distribution. The density is proportional to $B(\alpha, c - \alpha)^{-k} \theta_i^{*\alpha-1} (1-\theta_i^*)^{c-\alpha-1} \alpha^{\nu_1-1} (c-\alpha)^{\nu_2-1}$.

The Gibbs sampler described above was initialized with $s = (1, 2, \dots, 18)$, $\theta_i^* = X_i/n_i$, and $\alpha = .265c$. The sampler was run for 10,000 iterations, with the first 1,000 iterations discarded as burn-in. Table 1 contains a summary of the data and estimates, rounded to the traditional three decimal places. Conservative values for the standard errors of the estimates, calculated with the batch means method (batches of size 500) range from .00023 to

.00048. Consequently, the estimates are expected to be within .001 of the actual estimate under the model. Note that the strong prior used for the MDP analysis leads to considerably more shrinkage than does the empirical Bayes analysis. For the final criterion of predicting the batting average for the rest of the season, the MDP analysis results in a sum of squared error of prediction of 0.021388 while the empirical Bayes analysis results in a sum of squared error of 0.021611. The difference, on the order of one percent of the sum of squared error, is small, but favors the MDP analysis.

TABLE 1. The baseball data, the Stein estimate and an MDP estimate.

Player	X/z	Rest of Season	Stein	MDP
1	.400	.346	.290	.286
2	.378	.298	.286	.283
3	.356	.276	.281	.279
4	.333	.222	.277	.276
5	.311	.273	.273	.273
6	.311	.270	.273	.273
7	.289	.263	.268	.269
8	.267	.210	.264	.266
9	.244	.269	.259	.262
10	.244	.230	.259	.262
11	.222	.264	.254	.259
12	.222	.256	.254	.259
13	.222	.303	.254	.259
14	.222	.264	.254	.259
15	.222	.226	.254	.259
16	.200	.285	.249	.255
17	.178	.316	.244	.252
18	.156	.200	.239	.248

2.3.2 Blocking

The use of blocking in MDP models is more specialized than the use of the collapse. Its main application is with models that involve censored data, and it is this type of model that is used to illustrate the computational technique. To set up the technique, consider the case where all observations are right censored. Let $z_1 \geq \ldots \geq z_n$ denote the censoring times, ordered from greatest to least. In this instance the ith survival time, say θ_i, is known to exceed z_i, but its exact value is not known. Take the model to be $G \sim Dir(MG_0)$, with $\theta_i|G \sim G$ for $i = 1, \ldots, n$. The data part of the model, since all observations are right censored, is just a collection of indicators that guarantee $\theta_i > z_i$ for each i. The following algorithm generates $(\theta_1, \ldots, \theta_n)$ directly from its posterior distribution. See MacEachern

(1992) for a brief description.

- Let $G_{0;z}$ denote the conditional distribution function defined as $G_{0;z}(t) = 1 - G_0(t)/G_0(z)$ for $t \geq z$, and defined as $G_{0;z}(t) = 0$ for $t < z$. Generate the θ_i sequentially. For $i = 1, \ldots, n$, draw $\theta_i \sim G_{0;z_i}$ with probability proportional to $M(1 - G_0(z))$, and set $\theta_i = \theta_j, j < i$ with probability proportional to 1.

The algorithm can be extended to include uncensored observations as well. In this case, exact survival times are first absorbed into the base measure to create an effective base measure. If, for example, there are an additional m observations with exact survival times $\theta_{n+1}, \ldots, \theta_{n+m}$, the effective base measure is $MG_0 + \sum_{i=n+1}^{n+m} \delta(\theta_i)$, where $\delta(\theta_i)$ is a point mass of size one at the point θ_i. Generation from the posterior is accomplished using the algorithm described above, with the effective base measure substituted for the base measure.

The accuracy of the algorithm can be verified by writing the joint distribution of $(\theta_1, \ldots, \theta_n)$ as the mixture distribution induced by the Polya urn scheme, multiplied by the indicators due to the censoring times. An alternate, more intuitive derivation relies on a redistribution of mass argument.

As with the collapse of the state space, the fragment of an algorithm described above is worked into a larger problem. The problem described below, the bioassay problem, is one of the earliest proposed uses for the hierarchical MDP model, and one that I studied intensively while writing my dissertation (MacEachern, 1988). The problem may be viewed as one involving doubly censored data, where each observation is either left censored or right censored, or it may be considered the simplest example of a logistic regression problem, where there is a single covariate.

The bioassay problem

The goal of the bioassay problem is to learn about the dose-response relationship in a population. As an example, a drug to reduce hypertension could be administered in various doses. In order to determine an effective dose, one would need to collect data on the relationship between reduction in hypertension and dose of the drug. Each individual in a study would be assigned some dose of the drug. A sufficient reduction in blood pressure for an individual would be considered a success, and that person would be said to have responded to treatment. Individuals without a sufficient reduction in blood pressure would fail to respond.

The impact of the drug on members of the population is summarized by a cumulative distribution function, say $G(z)$. $G(z)$ gives the proportion of the population that would respond to dose z. This distribution, often called the dose-response curve, is used to find doses to which varying fractions of the population respond. At least half the population will respond to the median of G. The dose producing this response fraction is variously called the effective dose 50 (the ED50) or the lethal dose 50 (the LD50), depending on the nature of the response.

To fit the bioassay problem into the formal MDP model, one more quantity is needed: The smallest dose which would induce an individual to respond. This dose is called the tolerance of the individual. The tolerance is a latent variable and is not directly observed for any individual. However, some information is collected about the tolerance, in that after being assigned a dose, an individual will either be known to have a tolerance that does not exceed the dose (i.e., the individual responds, and the tolerance is known to be left censored) or a tolerance that exceeds the dose (i.e., the individual does not respond, and the tolerance is known to be right censored). Thus, the data are binary responses. The tolerances are introduced as latent variables, as an aid to simulation.

In the example to come, the unknown distribution function is modelled with an MDP model that is patterned after the logistic regression model. Formally, z_i represents the dose assigned to individual i. The individual's tolerance is θ_i. A right censored observation is denoted by $X_i = 0$ and a left censored observation is denoted by $X_i = 1$. The shape of the base measure is the logistic distribution, with $G_{0;a,b}(x) = (1+exp(-a(x-b)))^{-1}$, where b is the location parameter and $a > 0$ is the reciprocal of a scale parameter. The above discussion, coupled with a specific choice for the prior distribution for the coefficients of the logistic distribution, results in

$$\begin{aligned} X_i|(\theta_i, z_i) &= \begin{cases} 1 & \text{if } \theta_i \leq z_i \\ 0 & \text{if } \theta_i > z_i \end{cases} \\ \theta_i|G &\sim G \\ G|(M, a, b) &\sim Dir(MG_{0,a,b}) \\ b &\sim N(\mu_0, \tau^2) \\ a &\sim H(a) \end{aligned}$$

where $i = 1, \ldots, n$. $H(a)$ is a distribution chosen to introduce a uniform(a_0, b_0) distribution on the standard deviation of a variate with distribution G_0. As a consequence, a has a density proportional to a^{-2} over its range. To make the algorithm easier to describe, the data are reordered so that the first m observations have $X_i = 0$ and the last $n-m$ observations have $X_i = 1$. As a further convention, the first m observations have doses listed in decreasing order, $z_1 \geq \ldots \geq z_m$, and the last $n - m$ observations have doses listed in increasing order, $z_{m+1} \leq \ldots \leq z_n$. Since the θ_i are exchangeable, the only impact of this reordering is a relabelling of the cases. The essence of the posterior is not affected.

The Gibbs sampler uses blocking in two stages:

- Generate $(\theta_1, \ldots, \theta_m)|(\theta_{m+1}, \ldots, \theta_n, a, b, X)$. Absorb the left censored observations into the base measure, so that an effective base measure is defined for this step. The effective base measure is $\alpha + \sum_{i=m+1}^{n} \delta(\theta_i)$

and hence the effective M is $M + n - m$. The effective G_0 is proportional to the effective base measure. With this set-up, the algorithm fragment described above is used to generate the block of parameters $(\theta_1, \ldots, \theta_m)$.

- Generate $(\theta_{m+1}, \ldots, \theta_n)|(\theta_1, \ldots, \theta_m, a, b, X)$. Absorb the right censored observations into the base measure, so that an effective base measure is defined for this step. The effective base measure is $\alpha + \sum_{i=1}^{m} \delta(\theta_i)$ and hence the effective M is $M + m$. The effective G_0 is proportional to the effective base measure. With this set-up, the algorithm fragment described above is used, with the dose scale reversed, to generate the block of parameters $(\theta_{m+1}, \ldots, \theta_n)$.

- Generate (a, b) with a Metropolis step. To do so, a proposal value for the pair (a, b) is generated, and then either accepted or declined. The actual proposals used in the simulation below are a uniform proposal for a over its entire range and a standard normal proposal for b, centered at the current value of b. The proposal is accepted with the usual Metropolis acceptance probability.

- Generate $\theta^*|(a, b, s, X)$. The conditional distribution of θ^* is a truncated logistic distribution. For θ_i^*, examine cluster i, the set of observations for which $s_j = i$. Find the largest right censored observation in the cluster, call it r, and the smallest left censored observation in the cluster, call it l. If there are no right censored observations in the cluster, set $r = -\infty$. If there are no left censored observations in the cluster, set $l = \infty$. Since the value of θ_i^* must lie in the interval $(r, l]$, the conditional distribution of θ_i^* has $G_{\theta_i^*;a,b,s,z,X}(\theta^*) = 0$ for $\theta^* < r$ and $G_{\theta_i^*;a,b,s,z,X}(\theta^*) = 1$ for $\theta^* \geq l$. In between, the distribution function interpolates as a logistic, with $G_{\theta_i^*;a,b,s,z,X}(\theta^*) = (G_{0;a,b}(\theta^*) - G_{0;a,b}(r))/(G_{0;a,b}(l) - G_{0;a,b}(r))$

The first two generations described above make use of the blocking described in the previous subsection. For the right censored observations, the left censored observations are absorbed into the base measure of the Dirichlet process, and then blocking is used, exactly as described above. For the left censored observations, the dose axis is reversed, the right censored observations are absorbed into the base measure, and the algorithm as described above is applied. If the base-measure had a discrete component, and a left censoring time exactly matched an atom of mass from the discrete component, a slight modification to the algorithm would be needed, since the left censored observation could have a tolerance exactly matching its censoring time. The modification merely requires a slight change in the definition of the conditional distribution in the earlier fragment of an algorithm. The Metropolis step for generation of (a, b) can be replaced with a pair of Gibbs steps, generating a and then b from their full conditional distributions. These conditional distributions do not fall in any of the usual families of distributions.

The Ryanodine data

Dixon (1965) presents the results of a study which investigated the effect of Ryanodine on male mice. Ryanodine was administered intravenously, with dose adjusted for weight, in saline, under 12 experimental conditions.

The experimental conditions arise from a two-way table, with the margins of the table based on body weight and a time cut-off. The mice were grouped into three weight classes, so that the appropriateness of the dosage (defined as dose adjusted for body weight) to body weight scale could be investigated. The response of direct interest was time of last visible movement. Since it can be difficult to determine time of last visible movement, the data were collected on a coarser scale. Four time cut-offs, 64, 96, 144 and 216 seconds, were selected. For a particular time point, it was determined whether there was visible movement after that time point or not.

For each experimental condition, an up-and-down sequential design was conducted. Thus, data were collected on several mice, with the mice assigned varying doses and a binary response/non-response recorded for each mouse. The data for one experimental condition appear in Table 2. The sample sizes for the 12 experiments range from 5 to 8. The reader is referred to Dixon (1965) for the complete data set.

TABLE 2. The Ryanodine data for the weight class 18 - 20 and the time cut-off of 64 seconds. The data are presented in the order of experimentation.

z_i:	0	1	2	3	2	1	2
X_i:	0	0	0	1	1	0	1

The data are analyzed on the dosage scale, as twelve separate data sets, with a prior distribution for each chosen to reflect the basic MDP modelling strategy. The prior distribution relies on the information used to implement the up-and-down method. With the up-and-down method, the investigator attempts to guess the median of the tolerance distribution (also called the $LD50$). The first mouse is assigned this guess as his dosage. As the experiment progresses, steps of a constant size are taken, either up or down, depending on the result for the most recent mouse. For this experiment, the step size was targeted as the standard deviation of the distribution of tolerances. We take G_0 to be a logistic distribution with parameters a and b. The prior distribution on b is normal. The 12 prior means for the individual b's were chosen by fitting the two-way additive model to the doses assigned to the first mouse in each weight class by time cut-off pair. The two-way model was fit to smooth out the prior means. The standard deviation of the prior distribution is taken to be 1.61, a value which is consistent with Dixon's assessment of the investigator's prior beliefs. The prior distribution on a was chosen to produce a standard deviation of the tolerances near the dosage step size of 1. It induces a uniform$[0.1, 1.9]$ distribution on the standard deviation. a and b are a priori independent.

To complete the prior distribution, a value (or distribution) is needed for the mass of the base measure. This value was set to $M = 10$, a value that is reasonably large when compared to the sample sizes. Choice of an M that is either large in an absolute sense or large relative to the sample size is necessary when working with bioassay data in order to avoid the excessive clustering first described by Antoniak (1974).

A simulation was performed for each of the 12 data sets. For each data set, the algorithm described above was run, with a burn-in period of 1000 iterates and an estimation period of 5000 iterates. The results of the simulation were used to estimate the predictive distribution of a future tolerance, say $\tilde{\theta}$, via $\hat{G}(\tilde{\theta}) = R^{-1}\sum_{i=1}^{R}\hat{G}_i(\tilde{\theta})$ with the generated values at the end of iterate i yielding $\hat{G}_i(\tilde{\theta}) = (M+n)^{-1}\sum_{j=1}^{n} I(\theta_{s_j} \leq x) + MG_{0;a,b}(\tilde{\theta})/(M+n)$. This estimate, computed on a grid of values for $\tilde{\theta}$, was used to produce an estimate of the $LD50$. Table 3 briefly summarizes the results of the model fitting procedure. The fit also produces estimates of the tolerance distribution for each weight-time combination. Comparison of these estimates facilitates a visual assessment of the similarities and differences between weight groups and across the time cut-off points.

TABLE 3. Estimates of the LD50 for each of the 12 Ryanodine data sets. The top number in each cell is the prior predictive LD50; the middle number is the estimate from the MDP simulation described here; the bottom number is the estimate from Dixon's paper.

Wt. Time	64	96	144	216
18-20	-0.55	-3.66	-6.76	-8.10
	1.71	-0.48	-3.29	-6.74
	1.86	-0.25	-2.84	-6.35
21-23	0.95	-2.16	-5.26	-6.60
	3.78	-1.51	-5.38	-6.87
	4.18	-0.55	-5.06	-6.95
24-26	0.95	-2.16	-5.26	-6.60
	1.66	-2.07	-4.87	-6.26
	1.86	2.25	1.52	6.13

Estimates of the LD50 from the posterior are fairly close to Dixon's estimates. The great advantages of the MDP analysis are that it allows one to investigate many features of the posterior, and that it allows one to phrase decision problems that are appropriate for a nonparametric analysis. Estimates of other LD's are calculated in a fashion similar to that for the LD50. Once an estimate of an LD is found, traditional analyses, motivated by parametric models, attach a standard error to the estimate. The full posterior distribution described by the MDP model allows one to investigate the more relevant question of just what the impact of a proposed dose is. Such questions are about the distribution of $G(z)$ for a particular z rather than about how close z is to some particular LD.

2.3.3 Estimation

Estimation on the basis of MCMC output can be greatly improved by a variety of techniques. One of the most straightforward improvements, aimed at increasing the accuracy of estimates of marginal features of the posterior distribution, is the Rao-Blackwellization suggested by Gelfand and Smith (1990). With this technique, values generated as part of the simulation are replaced by conditional expectations, most often immediately before a parameter is generated. For example, in the easy Gibbs sampler described at the beginning of Section 3, the tabulation based estimate of $E[h(\theta_1)]$, $R^{-1}\sum_{i=1}^{R} h(\theta_1^{(i)})$, can be replaced with $R^{-1}\sum_{i=1}^{R} E[h(\theta_1^{(i)})|\theta_2^{(i-1)}, \theta_3^{(i-1)}]$. Such replacements are, of course, not limited to estimation within the Gibbs sampler or to use of conditional distributions used in the Markov chain. Were one to post-process the output of the simulation, one would find other estimators, including one which is a direct parallel to the usual Rao-Blackwellization, namely $R^{-1}\sum_{i=1}^{R} E[h(\theta_1^{(i)})|\theta_2^{(i)}, \theta_3^{(i-1)}]$.

The technique of Rao-Blackwellization can be used for the MDP model in a variety of ways. One natural use, developed by Bush and MacEachern (1996), is to create estimates at the stage where the θ_i^* are generated. To estimate $E[h(\theta_1)]$, one might use $R^{-1}\sum_{i=1}^{R} E[h(\theta_{s_1}^{(i)})|s^{(i)}, \nu^{(i)}, X]$, which often has a closed form expression. In the baseball example, the distribution of $\theta_{s_i}^*|(s, \nu, x)$ is a beta distribution, and so with polynomial h, the expectations are easily evaluated.

The bioassay model admits a wide range of Rao-Blackwellizations. To estimate the predictive density of the tolerance of an individual not included in the study, a particularly nice Rao-Blackwellization comes from Kuo's (1983) work. The distribution of $\theta_i^*|(s, \nu, X)$, used as part of the Gibbs sampler, is described in the last step of the algorithm of Section 3.2. It is a truncated logistic distribution. To estimate the predictive distribution of a future tolerance, say $\tilde{\theta}$, compute $\hat{G}(\tilde{\theta}) = R^{-1}\sum_{i=1}^{R} \hat{G}_i(\tilde{\theta})$ with $\hat{G}_i(\tilde{\theta}) = (M+n)^{-1}[\sum_{j=1}^{k^{(i)}} n_j^{(i)} G_{\theta_j^*;s^{(i)},\nu^{(i)},X}(\tilde{\theta}) + MG_{0;a,b}(\tilde{\theta})]$. The predictive density is the derivative of $\hat{G}$ which exists for all $\tilde{\theta}$ other than assigned doses. Since the expression for $\hat{G}$ is linear in the individual $G_{\theta_j^*}$ and G_0, the density can be computed directly, avoiding the need to differentiate $\hat{G}$.

This Rao-Blackwellization for the bioassay problem produces an estimate with a nice property. It matches two known features of the estimand. First, although the model assigns probability one to the event that the tolerance distribution is discrete, the predictive distribution of tolerances is known to be continuous. The estimator above, an average of continuous distributions, is continuous. There has been no need to artificially smooth the estimates in order to produce this continuity. Second, the estimand may be represented as a mixture over cluster structures and values of a and b. The pieces of this mixture are of the form $G_{\theta;s,\nu,X}$. The estimate has this same structure. The mixture estimate also allows one to pick up and display more subtle features of the estimand such as the discontinuity of the predictive density at the assigned dosages.

The variety of potential Rao-Blackwellizations in MDP models is enormous. My personal preference is to identify what seem to be key qualitative features of the posterior and to choose a Rao-Blackwellization that has the same features. There will typically be many such Rao-Blackwellizations. The final decision of which Rao-Blackwellization to choose involves a trade-off between the additional effort required to program the estimator, the time needed to compute the estimator, and the perceived benefits of use of the estimator.

2.4 Non-Conjugate Models

In many instances, the base measure, conditioned on the hyperparameters, will not have a conjugate form with the likelihood. In this instance, the integral to obtain q_0 may be difficult to compute, or even to approximate. In such a situation, one must turn to another method to perform the simulation. MacEachern and Müller (1998) describe an algorithm which can be routinely implemented. This algorithm relies on the creation of a model that underlies the Polya urn scheme. The model remains hidden throughout the simulation, and entails only small changes to the basic algorithm. It nevertheless enables one to avoid the integral needed to find q_0. The cost of avoiding the integration is an algorithm which appears to suffer to some extent from poorer convergence and mixing properties.

There is one situation, however, in which the basic algorithm can be rescued with a slight alteration. When the density $g_{0,\nu}(\theta_k^*) f_{\theta_k^*,z_i}(X_i)$ appearing in term **2** of step 2 of the basic algorithm is log-concave, one can make use of a variant on Gilks and Wild's (1992) adaptive rejection method to generate $\theta_i|(\nu, M, \theta_{-i}, X)$. The variant arises from the fact that the rejection envelope which those authors adaptively build can also be built for a piecewise log-concave (or log-convex!) density. The density at each of the cluster locations, θ_i^*, is trivially log-concave. So, to demonstrate piecewise log-concavity, we need only verify that the aforementioned density is log-concave. Evans and Swartz (1998) develop a more general version of this technique that allows for transformations more general than the log transform. Further work on generation from non-conjugate models can be found in Walker and Damien (1998).

2.5 Discussion

The development of the MDP model, as the development of the Dirichlet process itself, reflects the tension between devising a model that captures important features of the data and finding a model that can actually be fit to the data. Escobar's (1994) development of the first MCMC techniques for the Dirichlet process provided major impetus to the field, allowing the MDP modelling strategy to be implemented for an array of problems that had previously been beyond then current computational procedures.

Computational developments have included a focus on the cluster structure of θ, relying either on an integration (MacEachern, 1994) or an additional stage appended to the Gibbs sampler (Bush and MacEachern, 1996) to improve the mixing of the Markov chain used to fit the model. The material in Section 3 furthers these developments. It should also be mentioned that a number of alternative computational methods have also been developed for the MDP model. Notable among them are an MCMC algorithm developed by Doss (1994) which takes a different tack, essentially generating the random G rather than marginalizing it, and the sequential imputation method whose application to the MDP problem is due to Liu (1994). Some of the same modifications developed for the MCMC methods also pay dividends for these alternative strategies (MacEachern, Clyde and Liu, 1998; Doss and Huffer, 1998). Finally, techniques developed for finite mixture models, such as splitting and combining clusters (see Richardson and Green, 1997, and references therein), can also be used to improve MCMC methods for MDP models.

The models have also been refined, moving beyond a model that essentially describes one-sample problems but which can be used for problems that appear to require a more complex model (see, for a sterling example, Müller, West and Erkanli, 1996). The more refined models develop particular roles for the parametric and nonparametric portions of the model. Two directions are now well established: One direction, exemplified by the early paper of Newton, Czado and Chappell (1996) in the context of logistic regression, is to model the impact of covariates in a parametric fashion, but to replace the logistic distribution with a nonparametric model (in their case a variant on the Dirichlet process). The second direction is to draw a distinction between fixed effects–those effects for which we can model the individual effect–and random effects–those effects for which we most naturally model the distribution from which the effects arise. This distinction, first drawn for the MDP model in Bush and MacEachern's (1996) analysis of the randomized block design, and further developed in Bush (1994), suggests use of a parametric model for the fixed effects and a nonparametric model for the random effects. Further modelling developments are currently an active area of research. This author is pursuing research that will more sharply define the dividing line between the parametric and nonparametric portions of the model, and also the development of models that automatically display more robustness than do the current MDP models.

References

Antoniak, C.E. (1974). Mixtures of Dirichlet processes with applications to Bayesian nonparametric problems. *The Annals of Statistics,* 2, 1152-1174.

Blackwell, D. and MacQueen, J.B. (1973). Ferguson distributions via Polya urn schemes. *The Annals of Statistics,* 1, 353-355.

Bush, C.A. (1994). Semi-parametric Bayesian Linear Models. Unpublished Ph.D. dissertation, Ohio State University.

Bush, C.A. and MacEachern, S.N. (1996). A semi-parametric Bayesian model for randomized block designs. *Biometrika,* 83, 275-285.

Dixon, W.J. (1965). The up-and-down method for small samples *Journal of the American Statistical Association,* 60, 967-978.

Doss, H. (1994). Bayesian nonparametric estimation for incomplete data via successive substitution sampling. *The Annals of Statistics,* 22, 1763-1786.

Doss, H. and Huffer, F. (1998). Monte Carlo methods for Bayesian analysis of survival data using mixtures of Dirichlet priors. Technical Report, Department of Statistics, Ohio State University.

Efron, B. and Morris, C. (1975). Data analysis using Stein's estimator and its generalizations. *Journal of the American Statistical Association,* 70, 311-319.

Escobar, M.D. (1994). Estimating normal means with a Dirichlet process prior. *Journal of the American Statistical Association,* 89, 268-277.

Escobar, M.D. and West, M. (1995). Bayesian density estimation and inference using mixtures. *Journal of the American Statistical Association,* 90, 577-588.

Evans, M. and Swartz, T. (1998). Random variable generation using concavity properties of transformed densities. *Journal of Computational and Graphical Statistics,* to appear.

Ferguson, T.S. (1973). A Bayesian analysis of some nonparametric problems. *The Annals of Statistics,* 1, 209-230.

Gelfand, A.E. and Kuo, L. (1991). Nonparametric Bayesian bioassay including ordered polytomous response. *Biometrika,* 78, 657-666.

Gelfand, A.E. and Smith, A.F.M. (1990). Sampling-based approaches to calculating marginal densities. *Journal of the American Statistical Association,* 85, 398-409.

Gilks, W.R. and Wild, P. (1992). Adaptive rejection sampling for Gibbs sampling. *Jounal of Applied Statistics,* 41, 337-348.

Kuo, L. (1983). Bayesian bioassay design. *The Annals of Statistics,* 11, 886-895.

Kuo, L. (1986). Computations of mixtures of Dirichlet processes. *SIAM Journal of Scientific and Statistical Computing,* 7, 60-71.

Kuo, L. and Smith, A.F.M. (1992). "Bayesian computations in survival models via the Gibbs sampler" in *Survival Analysis: State of the Art* (eds: J.P. Klein and P.K. Goel), pp. 11-22.

Lindley, D.V. and Smith, A.F.M. (1972). Bayes estimates for the linear model (with discussion). *Journal of the Royal Statistical Society,* Ser. B, 34, 1-42.

Liu, J.S. (1994). The collapsed Gibbs sampler in Bayesian computations with application to a gene regulation problem. *Journal of the American Statistical Association,* 89, 958-966.

Liu, J.S., Wong, W.H. and Kong, A. (1994). Covariance structure of the Gibbs sampler with applications to the comparisons of estimators and augmentation schemes. *Biometrika,* 81, 27-40.

MacEachern, S.N. (1988). Sequential Bayesian bioassay design. Unpublished Ph.D. Dissertation. University of Minnesota.

MacEachern, S.N. (1992). Discussion of "Bayesian computations in survival models via the Gibbs sampler" by L. Kuo and A.F.M. Smith, in *Survival Analysis: State of the Art* (eds: J.P. Klein and P.K. Goel), pp. 22-23.

MacEachern, S.N. (1994). Estimating normal means with a conjugate style Dirichlet process prior. *Communications in Statistics: Simulation and Computation,* 23, 727-741.

MacEachern, S.N., Clyde, M. and Liu, J. (1998). Sequential importance sampling for nonparametric Bayesian models: The next generation. *Canadian Journal of Statistics,* to appear.

MacEachern, S.N. and Müller, P. (1998). Estimating mixtures of Dirichlet process models. *Journal of Computational and Graphical Statistics,* to appear.

Müller, P., Erkanli, A., and West, M. (1996). Bayesian curve fitting using multivariate normal mixtures. *Biometrika,* 83, 67-80.

Newton, M.A., Czado, C. and Chappell, R. (1996). Bayesian inference for semiparametric binary regression. *Journal of the American Statistical Association,* 91, 142-153.

Richardson, S. and Green, P.J. (1997). On Bayesian analysis of mixtures with an unknown number of components (with discussion). *Journal of the Royal Statistical Society,* Ser. B, 59, 731-792.

Walker, S. and Damien, P. (1998). "Sampling methods for Bayesian nonparametric inference involving stochastic processes," in *Practical Nonparametric and Semiparametric Bayesian Statistics* (eds: D. Dey, P. Müller, and D. Sinha), New York: Springer-Verlag. .

West, M., Müller, P. and Escobar, M.D. (1994). "Hierarchical priors and mixture models, with application in regression and density estimation," in *Aspects of Uncertainty: A tribute to D.V. Lindley* (eds: A.F.M. Smith and P. Freeman), pp. 363-368, New York: Wiley.

3

Nonparametric Bayes methods using predictive updating

Michael A. Newton
Fernando A. Quintana
Yunlei Zhang

ABSTRACT Approximate nonparametric Bayes estimates calculated under a Dirichlet process prior are readily obtained in a wide range of models using a simple recursive algorithm. This chapter develops the recursion using elementary facts about nonparametric predictive distributions, and applies it to an interval censoring problem and to a Markov chain mixture model. S-Plus code is provided.

3.1 Introduction

Sampling models that enforce relatively weak assumptions are naturally favored in many applications, but it is well known that the corresponding posterior computations can become very intensive when a Dirichlet process encodes prior uncertainty in the weakly specified part of the model. In all but the most simple models, posterior calculations involve a mixture of Dirichlet processes. As evidenced by companion chapters, advances in Markov chain Monte Carlo (MCMC) provide critical methodology for enabling these calculations, and have opened up a wide range of interesting applications to Dirichlet-process-based nonparametric Bayesian analysis.

Although MCMC provides the most effective computational solution, other algorithms can be advantageous in certain situations, and these deserve further consideration. In this chapter we discuss a simple recursive algorithm derived by approximating posterior predictive distributions. Specifically, we review the recent proposal of Newton and Zhang (1996). It yields approximate Bayes estimates through a simple and efficient algorithm, and may be particularly helpful when the standard MCMC algorithm runs on a very high dimensional space, or if a fast approximate solution is helpful prior to full MCMC implementation.

In the class of models under consideration, an unobserved random sample $\theta_1, \ldots, \theta_n$ from an unknown distribution G determines the conditional probability structure of observables $Y_1, \ldots, Y_n$. We think of θ_i as real or vector-valued; and observations can come in various types. The unobserved random sample may be missing data or it may represent parameters that fluctuate across experimental units. The interval censoring in Section 5 has θ_i as a survival time and Y_i an interval known to contain θ_i. The random effect θ_i encodes a person-specific transition matrix for a binary time series Y_i in the example in Section 6. Certainly the range of possibilities within this framework is broad.

Bayesian calculations are induced by placing a Dirichlet process prior on G. As reviewed in Escobar and West (1998), this prior is indexed by a positive finite measure defined on the range $\mathcal{X}$ of θ_i. Of course the prior measure, denoted m_0, contains the Bayesian's information about G, and may be factored into a probability measure G_0 representing a prior guess, and a mass $\alpha = m_0(\mathcal{X})$. Among other things, the prior mass α measures the extent to which new information will change the Bayesian's opinion.

Conditional on data $D = \{Y_i\}$, and for a measurable subset $B \subset \mathcal{X}$, the Bayes estimate $E\{G(B)|D\}$ coincides with the posterior predictive distribution $G_n(B) = P(\theta_{n+1} \in B|D)$ and it is this distribution that we are trying to calculate.

3.2 On $n = 1$

The most simple case motivates the general recursive approximation. The object of interest is the posterior predictive distribution

$$\begin{aligned} G_1(B) &= P(\theta_2 \in B|Y_1) \qquad (1)\\ &= E\left\{P(\theta_2 \in B|\theta_1)\,|Y_1\right\} \\ &= E\left\{\frac{\alpha}{1+\alpha}G_0(B) + \frac{1}{1+\alpha}1_B(\theta_1)\middle|\, Y_1\right\} \\ &= \frac{\alpha}{1+\alpha}G_0(B) + \frac{1}{1+\alpha}P(\theta_1 \in B|Y_1). \end{aligned}$$

Critical in the above development is the representation of $P(\theta_2 \in B|\theta_1)$ as a mixture of prior opinion G_0 with information from θ_1. A general argument supporting this claim comes from Polya sequence theory as developed in Blackwell and MacQueen (1973) and as discussed in Escobar and West (1998). A direct argument for the special case of $n = 1$ is to recall that the Dirichlet-process distributed G may be expressed $G(B) = \sum_k w_k 1_B(v_k)$ where $v_1, v_2, \ldots$ are independent and identically distributed from G_0 and where the w_k's arise from a simple stick-breaking exercise (Sethuraman, 1994). Specifically, $w_1 = b_1$, and for any $k \geq 2$, $w_k = b_k \prod_{j=1}^{k-1}(1-b_j)$ where $b_1, b_2, \ldots$ are independent and identically distributed Beta$(1, \alpha)$ random variates. If G_0 is non-atomic, then conditional on G, the probability of a tie, that is that $\theta_1 = \theta_2$, is $\sum_k w_k^2$. Averaging over G, the tie probability is readily calculated to be $1/(1+\alpha)$. On the other hand, if there is no tie, then θ_2 must be distributed as G_0, hence (1). This argument requires slight elaboration if G_0 has atoms, but (1) continues to hold.

Whereas the measure $m_0 = \alpha G_0$ represents uncertainty prior to observing data, the measure $m_1 = (\alpha + 1)G_1$ encodes updated uncertainties. We observe that Dirichlet-process based learning occurs by accumulating measure:

$$m_1(B) = m_0(B) + P_0(\theta_1 \in B|Y_1). \tag{2}$$

It is interesting to note that Bayes rule enters the second term in (2). The corresponding posterior calculations are driven by the prior G_0 on θ (hence the subscripting in P_0). If no information is lost going from θ_1 to Y_1, then the added measure is simply a point mass at θ_1, and we have the familiar Polya sequence rule; after sampling with replacement from an urn, add another bit of mass at the observed value. More generally, the rule then is to return an entire probability distribution to the urn. It is also interesting that the so-often criticized discreteness property of G coincides with the additive accumulation of measure.

3.3 A recursive algorithm

The idea that predictive uncertainty can be encoded in a single measure, and that learning occurs by adding measure lead us to the following recursion generalizing (2)

$$m_i(B) = m_{i-1}(B) + P_{i-1}(\theta_i \in B|Y_i) \tag{3}$$

for $i \geq 1$. In terms of probability distributions

$$G_i(B) = (1 - w_i)G_{i-1}(B) + w_i P_{i-1}(\theta_i \in B|Y_i) \tag{4}$$

where again Bayes rule enters the second term taking G_{i-1} as the updated prior distribution for θ_i. The nominal weights are $w_i = 1/(\alpha + i)$, though we consider some alternatives later.

As constructed, G_i is not the exact posterior predictive distribution for θ_{i+1} in general, even though it is so when $i = 1$. It will be only in the case of no information loss, that is when $Y_i = \theta_i$. Thus the value of G_i depends on the order in which $Y_1, \ldots, Y_i$ are processed. What gives some credence to the recursion is that the dependence on order can be relatively weak. The suggested algorithm then is to arrange in some order $Y_1, \ldots, Y_n$ and to process them through (4) to produce an approximate Bayes estimate G_n. Calculations being $O(n)$, we can easily re-evaluate G_n over a random sample of orderings and average the results.

That (4) is order-dependent and approximate for $n > 1$ may not be obvious though the calculations in the next sections bear this out. In Section 7, we dissect a particular example with $n = 2$ to study this phenomenon.

The proposed recursion has the form of a stochastic approximation algorithm (Kushner and Yin, 1997), one motivated by Dirichlet-process based learning and one formally residing in a function space.

An important special case of the proposed recursion is the quasi-Bayes sequential procedure discussed by Smith and Makov (1978) and elsewhere. The canonical quasi-Bayes problem concerns finite mixture models. In our notation, θ_i indicates a component population from which feature data Y_i are generated. The relevant sets B indicate the different component populations, and $G(B)$ is the mixing probability, i.e., the probability that an observation is from population B. Then, with a Dirichlet distribution prior for the mixing probabilities, a quasi-Bayes procedure arises by approximating the posterior distribution of these mixing probabilities in a certain way. The recursive approximation proposed by Newton and Zhang and reviewed here differs in several respects from the quasi-Bayes recursions, although they do coincide in the finite mixture case. Aside from a difference in scope, a general feature of the present recursive approximation is its emphasis on posterior predictive distributions rather than on posterior distributions over a parameter space. There are also some issues of implementation special to the recursion discussed here, as we see in Sections 5 and 6.

3.4 Interval Censoring

The form of information loss that we have studied most extensively is interval censoring. Rather than observing $\theta_i \in \mathcal{X}$ we observe an interval $Y_i \subset \mathcal{X}$ that is known to contain θ_i. Interval censoring arises frequently in statistical practice, and methods exist for obtaining nonparametric maximum likelihood estimates of G (e.g., Groeneboom and Wellner, 1992; Gentleman and Geyer, 1994). Less seems to have been done concerning nonparametric Bayesian approaches to this problem, although the MCMC methods in Doss (1994) certainly apply. As an illustrative example, we consider in the next section a much studied data set on cosmetic deterioration after radiotherapy of the breast. Each patient is monitored periodically, and a time to deterioration θ_i is known to occur either in an interval between hospital visits, or is right-censored and known to occur only beyond some maximum observed time. A second example concerns the weight at one year of age of calves on a large ranch in Brazil. At a time of round-up, calves ages are known and weights are obtained, but the ages at round-up vary around one year. Assuming that weight is nondecreasing during this period of growth, the weight at one year θ_i is known to exceed the measured weight if calf i is younger than one year, and is known to be smaller than the measured weight if calf i is older than one year. Data on this example are currently being compiled.

The recursive formula (4) is particularly simple for interval censored data:

$$G_i(B) = (1 - w_i)G_{i-1}(B) + w_i \frac{G_{i-1}(B \cap Y_i)}{G_{i-1}(Y_i)} \tag{5}$$

Newton and Zhang (1996) reported an interesting theoretical property of (5) in the restricted case that $\mathcal{X}$ is a finite set. (In his thesis, Zhang extended this result to the countable support case.) Suppose that $Y_1, Y_2, \ldots$ are independent and identically distributed random subsets of $\mathcal{X}$ (for example, a sequence of observed intervals), and that the prior guess G_0 has support $\mathcal{X}$. Then, with probability one, the sequence of approximate predictive distributions $G_1, G_2, \ldots$ converges weakly to a distribution G^*, say, which satisfies, for any $B \subset \mathcal{X}$,

$$G^*(B) = E\left\{\frac{G^*(B \cap Y)}{G^*(Y)}\right\}. \tag{6}$$

Randomness on the right comes through the generic random subset Y which is distributed as the other Y_i.

Note that (6) holds for any distribution of the random subsets and so further consideration of the problem is needed to connect this to samples from G. In what some authors call case I and case II interval censoring, random censoring times partition $\mathcal{X}$ into some number of subsets, and the observed Y_i is that subset which contains a random θ_i from G. To simplify the discussion, suppose that this censoring process partitions $\mathcal{X}$ into two random subsets $\mathcal{X} = A_i \cup A_i^c$, independently from θ_i. Then

$$Y_i = \begin{cases} A_i & \text{if } \theta_i \in A_i \\ A_i^c & \text{if } \theta_i \in A_i^c \end{cases} \tag{7}$$

and (6) becomes

$$\begin{aligned} G^*(B) &= E\left\{E\left[\frac{G^*(B \cap Y)}{G^*(Y)} \,\middle|\, A\right]\right\} \\ &= E\left\{E\left[\frac{G^*(B \cap A)}{G^*(A)} 1_A(\theta) + \frac{G^*(B \cap A^c)}{G^*(A^c)} 1_{A^c}(\theta) \,\middle|\, A\right]\right\} \\ &= E\left\{\frac{G^*(B \cap A)}{G^*(A)} G(A) + \frac{G^*(B \cap A^c)}{G^*(A^c)} G(A^c)\right\}. \end{aligned}$$

Certainly one solution to this system is $G^* = G$, owing to the independent censoring. Multiple solutions may exist depending on the censoring process, but if this is sufficiently rich, then G is the only solution, and hence the recursive approximations are consistent.

We present a simple numerical example to illustrate these points. Figure 1 shows the results of applying recursion (5) to random subsets Y_i of $\mathcal{X} = \{0, 1, \ldots, 49\}$ formed as follows. A random partition of $\mathcal{X}$ is formed by independent coin tossing on elements of $\mathcal{X}$, with heads going into a set A_i and tails into A_i^c. Then a random θ_i is sampled from a distribution G having masses $g(\theta) \propto \tan(c\theta)$, where $c = \pi/100$; the cumulative distribution function is indicated in dotted lines in Figure 1. Data $Y_1, \ldots, Y_n$ arise as in (7). We applied the recursive approximation (5) to these simulated data, obtaining approximate predictive distributions G_n shown in solid lines in Figure 1. The recursion was evaluated for sets $B = [0, \theta]$ with $\theta \in \mathcal{X}$. The reader may be surprised that so many samples are required for G_n to approach G, but remember that there is extreme information loss in this example. Also, cube-root rather than square-root asymptotics govern standard estimators with such information loss (Groeneboom and Wellner, 1992).

For the calculations reported in Figure 1 we did not use nominal weights $w_i = 1/(i+\alpha)$ suggested by (1). Inspection shows that these give relatively high weight to the first observations and quite low weight to later observations. It turns out that convergence holds as long as the positive weights satisfy $\sum_i w_i = \infty$ and $\lim_{i\to\infty} w_i = 0$. We use $w_i = .5i^{-1/3}$ in Figure 1.

3.5 Censoring Example

The theory outlined in the last section does not say how well the recursive approximations match the actual posterior predictive distributions of interest. We study in this section a small example concerning the effects of radiotherapy on cosmetic deterioration of the breast. This example has been considered by a number of authors. Our goal is not to provide further insight into this application, but rather to illustrate how the recursive approximations work in a typical problem. The data are reported in Table 1 of Finkelstein and Wolfe (1985), in Table 1 of Gentleman and Geyer (1994), and are available at the first author's web site through `http://www.stat.wisc.edu/`. There are $n = 46$ intervals in the data set. Associated with each woman in the study is a time θ_i (days) indicating the time after radiotherapy treatment that a defined amount of change occurs in the treated tissue, and the ith interval is known to contain θ_i.

Figure 2 shows the results of our recursive approximation (5) using the nominal weighting scheme. We take G_0 to be an exponential prior with mean equal to one half a year, and we consider two values of α; 1 and 5 (corresponding to the left and right sides in Figure 2). Panels (a) and (b) show G_{46} for 100 random orderings of the data. We use sets $B = [0, \theta]$ to drive the recursion, for θ in the grid $\{0, 1, 2, \ldots, 100, 1000\}$. The outlying grid point at 1000 days is included to account for mass beyond the range of data. (On a related point, the nonparametric MLE is typically a subdistribution function, with positive mass at an arbitrary point beyond the largest censoring time.) The lower panels (c) and (d) compare the pointwise

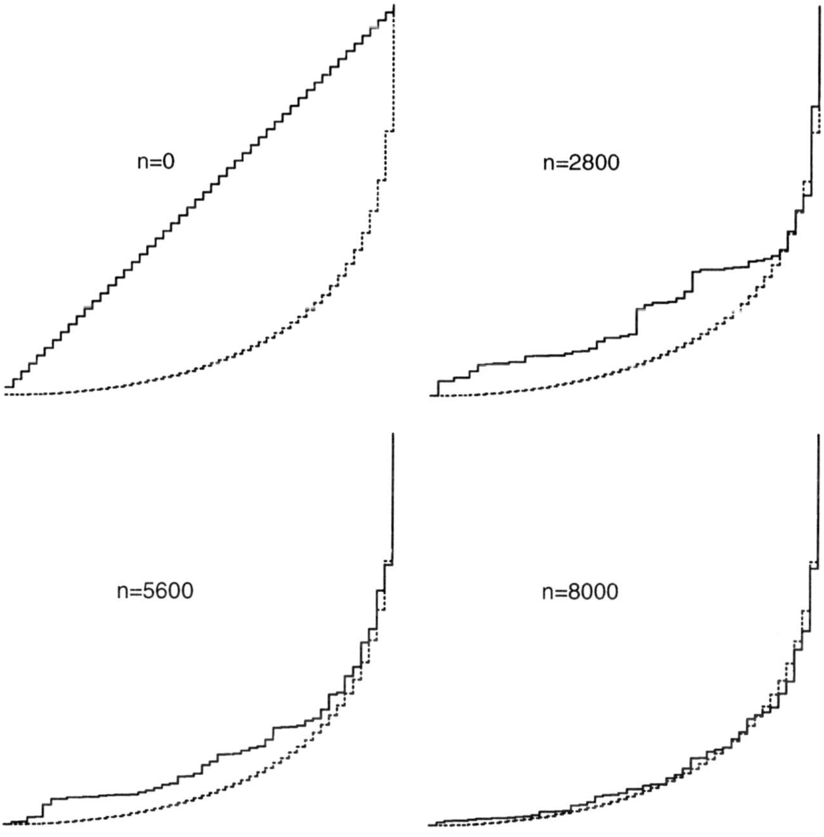

FIGURE 1. Extreme censoring: The true distribution function G is the dotted curve, and recursive approximations G_n are marked as solid curves after various sample sizes. Vertical axis is cumulative probability and horizontal axis is $\mathcal{X}$.

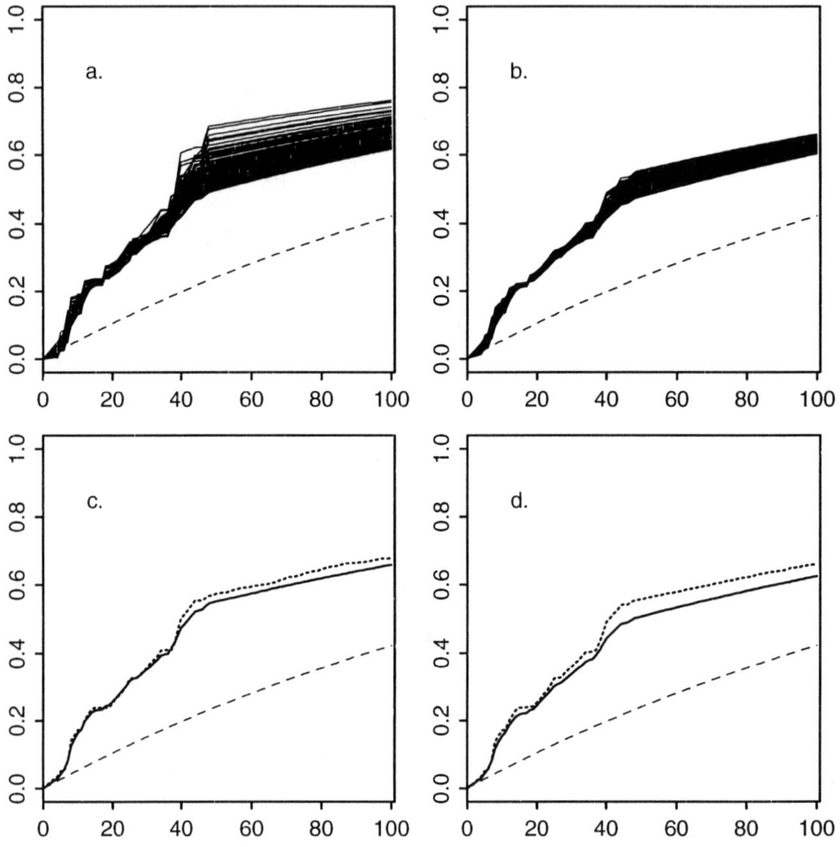

FIGURE 2. Interval censoring, nominal weights: Vertical axis is cumulative probability and horizontal axis is $\mathcal{X}$ in days in all panels. Panels (a) and (c) refer to calculations with prior mass $\alpha = 1$, and panels on the right correspond to $\alpha = 5$. Upper panels (a) and (b) show G_n calculated for 100 different random orderings of the $n = 46$ cases, with dashed line indicating the prior guess G_0. Solid lines in lower panels are pointwise averages of G_n, and dotted lines are Gibbs sampler output.

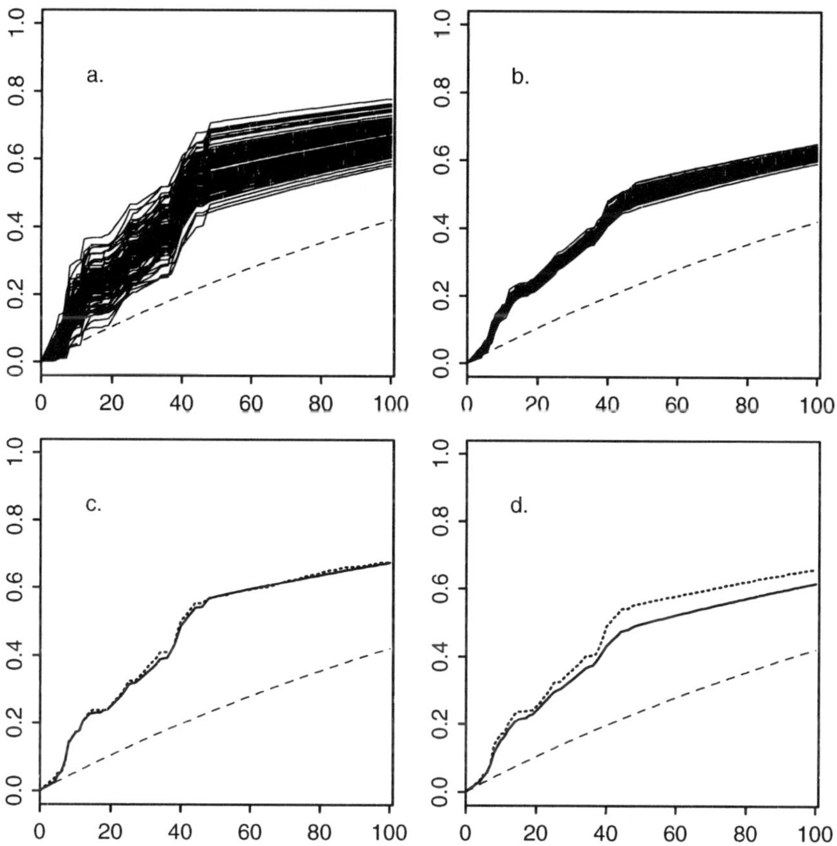

FIGURE 3. Interval censoring, square root weights: Descriptions are as in Figure 2.

average over orderings to a Gibbs sampler approximation. Similar results are shown in Figure 3, but here we use weights $w_i = 1/[(1+\alpha)\sqrt{i}]$. The average-over-orderings G_n is a very accurate approximation to the correct posterior predictive distribution in this example.

The Gibbs sampler calculation used for comparison here is an adaptation of the algorithm given by Escobar (1994) (see Escobar and West 1998; MacEachern 1998). It was run for 500^2 complete scans, subsampled to produce 500 vectors $(\theta_1, \ldots, \theta_{46})$ drawn approximately from their posterior distribution. Time series output analysis indicated that this represents an informative posterior sample. The dotted curves in panels (c) and (d) of Figures 2 and 3 show the marginal empirical distribution of the entire collection.

Appendix I provides some code to implement the recursion using built-in functions in S-Plus (e.g., Venables and Ripley, 1994).

3.6 Mixing Example

In a related class of models, the observed Y_i has some conditional density or mass function $p(y|\theta)$ given that $\theta_i = \theta$. The general recursion (4) may be expressed in terms of predictive densities instead of distribution functions:

$$g_i(\theta) = (1 - w_i)g_{i-1}(\theta) + w_i g_{i-1}(\theta)p(Y_i|\theta)/c_i \tag{8}$$

where c_i ensures that the posterior distribution on the right integrates to 1, that is $c_i = \int g_{i-1}(\theta)p(Y_i|\theta)\, d\theta$.

To illustrate (8), we consider modeling a set of survey data concerning employment status of youth in the United States. We confine attention to a sample of $n = 2390$ individuals from the National Longitudinal Survey of Youth, as in Quintana and Newton (1998). Briefly, annual employment history of each individual is recorded over time for up to thirteen years. For analysis here, data are summarized into binary indicators of employment during each year, yielding a binary time series for each individual. We account for positive correlation by supposing that each $Y_i = (Y_{i,1}, \ldots, Y_{i,k})$ is a binary Markov chain, with some person-specific transition matrix

$$\begin{pmatrix} p_i & 1 - p_i \\ 1 - q_i & q_i \end{pmatrix}.$$

determined by the parameter vector $\theta_i = (p_i, q_i)$. Now we take as prior assumptions that $\theta_i \sim G$, and that G is a Dirichlet process, so the recursive equations can be invoked to approximate $E(G(B)|D) = P(\theta_{n+1} \in B|D)$. We work with densities, noting that the Markov assumption implies that in (8),

$$p(Y_i|\theta_i) \propto p_i^{t_{0,0}}(1 - p_i)^{t_{0,1}} q_i^{t_{1,1}}(1 - q_i)^{t_{1,0}}$$

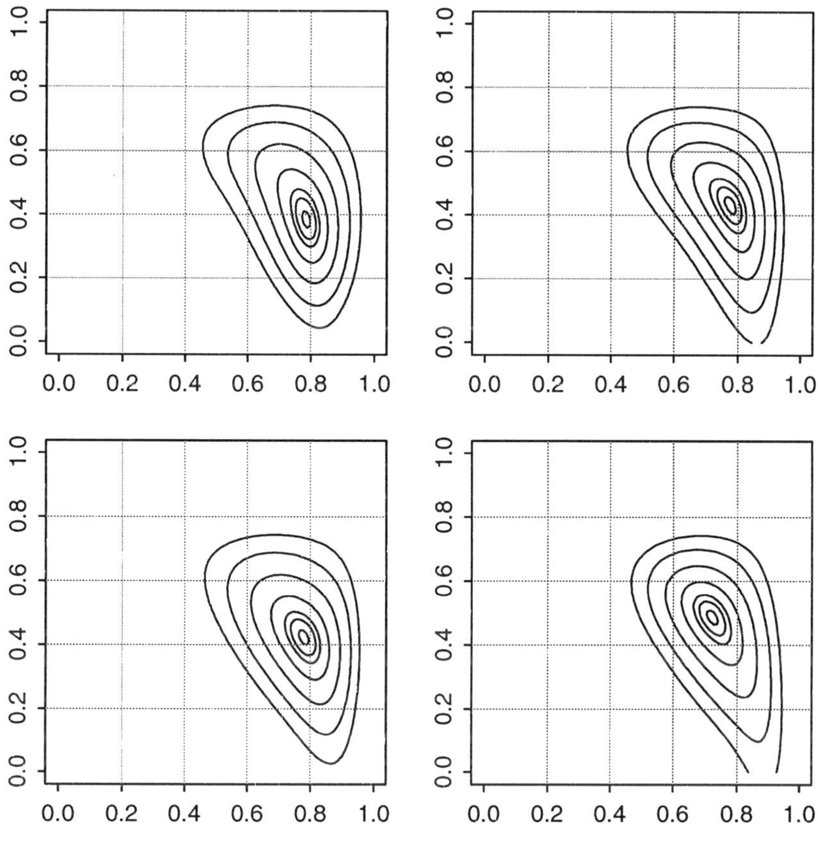

FIGURE 4. Markov chain mixtures: Shown are contours of recursive approximations to the posterior predictive distribution of $\theta_{n+1} = (p_{n+1}, q_{n+1})$, the *non*-transition probabilities. The four panels correspond to results from different random orderings of the $n = 2390$ binary sequences. Contours define probability content at the levels (.10, .25, .50, .75, .90, .95, .99).

where $t_{j,k}$ counts the transitions from j to k in the binary time series Y_i.

The move from one to two dimensions on θ creates no significant problems. A simple way to invoke the recursion (8) is to work on a grid in the unit square, which we do in the following calculations. (A speedier solution takes advantage of the smoothness, using Gauss-Legendre quadrature, as in Tao *et al.*, 1998.) Starting with a uniform prior guess G_0, so $g_0(\theta)$ is constant, we run (8) using the weight sequence $w_i = .5/\sqrt{i}$. We simply sum over the 100×100 grid after each step to calculate c_i. Figure 4 shows contours of G_n from four different random orderings of the data. Evidently, there is very little variation created by the processing order.

For comparison purposes, we also approximate the posterior predictive distribution of θ_{n+1} using MCMC. We adapted the Markov chain constructed in Bush and MacEachern (1996) (and discussed in MacEachern 1998), running it for 10,000 complete scans after a short burn in period. This chain moves through the $2n$ dimensional space of all θ values, nearly 5,000 dimensions in this example. Independent runs were performed for different values of the prior mass α, and we were encouraged to see reasonably rapid mixing as measured by simple time-series diagnostics on a few one-dimensional summaries. The posterior predictive distribution of interest is obtained from the Monte Carlo sample by collapsing all dimensions and recording the marginal empirical distribution of the θ's. Little information is lost if we simply accumulate counts in bins defined by the same 100×100 grid used in the recursive approximation. We smoothed by a very small bit of local averaging this empirical distribution before plotting contours, as shown in Figure 5.

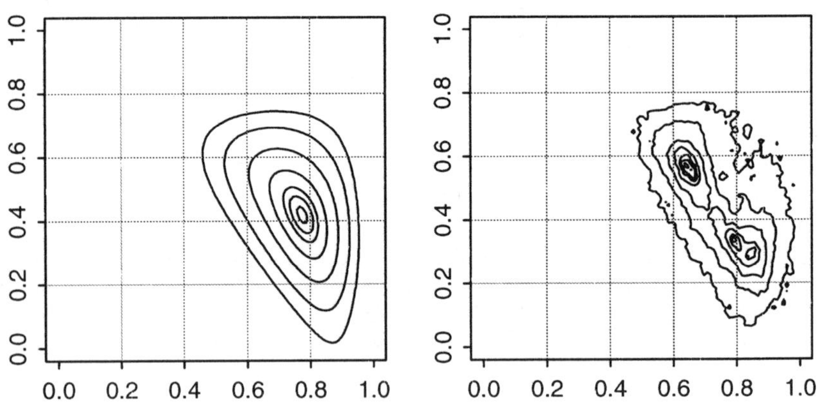

FIGURE 5. Comparing Methods: The left panel shows an average of distributions as in Figure 4 from 10 random orderings of the data. The right panel shows an MCMC approximation when $\alpha = 5$. Contours are defined as in Figure 4.

Generally, there is a close agreement between the MCMC approximation and the recursive approximation. There certainly are differences. For small values of α, the recursive approximation oversmooths slightly, missing what may be distinct modes in the true posterior predictive distribution. For larger values of α the two approximations agree quite well. We have reported in Figure 5 just an intermediate case, $\alpha = 5$. Of course an advantage of the recursive approximation is its computational simplicity.

A well-studied and somewhat simpler example has observations Y_i binomially distributed with success probability θ_i. Liu (1996) among others has studied the nonparametric Bayesian analysis of this problem, illustrating calculations on an interesting set of data on rolling tacks. In Appendix II we provide S-Plus code to implement the recursive approximation for this example.

3.7 On $n = 2$

It is difficult to make a direct comparison of the recursive approximation with the exact Bayes estimate unless we consider particular numerical examples or asymptotic properties. When $n = 2$, however, a comparison is quite feasible. To avoid technicalities, we work with densities as in Section 6. We can compute $P(\theta_3 \in B|Y_1, Y_2)$ exactly by noting that it equals the conditional expectation of the prior probability $P(\theta_3 \in B|\theta_1, \theta_2)$ given Y_1 and Y_2, and then by noting the mixture structure of the Polya sequence prior. Calculations reveal that this distribution is a mixture, with density

$$a_0 f_0(\theta) + a_1 f_1(\theta) + a_2 f_2(\theta) + a_{12} f_{12}(\theta). \tag{9}$$

Here, $f_0(\theta) = g_0(\theta)$, the prior guess, $f_1(\theta)$ is the posterior of θ if we were to observe Y_1 only, $f_2(\theta)$ is the same given Y_2, and $f_{12}(\theta)$ is the posterior of θ if Y_1 and Y_2 are independent and identically distributed from the common $p(y|\theta)$. Furthermore, the mixing proportions are

$$a_0 = \frac{\alpha}{\alpha+2}, \qquad a_1 = a_2 = \left[\frac{\alpha}{(\alpha+1)(\alpha+2)}\right] \frac{p(Y_1)p(Y_2)}{p(Y_1, Y_2)}$$

and $a_{12} = 1 - a_0 - a_1 - a_2$. These are prior predictive probabilities in the mixing weights. Consistent with our understanding of the role played by the prior mass α, we see that the f_0 component dominates for large α and the f_{12} component dominates for α near 0. Also, symmetry in Y_1 and Y_2 is evident in (9).

Now we turn to the recursive approximation (8). Obviously,

$$\begin{aligned} g_1(\theta) &= (1 - w_1)g_0(\theta) + w_1 g_0(\theta)p(Y_1|\theta)/c_1 \\ g_2(\theta) &= (1 - w_2)g_1(\theta) + w_2 g_1(\theta)p(Y_2|\theta)/c_2 \end{aligned}$$

and so by direct substitution, solving for g_2, we get

$$g_2(\theta) = b_0 f_0(\theta) + b_1 f_1(\theta) + b_2 f_2(\theta) + b_{12} f_{12}(\theta). \tag{10}$$

The recursive approximation produces a mixture of the same type as the correct answer (9), but with different mixing proportions. Using the nominal weights, $b_0 = a_0$, but

$$b_1 = \frac{\alpha + 1}{(\alpha + 2)^2} \text{ and } b_2 = b_1 c_2^* / c_2,$$

where we recall that c_2 is the normalizing constant above. Interestingly, c_2^* is a slightly different normalizing constant, being $\int p(Y_1|\theta)p(Y_2|\theta)g_0(\theta)\, d\theta$, i.e., the normalizer in f_{12}. Also the asymmetry in (10) with respect to Y_1 and Y_2 is clear.

3.8 Concluding Remarks

Recursive approximations are readily obtained for the posterior predictive distributions in Dirichlet-process-based nonparametric Bayesian analysis. A major advantage of this approach is that computations are extremely simple and thus can be deployed rapidly in a wide range of applications. They do not require keeping track of cluster structure among unobserved θ values, and thus have very low coding, storage, and CPU requirements. As a practical matter, the recursions might be useful as advance tools prior to full-blown implementation of MCMC. The accuracy of the recursive approximations is high in the examples studied here and in Newton and Zhang (1996), although in general this accuracy will depend on the particular data and prior, and further investigation is certainly warranted. The censoring examples we have studied exhibit quite high accuracy. In the mixing examples considered so far accuracy is also high but there is some indication of oversmoothing by the recursive approximation when the prior mass is small.

References

Blackwell, D., and MacQueen, J.B. (1973), "Ferguson distributions via Polya urn schemes," *The Annals of Statistics*, 1, 353-355.

Bush, C., and MacEachern, S. (1996), "A semiparametric Bayesian model for randomised block designs," *Biometrika*, 83, 275-286.

Doss, H. (1994), "Bayesian nonparametric estimation for incomplete data via successive substitution sampling," *The Annals of Statistics*, 22, 1763-1786.

Escobar, M.D. (1994), "Estimating normal means with a Dirichlet process prior," *Journal of the American Statistical Association*, 89, 268-277.

Escobar, M.D. and West, M. (1998), "Computing Bayesian nonparametric hierarchical models," in *Practical Nonparametric and Semiparametric Bayesian Statistics* (eds: Dey, D. and Müller, P. and Sinha, D.), New York: Springer-Verlag.

Finkelstein, D.M., and Wolfe, R.A. (1985), "A semiparametric model for regression analysis of interval-censored failure time data," *Biometrics*, 41, 933-945.

Gentleman, R., and Geyer, C.J. (1994), "Maximum likelihood estimation for interval censored data: consistency and computation," *Biometrika*, 81, 618-623.

Groeneboom, P., and Wellner, J.A. (1992), *Information bounds and nonparametric maximum likelihood estimation*, Basel: Birkhauser Verlag.

Kushner, H.J. and Yin, G.G. (1997), *Stochastic approximation algorithms and applications*, New York: Springer-Verlag.

Liu, J.S. (1996), "Nonparametric hierarchical Bayes via sequential imputations," *The Annals of Statistics*, 24, 911-930.

MacEachern, S. (1998). Computations for Mixture of Dirichlet Process models, in *Practical Nonparametric and Semiparametric Bayesian Statistics* (eds: Dey, D. and Müller, P. and Sinha, D.), New York: Springer-Verlag.

Newton, M.A., and Zhang, Y. (1996), "A recursive algorithm for nonparametric analysis with missing data," technical report 965, University of Wisconsin, Department of Statistics.

Quintana, F.A., and Newton, M.A. (1998), "Assessing the order of dependence for partially exchangeable binary data," *Journal of the American Statistical Association*, 93, to appear.

Sethuraman, J. (1994), "A constructive definition of Dirichlet priors," *Statistica Sinica*, 4, 639-650.

Smith, A.F.M., and Makov, U.E. (1978), "A quasi-Bayes sequential procedure for mixtures," *Journal of the Royal Statistical Society, Series B*, 40, 106-112.

Tao, H., Palta, M., Yandell, B.S., and Newton, M.A. (1998), "An estimation method for the semiparametric mixed effects model," *Biometrics*, in press.

Venables, W.N., and Ripley, B.D. (1994), *Modern Applied Statistics with S-Plus*, New York: Springer-Verlag.

Appendix I: Splus code for censoring example

```
# Have left endpoints of intervals in a vector
# ''lefts'' and right endpoints in ''rights''and
# set ''N'' equal to the common length.

# Partition support (0,infty) into sets B
grid <- c( seq(0,100,by=1), 1000 )
ngrid <- length(grid)
grt <- grid[2:ngrid]; glt <- grid[1:(ngrid-1)]

# Exponential prior guess and prior sample size
gg <- exp( -glt/(365/2) ) - exp( -grt/(365/2) )
alpha <- 1

# Identify partition elements that may contain
# each survival time
less <- function(x,y){ return( x<y ) }
ma <- t( outer(grt,rights,FUN="less") )
mb <- outer(lefts,glt,FUN="less")
dmat <- ma&mb

# Recursion yields approximate Bayes estimate gg
weight <- 1/(alpha+1:N) # Nominal weight sequence
ord <- sample( 1:N )    # Process in random order
for( i in 1:N )
 {
   ok <- dmat[ord[i],]        # A_i
   numer <- rep(0,ngrid-1)
   numer[ok] <- gg[ok]        # G(B and A_i)
   denom <- sum( numer )      # G(A_i)
   gg <- gg*( 1-weight[i] ) + weight[i]*numer/denom
 }
# Repeat loop to see variation over orderings.

# Approximate Bayes estimate of distribution function.
plot(grid, cumsum( c(0,gg) ), type="s", xlim=c(0,100),
     xlab="time (days)", ylab="cumulative probability")
```

Appendix II: Splus code for binomial mixture example

```
# Beckett Diaconis Tack Data
nsuccess <- c( rep(1,3), rep(2,13), rep(3,18), rep(4,48),
   rep(5,47), rep(6,67), rep(7,54), rep(8,51), rep(9,19) )
ntrials <- 9; N <- length( nsuccess )

# Support of mixing distribution
grid <- seq(0.01,.99,length=100); delta <- grid[2]-grid[1]

# Beta prior guess and prior sample size
gg <- dbeta(grid,shape1=.5,shape2=.5) ; alpha <- 1/3

# Binomial likelihood
db2 <- function(y,prob,n){return(dbinom(y,n,prob))}
lik <-  outer(nsuccess,grid,FUN="db2",n=ntrials)

# Recursion yields approximate Bayes estimate gg
weight <- 1/sqrt((alpha+1)*(alpha+1:N)) # A weight sequence
ord <- sample( 1:N )    # Process tacks in random order
for( i in 1:N )
 {
   post <- lik[ord[i],]*gg
   post <- ( post/sum(post) )/delta
   gg <- gg*( 1-weight[i] ) + weight[i]*post
 }
# Repeat loop to see variation over orderings.

# Estimated predictive density for success probability
# of a future tack.  Compare to Fig. 2, Liu (1996).
plot( grid, gg, type="l", xlab="tack success probability",
      ylab="posterior predictive density" )
```

4

Dynamic Display of Changing Posterior in Bayesian Survival Analysis

Hani Doss
B. Narasimhan

ABSTRACT We consider the problem of estimating an unknown distribution function F in the presence of censoring under the conditions that a parametric model is believed to hold approximately. We use a Bayesian approach, in which the prior on F is a mixture of Dirichlet distributions. A hyperparameter of the prior determines the extent to which this prior concentrates its mass around the parametric family. A Gibbs sampling algorithm to estimate the posterior distributions of the parameters of interest is reviewed. An importance sampling scheme enables us to use the output of the Gibbs sampler to very quickly recalculate the posterior when we change the hyperparameters of the prior. The calculations can be done sufficiently fast to enable the dynamic display of the changing posterior as the prior hyperparameters are varied.

4.1 Introduction and Summary

We consider the following model for censored data. There are random variables $X_1, \ldots, X_n$ independently and identically distributed from some distribution function F. The X_i's are not necessarily observed. Rather, for each i, there is a set A_i within which X_i is known to lie. This framework includes several forms of censoring. Right censoring corresponds to the case where the sets A_i are either singletons or right-infinite intervals. Interval-censored data corresponds to the case where the sets are intervals, and current-status data corresponds to the case where each A_i is either a right-infinite interval or a left-infinite interval. In this model there is a nonparametric maximum likelihood estimator, which in the general case is not available in closed form, but can be computed using the EM algorithm; see Turnbull (1974, 1976). There are often reasons for using a parametric model H_θ; $\theta \in \Theta \subset \mathbb{R}^p$. A parametric family may be indicated by a biological model, or simply may be suggested by previous data of a similar type. Such models give rise to estimators that can be substantially more efficient, especially in estimating the tail of the distribution. In addition, the parameters of these models sometimes have a physical or biological interpretation.

The most standard way to proceed in this case is to first carry out a goodness of fit test (see Hjort 1990 and the references cited therein). If the test rejects the parametric model, then one uses a nonparametric estimate; otherwise, one uses an estimate based on the parametric model. One problem with this approach is that if the parametric model holds only approximately, then it may be that neither estimator is satisfactory. If the parametric model does not hold exactly, the estimator based on a parametric model will in general not even be consistent. On the other hand, the nonparametric estimator does not make use of the information that the parametric model holds approximately.

A Bayesian nonparametric approach based on mixtures of Dirichlet priors (Ferguson 1973, 1974 and Antoniak 1974) offers a reasonable compromise between purely parametric and purely nonparametric models. Recall that if ν is some prior distribution on Θ and $M_\theta > 0$ for each θ, then if θ is chosen from ν, and then F is chosen from $\mathcal{D}_{M_\theta H_\theta}$, the Dirichlet prior with parameter measure $M_\theta H_\theta$, we say that the prior on F is a mixture of Dirichlets (with parameter $(\{M_\theta H_\theta\}_{\theta \in \Theta}, \nu)$). Often $M_\theta \equiv M$, i.e. the constants M_θ do not depend on θ. In this case, M can be interpreted as a precision parameter that indicates the degree of concentration of the prior on F around the parametric family $\{H_\theta;\, \theta \in \Theta\}$. For example, as $M \to \infty$ the distribution of F converges to $\int H_\theta\, \nu(d\theta)$, the standard Bayesian model for the parametric family $\{H_\theta;\, \theta \in \Theta\}$ in which θ has prior ν.

For the case where the data are not censored, there is a closed form expression for the posterior distribution of F, and this is given in Proposition 1 below. From this result it will be easily seen that (in the uncensored case) estimators based on mixtures of Dirichlet priors continuously interpolate between those based on the purely parametric and nonparametric models. For large values of M, the estimators are close to the Bayes estimator based on the parametric model. On the other hand, for small values of M, the estimators are essentially equal to the nonparametric maximum likelihood estimator. One therefore expects that the same will be true for the case where the data are censored.

For the censored case, there is no closed form expression for the posterior distribution of F given the data, and one has to use Monte Carlo methods. We review a Gibbs sampling algorithm given in Doss and Huffer (1998), that enables us to estimate posterior distributions of interest. Running this Markov chain for a single prior usually takes on the order of several minutes to a few hours. This is unfortunate, because in an interactive kind of analysis, one would like to see the posterior distributions for a variety of priors; in particular, one would like to change the hyperparameter M, seeing how the estimates vary continuously as one moves from the purely parametric to the nonparametric models.

In this paper we discuss an importance sampling scheme that enables us to use output from a Markov chain corresponding to a single prior $\pi^{(0)}$ to get estimates of posteriors corresponding to other priors. This does not work well when the other priors are far from $\pi^{(0)}$. We discuss an extension of the importance sampling scheme that applies when we run a Markov chain for each of m priors. This extension allows us to estimate posterior distributions for a wider range of priors. Reweighing the output of one or a few Markov chains works much faster than running a new chain. The main contribution of this paper is a computing software environment that enables one to interactively change the prior and immediately see the corresponding changes in the posterior. The software is built on top of `Lisp-Stat`, described in Tierney (1990). A graphical user-interface is provided for entering the data produced by the Markov chains. A set of sliders allow the user to vary prior parameters as well and see the corresponding change in a plot of the posterior distribution of $F(t)$.

This paper is organized as follows. In Section 4.2, we review the Gibbs sampling algorithm of Doss and Huffer (1998), and state a result (Proposition 1) which is needed to implement the algorithm. In Section 4.3 we give a proof of Proposition 1. In Section 4.4 we use the ideas in the proof of Proposition 1 to develop the formulas needed to carry out the importance sampling method. In Section 4.5 we describe our computing environment, and illustrate its use on a data set that compares the time to cosmetic deterioration for breast cancer patients under two treatment regimens. The data consist of times that are "interval censored."

4.2 A Gibbs Sampler for Censored Data

Recall that the prior on F is

$$F \sim \int \mathcal{D}_{\alpha_\theta}\, \nu(d\theta), \tag{1}$$

where for each $\theta \in \Theta, \alpha_\theta = M_\theta H_\theta$ and $M_\theta \in (0, \infty)$. Let $\boldsymbol{X} = (X_1, \dots, X_n)$. The X_i's are not necessarily observed. Let **data** represent the observed data, i.e. **data** is the event $\{X_i \in A_i;\, i = 1, \dots, n\}$. We will often find it convenient to use the convention that subscripting a probability measure signifies conditioning. For example, if U and V are random variables, then $\mathcal{L}_V(U)$ denotes $\mathcal{L}(U \mid V)$, the conditional distribution of U given V. Also, if U and V are random variables defined on the same space, and Q is a probability measure such that $U \sim Q$, then Q_V denotes the conditional distribution of U given V.

We are interested in $\mathcal{L}_{\mathbf{data}}(F)$. We will use the fact that

$$\begin{aligned}\mathcal{L}_{\mathbf{data}}(F) &= \int \mathcal{L}_{\mathbf{data}}(F \mid \boldsymbol{X}, \theta)\mathcal{L}_{\mathbf{data}}(\boldsymbol{X}, \theta) \\ &= \int \mathcal{D}_{\alpha_\theta + \sum_{i=1}^n \delta_{X_i}} \mathcal{L}_{\mathbf{data}}(\boldsymbol{X}, \theta). \qquad (2)\end{aligned}$$

From (2), we see that if we can generate a sample $(\boldsymbol{X}^{(1)}, \theta^{(1)}), \ldots,$ $(\boldsymbol{X}^{(G)}, \theta^{(G)}) \overset{iid}{\sim} \mathcal{L}_{\mathbf{data}}(\boldsymbol{X}, \theta)$, then we can estimate $\mathcal{L}_{\mathbf{data}}(F)$. Thus, for example, for any fixed t, we can estimate $\mathcal{L}_{\mathbf{data}}(F(t))$ by the average of Beta distributions

$$\frac{1}{G}\sum_{g=1}^{G} \mathrm{Be}\Big(\alpha_{\theta^{(g)}}(t) + \textstyle\sum_{i=1}^n \delta_{X_i^{(g)}}(t),\ M_{\theta^{(g)}} + n - \alpha_{\theta^{(g)}}(t) - \sum_{i=1}^n \delta_{X_i^{(g)}}(t)\Big). \qquad (3)$$

So we need to be able to generate samples from $\mathcal{L}_{\mathbf{data}}(X_1, \ldots, X_n, \theta)$. We will do this using the Gibbs sampler, using two well-known facts about the Dirichlet prior. The first of these states that if $F \sim \mathcal{D}_\beta$, where $\beta = M\beta_0$, and if $X_1, \ldots, X_n \overset{iid}{\sim} F$, then for any j,

$$\mathcal{L}(X_j \mid X_i,\, i \neq j) = \frac{\beta + \sum_{i\neq j} \delta_{X_i}}{M + n - 1}. \qquad (4)$$

The second fact is the proposition below, which enables the generation of θ in the Gibbs sampler. Several versions of this result are already known (e.g. Lemma 1 of Antoniak 1974, Lemmas 2.1 and 3.1 of Diaconis and Freedman 1986b) and a number of different proofs are possible. In Section 4.3 we give a new proof, the ideas of which can be used to obtain formulas for the importance sampling that underlies the dynamic graphics discussed in Section 4.5.

If $\boldsymbol{v}$ is a vector in R^n, we use $\#(\boldsymbol{v})$ to denote the number of distinct values of $v_1, \ldots, v_n$.

Proposition 1 *Assume that for each $\theta \in \Theta$, H_θ is absolutely continuous, with a density h_θ that is continuous on R. If the prior on F is given by (1), then the posterior distribution of F given $X_1, \ldots, X_n$ is*

$$\int \mathcal{D}_{\alpha_\theta + \sum_{i=1}^n \delta_{X_i}} \nu_{\boldsymbol{X}}(d\theta), \qquad (5)$$

where $\nu_{\boldsymbol{X}}$ is the measure which is absolutely continuous with respect to ν and is defined by

$$\nu_{\boldsymbol{X}}(d\theta) = c(\boldsymbol{X})\Bigg(\prod_{\text{dist. } j} h_\theta(X_i)\Bigg)\left[\frac{(M_\theta)^{\#(\boldsymbol{X})}\Gamma(M_\theta)}{\Gamma(M_\theta + n)}\right]\nu(d\theta). \qquad (6)$$

In (6) the "dist." in the product indicates that the product is taken over distinct values only, Γ is the gamma function, and $c(\boldsymbol{X})$ is a normalizing constant.

The Gibbs sampler described in Doss and Huffer (1998) is very easy to describe, and relies on (4) and (6). We pick starting values $X_1^{(0)}, \ldots, X_n^{(0)}$, $\theta^{(0)}$, with $X_i^{(0)} \in A_i$. Suppose that the current value of this vector is $(X_1^{(k-1)}, \ldots, X_n^{(k-1)}, \theta^{(k-1)})$. Generate

$$X_i^{(k)} \sim \left(\frac{\alpha_{\theta^{(k-1)}} + \sum_{j<i} \delta_{X_j^{(k)}} + \sum_{j>i} \delta_{X_j^{(k-1)}}}{M_\theta + n - 1} \right)_{A_i} \qquad i = 1, \ldots, n, \tag{7}$$

and then

$$\theta^{(k)} \sim \nu_{\boldsymbol{X}^{(k)}}, \tag{8}$$

where $\nu_{\boldsymbol{X}^{(k)}}$ is given by (6). In (7) we are using the following notation. If Λ is a probability measure, B is a set, and U is distributed according to Λ, then Λ_B denotes the conditional distribution function of U given that $U \in B$; that is

$$\Lambda_B(A) = \Lambda(A \cap B)/\Lambda(B) \tag{9}$$

when $\Lambda(B) > 0$. Thus, in (7), the subscript A_i indicates that X_i is generated from the probability distribution inside the parentheses, conditional on its lying in the set A_i. (When conditioning on X_j $(j \neq i)$ and **data**, conditioning on the events $X_j \in A_j$ $(j \neq i)$ is superfluous.)

Consider now (6). If M_θ is constant in θ (which we shall be assuming in our examples), then the expression in brackets in (6) is a constant, and can be absorbed into $c(\boldsymbol{X})$. In this case, except for the fact that the product is over the distinct X_j's only, (6) is the posterior distribution of θ in the standard finite-dimensional parametric model in which $\theta \sim \nu$, and $X_i \overset{iid}{\sim} H_\theta$ given θ, for which the posterior is proportional to the likelihood times the prior.

This Gibbs sampler is distinct from the one described in Doss (1994). Doss and Huffer (1998) describe other Monte Carlo schemes for generating observations from $\mathcal{L}_{\mathbf{data}}(X_1, \ldots, X_n, \theta)$. For the case where the data are right censored, one of these, based on the idea of sequential importance sampling (Kong, Liu, and Wong 1994), yields independent observations that are distributed exactly according to $\mathcal{L}_{\mathbf{data}}(X_1, \ldots, X_n, \theta)$.

4.3 Proof of Proposition 1

As mentioned earlier, there are a number of different proofs of Proposition 1 that have been given in the literature. Below we outline a simple proof based on a straightforward calculation similar in spirit to one in Doss (1985a) (the problem considered in Doss (1985a) is simpler than the one we consider here since it involves only a location mixture; there are technical problems that arise when one uses more general mixtures). This method of proof has the advantage that it generalizes very easily to variants of the Dirichlet such as the "symmetrized Dirichlets" and "conditional Dirichlets" that were studied by Diaconis and Freedman (1986a,b) and Doss (1984, 1985a,b). More importantly, a certain equation arising in the proof of the theorem will be needed to give the formulas that are necessary to carry out the importance sampling.

We begin by writing

$$\begin{aligned} \mathcal{L}_{\boldsymbol{X}}(F) &= \int \mathcal{L}_{\boldsymbol{X}}(F \mid \theta)\, \nu_{\boldsymbol{X}}(d\theta) \\ &= \int \mathcal{D}_{\alpha_\theta + \sum_{i=1}^n \delta_{X_i}}\, \nu_{\boldsymbol{X}}(d\theta), \end{aligned} \tag{10}$$

where $\nu_{\boldsymbol{X}}$ is the conditional distribution of θ given $\boldsymbol{X}$. The first equality in (10) is in effect the "law of total probability" in this situation, while the second follows from the standard formula that gives the posterior distribution of F given a sample of size n for a single Dirichlet.

Our goal now is to obtain $\nu_{\boldsymbol{X}}$, the difficulty being that we are conditioning on an event of probability 0. Let B be a Borel subset of R. To calculate $\nu_{\boldsymbol{X}}(B)$ we consider, for $\boldsymbol{x} \in R^n$ and $\eta > 0$, the probability measure $\nu_{\boldsymbol{x}}^{\eta}$ defined by

$$\nu_{\boldsymbol{x}}^{\eta}(B) = \frac{P\Big\{\theta \in B,\, \boldsymbol{X} \in C_{\boldsymbol{x}}^{(\eta)}\Big\}}{P\Big\{\boldsymbol{X} \in C_{\boldsymbol{x}}^{(\eta)}\Big\}}, \tag{11}$$

where $C_{\boldsymbol{x}}^{\eta}$ is the cube $(x_1 - \eta/2, x_1 + \eta/2) \times \cdots \times (x_n - \eta/2, x_n + \eta/2)$. Here, P denotes the joint distribution of $(\theta, F, \boldsymbol{X})$. We will find the limit as $\eta \to 0$, which we shall denote by by $\nu_{\boldsymbol{x}}$, and in the appendix complete the proof of the proposition by showing that the family of limits $\{\nu_{\boldsymbol{x}};\, \boldsymbol{x} \in R^n\}$ defines a "regular conditional distribution of θ given $\boldsymbol{X}$" (a definition of this is given in the appendix). Our definition of $\nu_{\boldsymbol{x}}^{\eta}(B)$ involves a slight abuse of our convention for subscripting probability measures, but this will not cause any problems.

To calculate $\nu_{\boldsymbol{x}}^{\eta}(B)$ we write the right side of (11) as

$$\frac{P\Big\{\theta \in B,\, F \in \mathcal{P},\, \boldsymbol{X} \in C_{\boldsymbol{x}}^{(\eta)}\Big\}}{P\Big\{F \in \mathcal{P},\, \boldsymbol{X} \in C_{\boldsymbol{x}}^{(\eta)}\Big\}}$$

$$= \frac{\displaystyle\int_B \int_{\mathcal{P}} \prod_{i=1}^{n} \big[F(x_i + \eta/2) - F(x_i - \eta/2)\big] \mathcal{D}_{\alpha_\theta}(dF)\nu(d\theta)}{\displaystyle\int_\Theta \int_{\mathcal{P}} \prod_{i=1}^{n} \big[F(x_i + \eta/2) - F(x_i - \eta/2)\big] \mathcal{D}_{\alpha_\theta}(dF)\nu(d\theta)}$$

$$= \frac{\displaystyle\int_B \frac{\int_{\mathcal{P}} \prod_{i=1}^{n} \big[F(x_i + \eta/2) - F(x_i - \eta/2)\big] \mathcal{D}_{\alpha_\theta}(dF)}{\eta^r \prod_{j=1}^{r}(n_j - 1)!} \nu(d\theta)}{\displaystyle\int_\Theta \frac{\int_{\mathcal{P}} \prod_{i=1}^{n} \big[F(x_i + \eta/2) - F(x_i - \eta/2)\big] \mathcal{D}_{\alpha_\theta}(dF)}{\eta^r \prod_{j=1}^{r}(n_j - 1)!} \nu(d\theta)} \tag{12}$$

$$= \frac{\displaystyle\int_B f_{\boldsymbol{x}}^{\eta}(\theta)\nu(d\theta)}{\displaystyle\int_\Theta f_{\boldsymbol{x}}^{\eta}(\theta)\nu(d\theta)},$$

where

$$f_{\boldsymbol{x}}^{\eta}(\theta) = \frac{\int_{\mathcal{P}} \prod_{i=1}^{n} \big[F(x_i + \eta/2) - F(x_i - \eta/2)\big] \mathcal{D}_{\alpha_\theta}(dF)}{\eta^r \prod_{j=1}^{r}(n_j - 1)!}. \tag{13}$$

Let $x^{(1)} < x^{(2)} < \cdots < x^{(r)}$ be the distinct values of $x_1, \ldots, x_n$, and let $n_1, \ldots, n_r$ be their multiplicities (we will occasionally use r instead of $\#(\boldsymbol{x})$ for typographical reasons). Then we may rewrite (13) as

$$f_{\boldsymbol{x}}^{\eta}(\theta) = \frac{\int_{\mathcal{P}} \prod_{j=1}^{r} \big[F(x^{(j)} + \eta/2) - F(x^{(j)} - \eta/2)\big]^{n_j} \mathcal{D}_{\alpha_\theta}(dF)}{\eta^r \prod_{j=1}^{r}(n_j - 1)!}. \tag{14}$$

Let $A_j(\eta) = \alpha_\theta\big\{(x^{(j)} - \eta/2, x^{(j)} + \eta/2)\big\}$, $j = 1, \ldots, r$, and also define $A_{r+1}(\eta) = M_\theta - \sum_{j=1}^{r} A_j(\eta)$. (Recall that $M_\theta = \alpha_\theta(R)$.) Assume that η is so small that the sets $A_j(\eta)$, $j = 1, \ldots, r$ are disjoint. Note that calculation of (14) is routine since it involves only the finite-dimensional Dirichlet distribution. The integral on the right side of (14) is $E(D_1^{n_1} \cdots D_r^{n_r})$ where $(D_1, \ldots, D_r, D_{r+1}) \sim \text{Dirichlet}\big(A_1(\eta), \ldots, A_{r+1}(\eta)\big)$, and a simple calculation shows that this expectation is equal to

$$\frac{\Gamma(M_\theta)}{\Big(\prod_{j=1}^{r} \Gamma(A_j(\eta))\Big)\Gamma(A_{r+1}(\eta))} \frac{\Big(\prod_{j=1}^{r} \Gamma(A_j(\eta) + n_j)\Big)\Gamma(A_{r+1}(\eta))}{\Gamma(M_\theta + n)}.$$

Let

$$f_{\boldsymbol{x}}(\theta) = \left(\prod_{j=1}^{r} h_\theta(x^{(j)})\right) \frac{M_\theta^r \Gamma(M_\theta)}{\Gamma(M_\theta + n)}. \tag{15}$$

Using the recursion $\Gamma(x+1) = x\Gamma(x)$ and the definition of the derivative, we see that for each $\theta \in \Theta$, $f_{\boldsymbol{x}}^\eta(\theta) \to f_{\boldsymbol{x}}(\theta)$ as $\eta \to 0$. If we operate heuristically and take limits under the integral signs, we obtain

$$\nu_{\boldsymbol{x}}^\eta(B) \to \frac{\int_B f_{\boldsymbol{x}}(\theta)\nu(d\theta)}{\int_\Theta f_{\boldsymbol{x}}(\theta)\nu(d\theta)} \quad \text{as } \eta \to 0, \tag{16}$$

which is the statement of the proposition.

4.4 Importance Sampling

Let $\pi^{(0)}$ denote the prior on F induced by $\left(\{\alpha_\theta^{(0)}\}, \nu^{(0)}\right)$ where $\alpha_\theta^{(0)} = M_\theta^{(0)} H_\theta^{(0)}$; i.e. the statement $F \sim \pi^{(0)}$ is equivalent to $F \sim \int \mathcal{D}_{\alpha_\theta^{(0)}} \nu^{(0)}(d\theta)$. This prior induces a distribution on the triple $(F, \theta, \boldsymbol{X})$, and by a slight abuse of notation, we shall use $\pi^{(0)}$ to denote this distribution also. We run the Markov chain described in Section 4.2 corresponding to this prior, and obtain as output $\left((\theta^{(1)}, \boldsymbol{X}^{(1)}), \ldots, (\theta^{(G)}, \boldsymbol{X}^{(G)})\right)$. We shall be interested in reweighing the G vectors in this output in order to estimate distributions and expectations corresponding to many different priors $\pi^{(\kappa)} = \int \mathcal{D}_{\alpha_\theta^{(\kappa)}} \nu^{(\kappa)}(d\theta)$, where $\alpha_\theta^{(\kappa)} = M_\theta^{(\kappa)} H_\theta^{(\kappa)}$ and $H_\theta^{(\kappa)}$ has density $h_\theta^{(\kappa)}$, κ ranging over some set. Since the reweighing will be computationally intensive, and has to be redone for every specification of κ, it will be important that the G vectors in this output be independent or at least nearly so. This can be accomplished by running the Markov chain for a very large number of iterations and taking every k^{th} value, where k is some large number, or by running G independent chains and keeping only the last iteration for each chain. Here, we assume that we have devised a satisfactory way of obtaining G nearly independent vectors, and do not discuss this issue any further.

Let $\pi_{\mathbf{data}}^{(\kappa)}$ denote the conditional distribution of $(F, \theta, \boldsymbol{X})$ given the data, when the prior is $\pi^{(\kappa)}$. Also, let $\lambda^{(\kappa)}$ be the marginal distribution of $(\theta, \boldsymbol{X})$ when the prior is $\pi^{(\kappa)}$, and let $\lambda_{\mathbf{data}}^{(\kappa)}$ be the conditional distribution of $(\theta, \boldsymbol{X})$ given the data, when the prior is $\pi^{(\kappa)}$. We wish to calculate the Radon-Nikodym derivative of $\lambda_{\mathbf{data}}^{(\kappa)}$ with respect to $\lambda_{\mathbf{data}}^{(0)}$. It will be very helpful to first consider the case where the constants $C(\kappa) = \pi^{(\kappa)}\{X_i \in A_i;\, i = 1, \ldots, n\}$ and $C(0) = \pi^{(0)}\{X_i \in A_i;\, i = 1, \ldots, n\}$ are both positive. In this case,

$$\left[\frac{d\lambda^{(\kappa)}_{\mathbf{data}}}{d\lambda^{(0)}_{\mathbf{data}}}\right](\theta,\boldsymbol{x}) = \left[\frac{d\lambda^{(\kappa)}/C(\kappa)}{d\lambda^{(0)}/C(0)}\right](\theta,\boldsymbol{x}) =$$

$$\frac{C(0)\Big\{\prod_{j=1}^{\#(\boldsymbol{x})} h_\theta^{(\kappa)}(x^{(j)})\Big\}\dfrac{\big(M_\theta^{(\kappa)}\big)^{\#(\boldsymbol{x})}\Gamma\big(M_\theta^{(\kappa)}\big)}{\Gamma\big(M_\theta^{(\kappa)}+n\big)}}{C(\kappa)\Big\{\prod_{j=1}^{\#(\boldsymbol{x})} h_\theta^{(0)}(x^{(j)})\Big\}\dfrac{\big(M_\theta^{(0)}\big)^{\#(\boldsymbol{x})}\Gamma\big(M_\theta^{(0)}\big)}{\Gamma\big(M_\theta^{(0)}+n\big)}}\left[\frac{d\nu^{(\kappa)}}{d\nu^{(0)}}\right](\theta), \qquad (17)$$

where the second equality in (17) follows from the development in Section 4.3 dealing with the calculation of $P\{\theta \in B,\ F \in \mathcal{P},\ \boldsymbol{X} \in C_{\boldsymbol{x}}^{(\eta)}\}$, i.e. the calculation of the numerator of the left side of (12) (see the material from equations (11) to (16), and the final equation (15)).

Formula (17) is still valid when $C(\kappa)$ and $C(0)$ are both zero (for example when some of the sets A_i are singletons), but in this case, $C(\kappa)$ and $C(0)$ must be interpreted as limits of probabilities of the set $\{X_i \in A_i;\ i = 1,\dots,n\}$, with the A_i's "fattened," as in the proof of Proposition 1.

Formula (17) appears to be too complicated to be usable, but in fact, this is not true. Suppose that the families $\{H_\theta^{(\kappa)}\}$ do not depend on κ. That is, we have in mind a single parametric family of distributions (in the example in Section 4.5, this is taken to be the family of exponential distributions). Suppose in addition that for each κ, including $\kappa = 0$, the $M_\theta^{(\kappa)}$'s do not depend on θ, i.e. $M_\theta^{(\kappa)} \equiv M^{(\kappa)}$. Thus, as κ varies, $M^{(\kappa)}$ and ν_κ are the only things that change. In this case, the terms in braces in the numerator and denominator of (17) cancel out exactly; moreover, the term $\Gamma\big(M_\theta^{(\kappa)}\big)/\Gamma\big(M_\theta^{(\kappa)}+n\big)$ can be absorbed into the constant $C(\kappa)$ and similarly for the term $\Gamma\big(M_\theta^{(0)}\big)/\Gamma\big(M_\theta^{(0)}+n\big)$. Therefore, assuming that $\nu^{(\kappa)}$ is absolutely continuous with respect to $\nu^{(0)}$ and that they have densities $(\nu^{(\kappa)})'$ and $(\nu^{(0)})'$ with respect to Lebesgue measure, we are left with

$$\left[\frac{d\lambda^{(\kappa)}_{\mathbf{data}}}{d\lambda^{(0)}_{\mathbf{data}}}\right](\theta,\boldsymbol{x}) = \frac{\widetilde{C}(0)}{\widetilde{C}(\kappa)}\left(\frac{M^{(\kappa)}}{M^{(0)}}\right)^{\#(\boldsymbol{x})}\frac{(\nu^{(\kappa)})'(\theta)}{(\nu^{(0)})'(\theta)}. \qquad (18)$$

From now on, we shall assume the conditions on the priors that yield a Radon-Nikodym derivative of the simple form (18). Note that (18) gives $[d\lambda^{(\kappa)}_{\mathbf{data}}/d\lambda^{(0)}_{\mathbf{data}}]$ *only up to a multiplicative constant.* We shall now review how this is enough to enable us to estimate the expectation under $\lambda^{(\kappa)}_{\mathbf{data}}$ of a function of θ and $\boldsymbol{x}$, when we have a sample from $\lambda^{(0)}_{\mathbf{data}}$. Important examples of such functions arise when we want to estimate the posterior distribution of $F(t)$ for fixed t, when the prior is $\pi^{(\kappa)}$. Specifically, let $t \in R$ and $u \in (0,1)$ be fixed, and suppose we wish to calculate $\pi^{(\kappa)}_{\mathbf{data}}\{F(t) \le u\}$.

We write

$$\pi^{(\kappa)}_{\mathbf{data}}\{F(t) \le u\} = \int \pi^{(\kappa)}\{F(t) \le u \mid \theta, \boldsymbol{X}\}\, d\lambda^{(\kappa)}_{\mathbf{data}}(\theta, \boldsymbol{X}), \tag{19}$$

and note that the integrand is a function of θ and $\boldsymbol{X}$, which has the explicit form

$$\mathrm{Be}\Big(\alpha^{(\kappa)}_\theta(t) + \textstyle\sum_{i=1}^n \delta_{X_i}(t),\ M^{(\kappa)}_\theta + n - \alpha^{(\kappa)}_\theta(t) - \sum_{i=1}^n \delta_{X_i}(t)\Big)(u),$$

i.e. a Beta distribution evaluated at the point u.

To evaluate the expectation under $\lambda^{(\kappa)}_{\mathbf{data}}$ of $g(\theta, \boldsymbol{x})$ we write

$$\begin{aligned}
\int g(\theta,\boldsymbol{x}) d\lambda^{(\kappa)}_{\mathbf{data}}(\theta,\boldsymbol{x}) &= \int g(\theta,\boldsymbol{x}) \left[\frac{d\lambda^{(\kappa)}_{\mathbf{data}}}{d\lambda^{(0)}_{\mathbf{data}}}\right](\theta,\boldsymbol{x})\, d\lambda^{(0)}_{\mathbf{data}}(\theta,\boldsymbol{x}) \\
&= \int g(\theta,\boldsymbol{x}) \frac{\widetilde{C}(0)}{\widetilde{C}(\kappa)} \left(\frac{M^{(\kappa)}}{M^{(0)}}\right)^{\#(\boldsymbol{x})} \frac{(\nu^{(\kappa)})'(\theta)}{(\nu^{(0)})'(\theta)}\, d\lambda^{(0)}_{\mathbf{data}}(\theta,\boldsymbol{x}) \\
&= \frac{\displaystyle\int g(\theta,\boldsymbol{x}) \frac{\widetilde{C}(0)}{\widetilde{C}(\kappa)} \left(\frac{M^{(\kappa)}}{M^{(0)}}\right)^{\#(\boldsymbol{x})} \frac{(\nu^{(\kappa)})'(\theta)}{(\nu^{(0)})'(\theta)}\, d\lambda^{(0)}_{\mathbf{data}}(\theta,\boldsymbol{x})}{\displaystyle\int \frac{\widetilde{C}(0)}{\widetilde{C}(\kappa)} \left(\frac{M^{(\kappa)}}{M^{(0)}}\right)^{\#(\boldsymbol{x})} \frac{(\nu^{(\kappa)})'(\theta)}{(\nu^{(0)})'(\theta)}\, d\lambda^{(0)}_{\mathbf{data}}(\theta,\boldsymbol{x})} \qquad (20) \\
&= \frac{\displaystyle\int g(\theta,\boldsymbol{x}) \left(\frac{M^{(\kappa)}}{M^{(0)}}\right)^{\#(\boldsymbol{x})} \frac{(\nu^{(\kappa)})'(\theta)}{(\nu^{(0)})'(\theta)}\, d\lambda^{(0)}_{\mathbf{data}}(\theta,\boldsymbol{x})}{\displaystyle\int \left(\frac{M^{(\kappa)}}{M^{(0)}}\right)^{\#(\boldsymbol{x})} \frac{(\nu^{(\kappa)})'(\theta)}{(\nu^{(0)})'(\theta)}\, d\lambda^{(0)}_{\mathbf{data}}(\theta,\boldsymbol{x})},
\end{aligned}$$

where the third equality in (20) follows because the denominator is simply the expectation of the constant 1, and the fourth equality follows because the constant $\widetilde{C}(0)/\widetilde{C}(\kappa)$ cancels from the numerator and denominator. The last line in (20) is the ratio of two expectations under $\lambda^{(0)}_{\mathbf{data}}(\theta, \boldsymbol{x})$, each of which can be estimated by an average. Let $\big(\theta^{(1)}, \boldsymbol{X}^{(1)}\big), \ldots, \big(\theta^{(G)}, \boldsymbol{X}^{(G)}\big)$ be a sample from $\lambda^{(0)}_{\mathbf{data}}(\theta, \boldsymbol{x})$, and define

$$v^{(\kappa)}_g = \left(\frac{M^{(\kappa)}}{M^{(0)}}\right)^{\#(\boldsymbol{X}^{(g)})} \frac{(\nu^{(\kappa)})'(\theta^{(g)})}{(\nu^{(0)})'(\theta^{(g)})} \qquad g = 1, \ldots, G.$$

Then the numerator and the denominator of the right side of (20) may be estimated by

$$\frac{1}{G}\sum_{g=1}^{G} g(\theta^{(g)}, \boldsymbol{X}^{(g)}) v^{(\kappa)}_g \quad \text{and} \quad \frac{1}{G}\sum_{g=1}^{G} v^{(\kappa)}_g,$$

respectively. Defining the weights $w_g^{(\kappa)} = v_g^{(\kappa)} / \sum_{g=1}^G v_g^{(\kappa)}$, we see that $\int g(\theta, x) d\lambda_{\mathbf{data}}^{(\kappa)}(\theta, x)$ may be estimated by

$$\sum_{g=1}^{G} g(\theta^{(g)}, \boldsymbol{X}^{(g)}) w_g^{(\kappa)}. \tag{21}$$

This is the well-known "ratio estimate" in importance sampling theory; see Hastings (1970).

As an illustration, to estimate $\pi_{\mathbf{data}}^{(\kappa)}\{F(t) \le u\}$, we use

$$\sum_{g=1}^{G} w_g^{(\kappa)} \mathrm{Be}\Big(\alpha_{\theta^{(g)}}^{(\kappa)}(t) + \textstyle\sum_{i=1}^n \delta_{X_i^{(g)}}(t), M_{\theta^{(g)}}^{(\kappa)} + n - \alpha_{\theta^{(g)}}^{(\kappa)}(t) - \sum_{i=1}^n \delta_{X_i^{(g)}}(t)\Big). \tag{22}$$

instead of the unweighted average (3). Of course, if $\kappa = 0$, this formula reduces to (3).

We remark that the specifications that give rise to the simplified formula (18) include the situation discussed in Section 4.1, in which $M^{(\kappa)}$ acts as a tuning parameter that allows us to interpolate between the purely parametric and nonparametric models. In this case, the ratio of the densities of θ in (18) is just 1, and consequently

$$w_g^{(\kappa)} \propto \left(\frac{M^{(\kappa)}}{M^{(0)}}\right)^{\#(\boldsymbol{X}^{(g)})}. \tag{23}$$

Suppose for example that $M^{(\kappa)} > M^{(0)}$. This formula makes precise the intuitive notion that samples $(\theta^{(g)}, \boldsymbol{X}^{(g)})$ for which $\#(\boldsymbol{X}^{(g)})$ is large get counted more heavily than those for which $\#(\boldsymbol{X}^{(g)})$ is small.

We now discuss a important practical limitation of the importance sampling method we have described. When the distribution $\pi^{(\kappa)}$ is far from $\pi^{(0)}$, it will turn out in general that a few of the observations have large weights, while the rest have very small weights. Therefore, the effective sample size may be small even if G is large. We deal with this problem by using an importance sampling distribution $\lambda^{\text{I.S.}}$ which is a mixture of distributions $\lambda_{\mathbf{data}}^{(\kappa_m)}$ for several different values of κ_m. If the hyperparameters κ_m are chosen appropriately, there will be a wide range of κ's for which $\lambda_{\mathbf{data}}^{(\kappa)}$ is close to $\lambda^{\text{I.S.}}$. Thus, having fixed $\kappa_1, \ldots, \kappa_M$, we form

$$\lambda^{\text{I.S.}} = \frac{1}{M} \sum_{m=1}^{M} \lambda_{\mathbf{data}}^{(\kappa_m)}.$$

For each $m = 1, \ldots, M$, we run a Markov chain corresponding to the prior $\pi^{(\kappa_m)}$. So for each m, we have $\{(\theta^{(\kappa_m, g)}, \boldsymbol{X}^{(\kappa_m, g)}); g = 1, \ldots, G\}$, which is a sample from $\lambda_{\mathbf{data}}^{(\kappa_m)}$.

Carrying out importance sampling with respect to $\lambda^{\text{I.S.}}$ is not entirely trivial. The problem is that for each m, $\lambda_{\mathbf{data}}^{(\kappa_m)}$ is known only up to an unknown constant, and there are m unknown constants involved, precluding a simple application of the trick that was used in (20). There are two ways of dealing with this problem. The first, which is the method we implement in this paper, is to use a technique due to Geyer (1994), for which we now give an explanation adapted to the present context.

Consider the set

$$\Big\{\big(\theta^{(\kappa_m,g)}, \boldsymbol{X}^{(\kappa_m,g)}\big);\ g = 1,\dots,G;\ m = 1,\dots,M\Big\}. \tag{24}$$

If we select an element at random from this set, ignore the Markov chain from which it came, and note only its value, which we denote by $(\theta, \boldsymbol{X})$, then the probability that it came from chain m is

$$\frac{\big(\widetilde{C}(\kappa_m)\big)^{-1}\big(M^{(\kappa_m)}\big)^{\#(\boldsymbol{X})}\big(\nu^{(\kappa_m)}\big)'(\theta)}{\sum\limits_{q=1}^{M}\big(\widetilde{C}(\kappa_q)\big)^{-1}\big(M^{(\kappa_q)}\big)^{\#(\boldsymbol{X})}\big(\nu^{(\kappa_q)}\big)'(\theta)}. \tag{25}$$

This probability involves the m unknown constants $\widetilde{C}(\kappa_m)$. Thus, we think of (25) as the "likelihood" of the point $(\theta, \boldsymbol{X})$. If we now proceed blindly and act as if the MG values that we select out of the set (24) are independent, then we get the overall "likelihood"

$$l\big(\widetilde{C}(\kappa_1),\dots,\widetilde{C}(\kappa_M)\big) = \prod_{m=1}^{M}\prod_{g=1}^{G}\frac{\big(\widetilde{C}(\kappa_m)\big)^{-1}\big(M^{(\kappa_m)}\big)^{\#(\boldsymbol{X}^{(\kappa_m,g)})}\big(\nu^{(\kappa_m)}\big)'\big(\theta^{(\kappa_m,g)}\big)}{\sum_{q=1}^{M}\big(\widetilde{C}(\kappa_q)\big)^{-1}\big(M^{(\kappa_q)}\big)^{\#(\boldsymbol{X}^{(\kappa_m,g)})}\big(\nu^{(\kappa_q)}\big)'\big(\theta^{(\kappa_m,g)}\big)}. \tag{26}$$

In this "likelihood," the vector $\big(\widetilde{C}(\kappa_1),\dots,\widetilde{C}(\kappa_M)\big)$ is identifiable only up to an overall constant, i.e. if a is a constant, then $l\big(\widetilde{C}(\kappa_1),\dots,\widetilde{C}(\kappa_M)\big) = l\big(a\widetilde{C}(\kappa_1),\dots,a\widetilde{C}(\kappa_M)\big)$. We fix $\widetilde{C}(\kappa_1) = 1$, maximize (26) with respect to $\widetilde{C}(\kappa_2),\dots,\widetilde{C}(\kappa_M)$, and take these as our estimates of the $\widetilde{C}(\kappa_m)$'s.

Now on the surface, there is no reason to expect that estimates obtained in this way have good statistical properties: Even if we ignore the unidentifiability problem, the actual likelihood is not (26), since (26) assumes that the MG draws from the set (24) are independent. Nevertheless, Geyer (1994) shows that the estimates obtained in this way behave very much like ordinary maximum likelihood estimates, even when the G points from each Markov chain are not independent. Under some regularity conditions, he shows that they are consistent and asymptotically normal.

Having obtained an estimate of the vector $\big(\widetilde{C}(\kappa_1),\dots,\widetilde{C}(\kappa_M)\big)$ up to an overall normalizing constant, we then have (an estimate of) the importance sampling distribution $\lambda^{\text{I.S.}}$ up to a normalizing constant. We are then in a position to use the trick in (20), and arrive at importance sampling weights as before. Specifically, we have

$$w_g^{(\kappa)} \propto \frac{(\nu^{(\kappa)})'(\theta^{(g)})}{\sum_{q=1}^{M} \Big\{ \Big[\Big(\frac{M^{(\kappa_q)}}{M^{(\kappa)}} \Big)^{\#(\boldsymbol{X}^{(g)})} \Big] \Big/ \hat{\tilde{C}}(\kappa_q) \Big\} (\nu^{(\kappa_q)})'(\theta^{(g)})},$$

where $\big(\hat{\tilde{C}}(\kappa_1), \dots, \hat{\tilde{C}}(\kappa_M)\big)$ is the estimate of $\big(\tilde{C}(\kappa_1), \dots, \tilde{C}(\kappa_M)\big)$ up to an overall normalizing constant.

The second method of dealing with the unknown constants $\widetilde{C}(\kappa_1), \dots, \widetilde{C}(\kappa_M)$ is to calculate them directly. Note that the $\widetilde{C}(\kappa_m)$'s differ from the $C(\kappa_m)$'s only by a factor involving the gamma function, so it is sufficient to calculate the $C(\kappa_m)$'s. If we write

$$C(\kappa_m) = \int \pi^{(\kappa_m)}\{X_i \in A_i;\, i = 1, \dots, n \mid \theta\} \nu^{(\kappa_m)}(d\theta),$$

we can evaluate the integrand using known distributional properties of samples from a single Dirichlet. This may work faster than the method using Geyer's idea, but we have not tried it.

Our experience has shown that if instead of running a single chain, we run several chains and use a mixture importance sampling distribution, then for the same total sample size, we can vary the hyperparameters over a *far* greater range than if we run a single chain. It is difficult to quantify the gain, because there are so many variables involved (one of which being the specific data set that is used), but the reader can do his or her own experimentation. It is useful to keep in mind that if we wish to vary only $M^{(\kappa)}$, then the formula for the weights if we run only one chain (equation (23)) indicates that if $M^{(\kappa)}$ is different from $M^{(0)}$ by a factor of, say, 10, then the two posterior distributions may be nearly orthogonal.

4.5 The Environment for Dynamic Graphics

The software for dynamic graphics is written in the lisp dialect implemented in `Lisp-Stat`. It is available in Doss and Narasimhan (1998) as a literate program in the sense of Knuth (1992). Briefly put, a literate program is a program written in a style that makes it easy for humans to read, understand and modify. All our programs can be obtained electronically from `http://www-stat.stanford.edu/~naras/dirichlet-dynamic-graphics`. It must be noted that the software only does sensitivity analysis—no general facility is provided for generating observations from Markov chains. We have, however, included the Fortran programs used in generating the output for our example so that others may use it for running Gibbs samplers in models like ours. The software can handle output from any source provided the samples are available in the proper format. The specific requirements are spelled out in detail in the literate program. We illustrate the capabilities of the software with an example.

Klein and Moeschberger (1997) describe a data set from a retrospective study carried out to compare the cosmetic effects of radiotherapy alone versus radiotherapy and adjuvant chemotherapy on women with early breast cancer. The use of excision biopsy followed by radiation therapy has been suggested as an alternative to mastectomy. This therapy preserves the breast and hence has the benefit of an improved cosmetic effect. The use of adjuvant chemotherapy is indicated to prevent reoccurrence of the cancer, but there is some clinical evidence that it enhances the effects of radiation on normal tissue, thus offsetting the cosmetic benefit of this procedure.

To compare the two treatment regimes, a retrospective study of 46 radiation only and 48 radiation plus chemotherapy patients was made. Patients were observed initially every 4–6 months, but as their recovery progressed the intervals between visits lengthened. At each visit the clinician recorded a measure of breast retraction on a three-point scale (none, moderate, severe). The event of interest was the time to first appearance of moderate or severe breast retraction. Due to the fact that patients were only observed at these random times, the exact time of breast retraction is known only to fall in the interval between visits. The data consist of the interval, in months, in which deterioration occurred (interval-censored observations) or the last time a patient was seen without deterioration having occurred as yet (right-censored observations).

For treatment j ($j = 1, 2$), we assume that the time until deterioration is exponentially distributed, with parameter λ_j. We also assume that λ_1 and λ_2 are independent, with Gamma distribution Gamma(a, b) (the same Gamma for the two λ's).

Several Gibbs samplers were run to estimate the posterior distribution of λ_1 and λ_2 given the data. For the purpose of carrying out the reweighing, eight chains were run for each of the treatments at settings of $M = 1, 4, 16, 64$, $a = 4$, and $b = 120, 240$, with each producing 500 data points.

Figures 1 and 2 show results of feeding the Gibbs sampler outputs to the software. It must be remarked that there are a number of inputs the program needs, such as the names of the data files containing the Gibbs sampler output, the priors corresponding to them, and so on. We have designed the software so that this tedium is encountered at most once; after the first run, the program saves these inputs, thus making subsequent runs easier. When entering the inputs, the user is asked to provide an initial guess for maximizing the "likelihood" given by equation (26). The program then uses a Nelder-Mead maximization routine built into `Lisp-Stat` for estimating the constants of proportionality. All calculations are done on a logarithmic scale when possible, to avoid numeric precision problems.

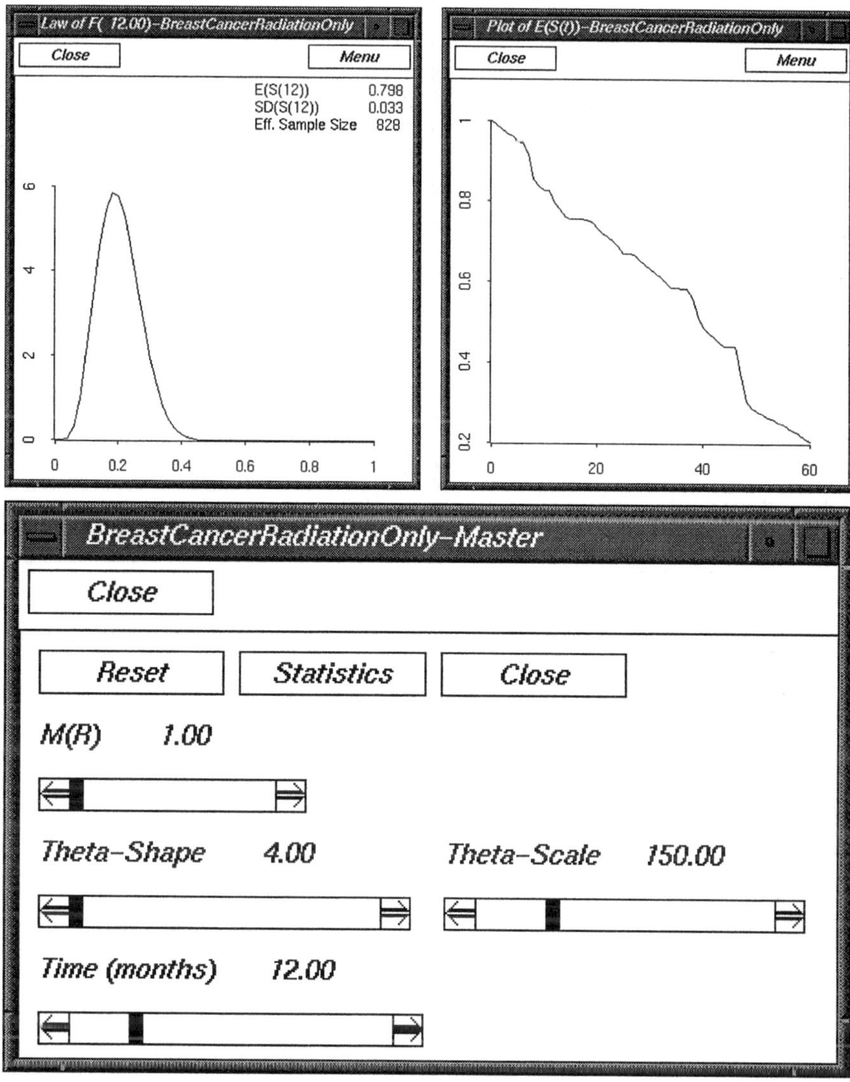

FIGURE 1. Dynamic exploration of posteriors (Breast Cancer Data, radiation only).

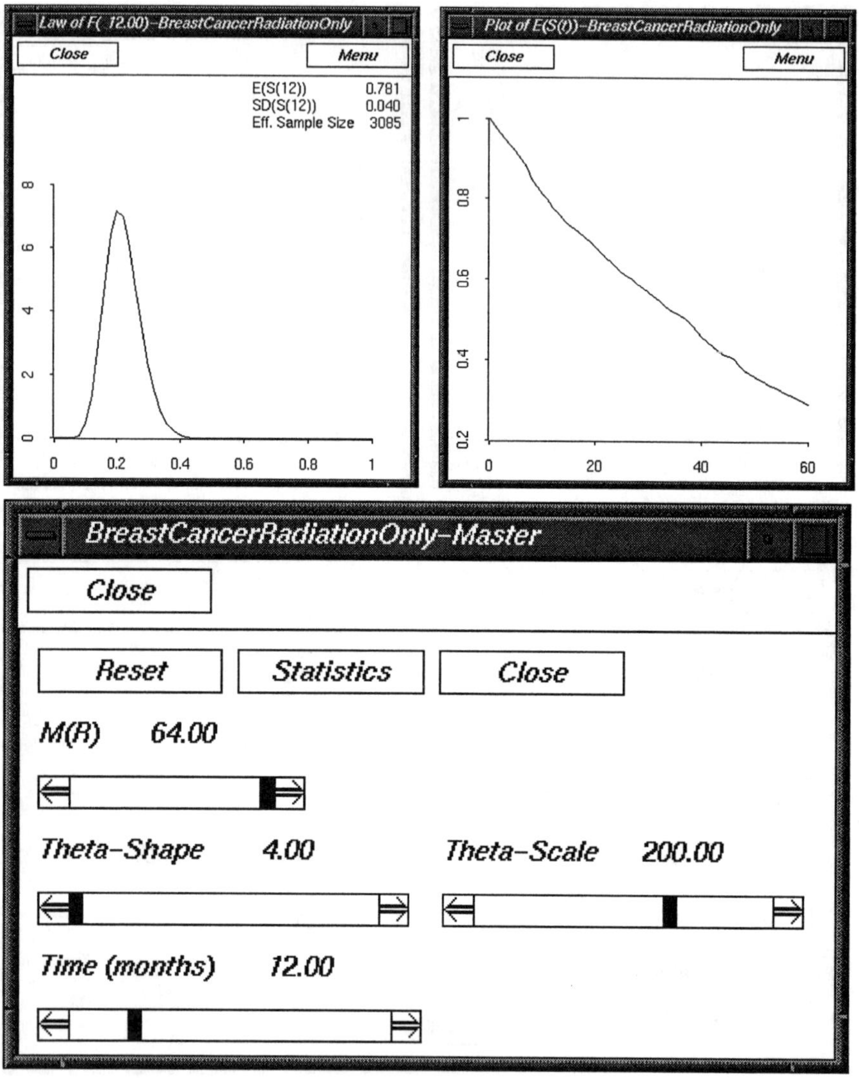

FIGURE 2. Effect of changing hyperparameters (Breast Cancer Data, radiation only).

Once the constants are estimated, a mixture density is computed. The mixture of Beta densities corresponding to equation (22) is plotted, along with $E(S(t))$, the posterior expectation of $S(t)$, for several values of t. Here, $S(t) = 1 - F(t)$. A dialog box contains sliders that allow the user to manipulate the prior hyperparameters. The quantity M is varied by the slider labeled **M(R)**. The bottom slider relates to t, the time in months, and as this is varied, the distribution of $F(t)$ is replotted. Note that varying t alone has no effect on the plot of $E(S(t))$. The values of the parameters are changed by dragging or clicking the mouse on the slider. Several statistics are displayed in the margin of the plot of the density of $F(t)$. The "effective sample size" is calculated via the formula $n/(1 + cv_t^2)$, where n is the total sample size from all Gibbs samplers, and cv_t^2 is the coefficient of variation of the importance weights. This formula is often used when carrying out importance sampling, to give a rough idea of the accuracy of the estimates.

Several buttons are also available in the dialog box. The `Reset` button begins the exploration anew from the initial settings of the hyperparameters, and the `Close` button destroys all windows. The button labeled `Statistics` is important. Clicking on this produces a summary of the hyperparameter settings, the mean of $S(t)$ and the standard deviation of $S(t)$. This is followed by a table of $E(S(t))$ and the standard deviation of $S(t)$ for several values of t. An edited version of the summary is shown below.

```
Hyperparameter Settings:
   M(R)                   =    64.00
   Theta-Shape            =     4.00
   Theta-Scale            =   200.00
   Time (months)          =    60.00
Total Sample Size         =   4000.
Effective Sample Size =   3085.
Mean of S(60):                  0.295
Std. Dev. of S(60):             0.098
```

Table of E(S(t))

Time (months)	E(S(t))	SD(S(t))
0	1.000	0.000
1	0.983	0.008
2	0.966	0.012
3	0.950	0.015
4	0.935	0.018
5	0.918	0.021
6	0.902	0.024
7	0.880	0.029
8	0.851	0.034

Figures 1 and 2 show two instances of the plots with different settings of the parameters. The effect of changing M is evident in the plots of $E(S(t))$: As M increases, the posterior expectation of $S(t)$ resembles a power law, which is what one would get from fitting the purely parametric exponential model with a Gamma prior. Figures 3 and 4 show the plots for the second treatment at the same hyperparameter settings. We note that the survival curves $E(S(t))$ in Figures 1 and 3, in which $M = 1$, look remarkably like the frequentist nonparametric maximum likelihood estimates of Turnbull (1974) (see page 133 of Klein and Moeschberger (1997)).

As we vary the hyperparameters using our software (Figures 1–4 showing snapshots), we conclude that retraction tends to occur earlier for the radiation and chemotherapy group, for a wide range of the hyperparameters M, a, and b. In particular, we draw this conclusion whether our model is reproducing the standard frequentist estimates, or estimates based on a purely parametric fit. (There is one exception, which is that $S(t)$ is actually smaller for the radiation only group for low values of t, such as $t = 12$; this phenomenon had been observed by Klein and Moeschberger (1997).)

In our software, the range within which one can vary M, shape and scale of the Gamma distribution is determined by the minimum and maximum values of these parameters used in running the Gibbs samplers. In our case, since the shape of the Gamma prior was fixed at 2, this is not a parameter that can be changed despite its appearance on a slider.

Table 1 gives a comparison of estimates obtained by reweighting with estimates of the mean and variance of $F(t)$ obtained by actually running a Gibbs sampler at several hyperparameter values. For each month, two rows of estimates are given. Each entry is a pair, the first being the estimate of $E(S(t))$ and the number in parentheses is the estimate of the standard deviation of $S(t)$. For each time, the first row shows the estimate obtained by reweighting and the following row shows the actual estimates obtained by running a Gibbs Sampler at the hyperparameter settings. The agreement is generally quite good, and discrepancies in some cases seem to be due to effective sample sizes being smaller than the sample size used for the actual run of the Gibbs sampler at those settings.

For the breast cancer data, varying the scale of the Gamma prior seems to have little effect, as a direct Gibbs sampler run confirms. However, changing M does affect the distribution of $F(t)$ for any t (although this does not change our overall conclusions in the analysis of this data set).

The software has been designed with extensibility in mind—it is easy to add capabilities so that other quantities that vary as the prior is varied are also shown changing dynamically. Several improvements are planned, including ways of saving snapshots along a given path in the hyperparameter space. Details are given in the literate program. The software, including all data files, may be obtained from the same location as the literate program.

We conclude with some information on the speed of the dynamic graphics. The calculations involved in reweighting the data points produced by

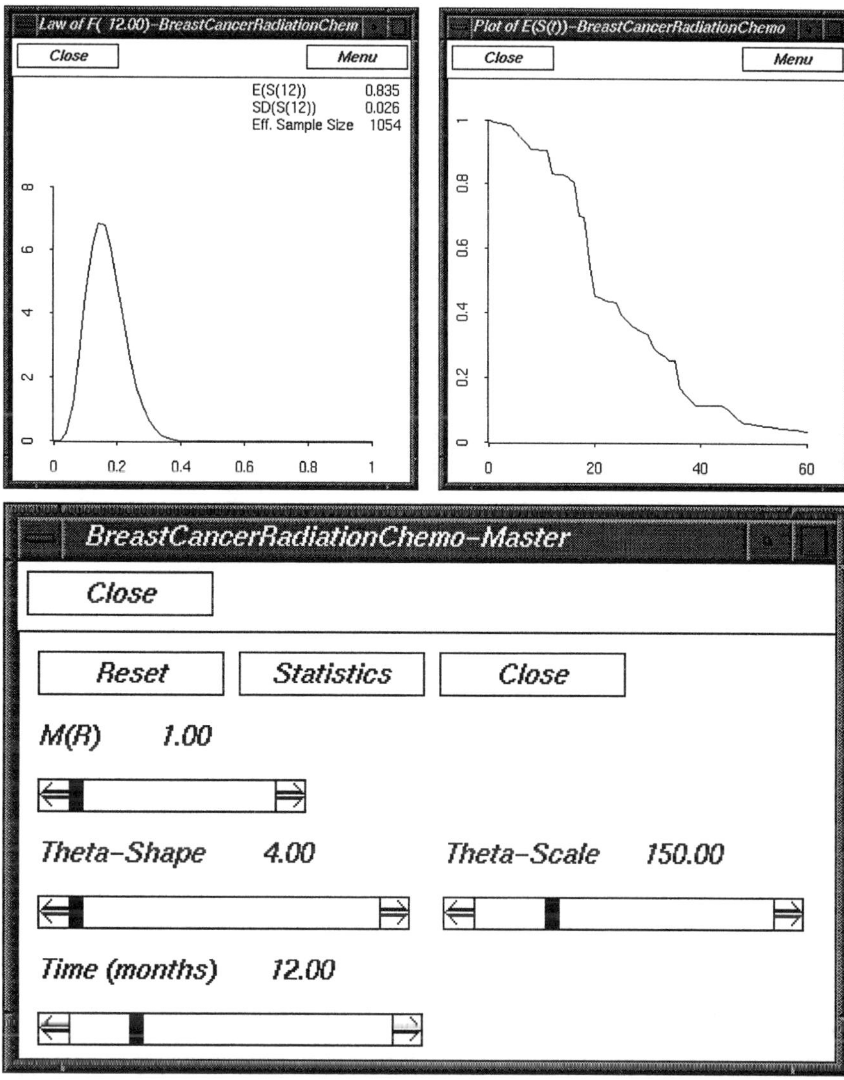

FIGURE 3. Dynamic exploration of posteriors (Breast Cancer Data, radiation plus chemotherapy).

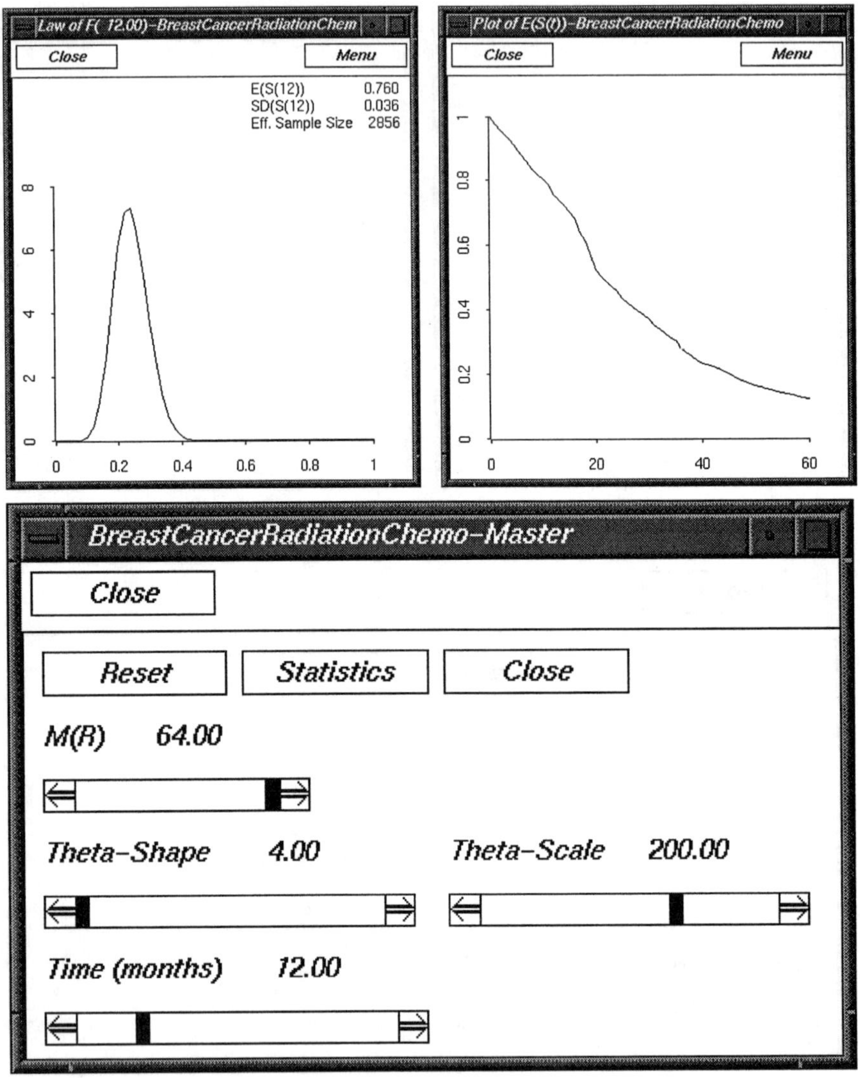

FIGURE 4. Effect of changing hyperparameters Dynamically (Breast Cancer Data, radiation plus chemotherapy).

TABLE 1. Comparison of estimates obtained by reweighting with actual estimates.

	Breast Cancer Data Radiation Only							
	$M = 1$		$M = 20$		$M = 40$		$M = 60$	
	Gamma Shape		Gamma Shape		Gamma Shape		Gamma Shape	
Time	150	200	150	200	150	200	150	200
12	.798(.033)	.798(.033)	.764(.032)	.774(.031)	.748(.040)	.763(.037)	.753(.046)	.771(.041)
	.796(.032)	.796(.032)	.772(.029)	.772(.030)	.755(.039)	.756(.039)	.745(.046)	.745(.046)
24	.689(.036)	.691(.036)	.626(.040)	.640(.039)	.594(.052)	.616(.051)	.594(.063)	.620(.059)
	.690(.035)	.688(.034)	.631(.038)	.632(.039)	599(.052)	.600(.053)	.579(.063)	.579(.063)
36	.586(.029)	.586(.029)	.512(.037)	.526(.038)	474(.052)	.496(.053)	.471(.067)	.499(.065)
	.588(.027)	.587(.027)	.518(.040)	.518(.040)	478(.055)	.478(.055)	.454(.066)	.454(.066)
48	.305(.196)	.318(.190)	.301(.136)	.324(.129)	298(.109)	.325(.104)	.322(.098)	.355(.093)
	.354(.164)	.360(.159)	.371(.056)	.374(.053)	.350(.058)	.351(.057)	.334(.065)	.335(.066)
60	.205(.211)	.224(.211)	.211(.150)	.236(.148)	.212(.120)	.241(.119)	.237(.106)	.271(.103)
	.258(.196)	.262(.192)	.286(.071)	.293(.069)	.272(.062)	.274(.060)	.258(.064)	.260(.064)

	Breast Cancer Data, Radiation plus Chemotherapy							
	$M = 1$		$M = 20$		$M = 40$		$M = 60$	
	Gamma Shape		Gamma Shape		Gamma Shape		Gamma Shape	
Time	150	200	150	200	150	200	150	200
12	.835(.026)	.836(.026)	.792(.028)	.800(.027)	.765(.035)	.777(.034)	.749(.038)	.763(.037)
	.840(.028)	.840(.029)	.804(.029)	.804(.029)	.776(.037)	.776(.036)	.757(.043)	.757(.042)
24	.435(.039)	.436(.039)	.436(.041)	.448(.041)	.439(.048)	.457(.048)	.441(.052)	.460(.052)
	.434(.039)	.435(.039)	.440(.041)	.441(.042)	.448(.051)	.448(.050)	.452(.058)	.454(.057)
36	.170(.075)	.172(.076)	.213(.059)	.225(.060)	.235(.059)	.254(.060)	.255(.059)	.275(.059)
	.178(.072)	.177(.072)	.245(.045)	.246(.045)	.268(.052)	.269(.052)	.280(.060)	.282(.059)
48	.062(.036)	.062(.037)	.107(.038)	.119(.039)	.132(.045)	.149(.048)	.152(.051)	.171(.053)
	.064(.037)	.065(.037)	.136(.039)	.140(.040)	.165(.048)	.165(.049)	.178(.056)	.180(.055)
60	.038(.038)	.041(.039)	.070(.038)	.081(.040)	.089(.044)	.105(.047)	.104(.046)	.121(.050)
	.043(.039)	.045(.039)	.093(.038)	.098(.038)	.113(.043)	.115(.044)	.124(.049)	.126(.049)

several Gibbs samplers are quite substantial even with as few as 500 points from each chain. Acceptable performance requires a reasonably powerful computer. On a machine that does about 75 million floating point operations per second, it takes about 2–3 seconds to update the plots, once any hyperparameter is changed. On any platform, the efficiency may be dramatically improved by dynamically loading C programs that perform the reweighting. Our experience has shown about a 5-fold gain in speed. In order to avoid machine-specific issues that dynamic loading entails, the current version of our software does not use it. The only requirement for using our software is a properly installed version of `Lisp-Stat`. The C programs for dynamic loading may be obtained from the authors, along with instructions on how to use them. Further optimizations are currently being developed to improve performance for those users who do not wish to use dynamic loading.

4.6 Appendix: Completion of the Proof of Proposition 1

We now complete the proof of Proposition 1 by giving a technical argument to justify the interchange of order of limits and integration at the end of Section 4.3. Formally, we wish to show that the family of probability measures $\{\nu_{\boldsymbol{x}};\, \boldsymbol{x} \in R^n\}$ given by (6) defines a "regular conditional distribution of θ given $\boldsymbol{X}$," i.e.

(i) for each Borel subset B of Θ, $\nu_{.}(B)$ is a measurable function of $\boldsymbol{x}$;

(ii) for each Borel subset B of Θ, $\nu_{.}(B)$ is a version of the conditional probability of B given $\boldsymbol{X}$, i.e. for each Borel subset C of R^n, $\int_C \nu_{\boldsymbol{X}}(B)dP = P\{\theta \in B,\, \boldsymbol{X} \in C\}$.

Consider for each $r = 1, \dots, n$, the probability model in which θ has the prior distribution $\nu^{(r)}$ given by

$$\nu^{(r)}(d\theta) \propto \frac{M_\theta^r \Gamma(M_\theta)}{\Gamma(M_\theta + n)}\, \nu(d\theta),$$

and conditional on θ, $X_1, \dots, X_r$ are i.i.d. $\sim H_\theta$. Let $\mu^{(r)}$ be the marginal distribution of $(X_1, \dots, X_r)$ in this model. It is well known that in this model, there exists a regular conditional distribution of θ given $\boldsymbol{X}^{(r)}$, and it is given by

$$\nu^{(r)}_{\boldsymbol{X}^{(r)}}(d\theta) \propto \left(\prod_{j=1}^{r} h_\theta(X_j)\right) \nu^{(r)}(d\theta), \tag{27}$$

where $\boldsymbol{X}^{(r)} = (X_1, \dots, X_r)$. According to a theorem of Pfanzagl (1979), the measures (27) can be obtained as the weak limits as $\eta \to 0$ of the measures $\{\nu^{(r)\eta}_{\boldsymbol{X}^{(r)}}\}$ defined by

$$\nu^{(r)\eta}_{\boldsymbol{X}^{(r)}}(B) = \frac{\int_B \left(\prod_{j=1}^r H_\theta\{(X_j - \eta/2, X_j + \eta/2)\}\right)\nu^{(r)}(d\theta)}{\int_\Theta \left(\prod_{j=1}^r H_\theta\{(X_j - \eta/2, X_j + \eta/2)\}\right)\nu^{(r)}(d\theta)},$$

i.e.

$$\nu^{(r)\eta}_{\boldsymbol{X}^{(r)}} \xrightarrow{d} \nu^{(r)}_{\boldsymbol{X}^{(r)}} \quad \text{for } [\mu^{(r)}] \text{ a.e. } \boldsymbol{X}^{(r)}. \tag{28}$$

Returning now to $\nu^\eta_{\boldsymbol{x}}$, let us write

$$P\{\boldsymbol{X} \in R^n;\, \nu^\eta_{\boldsymbol{X}} \xrightarrow{d} \nu_{\boldsymbol{X}}\} = \sum_{r=1}^n P\{\boldsymbol{X} \in R^n;\, \nu^\eta_{\boldsymbol{X}} \xrightarrow{d} \nu_{\boldsymbol{X}} \mid \#(\boldsymbol{X}) = r\} P\{\#(\boldsymbol{X}) = r\}, \tag{29}$$

and let us consider the conditional probabilities in (29). The conditional distribution of the distinct values of $X_1, \dots, X_n$ given that $\#(\boldsymbol{X}) = r$ (call these $Y_1, \dots, Y_r$) is obtained as follows. For a Borel subset B of R^r,

$$P\{(Y_1, \dots, Y_r) \in B \mid \#(\boldsymbol{X}) = r\} = \int P\{(Y_1, \dots, Y_r) \in B \mid \#(\boldsymbol{X}) = r,\, \theta\}\, \nu\big(d\theta \mid \#(\boldsymbol{X}) = r\big). \tag{30}$$

Now if $F \sim \mathcal{D}_{\alpha_\theta}$ for some $\theta \in \Theta$ and if $X_1, \dots, X_n$ are i.i.d. $\sim F$, then, as is well known, the distinct values $Y_1, \dots, Y_r$ are i.i.d. $\sim H_\theta$. This implies that the right side of (30) is equal to

$$\int H^{(r)}_\theta(B)\, \nu(d\theta \mid \#(\boldsymbol{X}) = r).$$

Here, the measure $H^{(r)}_\theta$ denotes the product of H_θ with itself r times. From this we see that, conditional on $\#(\boldsymbol{X}) = r$, the distinct values $Y_1, \dots, Y_r$ have a distribution which is absolutely continuous with respect to the measure on R^r given by

$$\int H^{(r)}_\theta(\cdot)\, \nu(d\theta),$$

and this in turn is absolutely continuous with respect to the measure

$$\int H^{(r)}_\theta(\cdot)\, \nu^{(r)}(d\theta).$$

Hence, (28) implies that each of the conditional probabilities in (29) is 1, and this proves the proposition.

Acknowledgments: This work was supported by Air Force Office of Scientific Research Grant F49620-94-1-0028.

References

Antoniak, C. (1974). Mixtures of Dirichlet processes with applications to Bayesian nonparametric problems. *Ann. Statist.* **2**, 1152–1174.

Diaconis, P. and Freedman, D.A. (1986a). On the consistency of Bayes estimates. *Ann. Statist.* **14**, 1–26.

Diaconis, P. and Freedman, D.A. (1986b). On inconsistent Bayes estimates of location. *Ann. Statist.* **14**, 68–87.

Doss, H. (1984). Bayesian estimation in the symmetric location problem. *Z. Wahrsch. verw. Gebiete* **68**, 127–147.

Doss, H. (1985a). Bayesian nonparametric estimation of the median; Part I: computation of the estimates. *Ann. Statist.* **13**, 1432–44.

Doss, H. (1985b). Bayesian nonparametric estimation of the median; Part II: asymptotic properties of the estimates. *Ann. Statist.* **13**, 1445–64.

Doss, H. (1994). Bayesian nonparametric estimation for incomplete data via successive substitution sampling. *Ann. Statist.* **22**, 1763–1786.

Doss, H. and Huffer, F. (1998). Monte Carlo methods for Bayesian analysis of survival data using mixtures of Dirichlet priors. Technical Report, Department of Statistics, Ohio State University.

Doss, H. and Narasimhan, B. (1998). Dynamic display of changing posterior in Bayesian survival analysis: the software. Technical Report, Department of Statistics, Stanford University.

Ferguson, T.S. (1973). A Bayesian analysis of some nonparametric problems. *Ann. Statist.* **1**, 209–230.

Ferguson, T.S. (1974). Prior distributions on spaces of probability measures. *Ann. Statist.* **2**, 615–629.

Geyer, C.J. (1994). Reweighting Monte Carlo mixtures. Technical Report 568r, Department of Statistics, University of Minnesota.

Hastings, W.K. (1970). Monte Carlo sampling methods using Markov chains and their applications. *Biometrika* **57**, 97–109.

Hjort, N.L. (1990). Goodness of fit tests in models for life history data based on cumulative hazard rates. *Ann. Statist.* **18**, 1221–1258.

Klein, J. and Moeschberger, M. (1997). *Survival Analysis.* Springer-Verlag, New York.

Knuth, D.E. (1992). *Literate Programming.* Center for Study of Language and Information, Stanford University.

Kong, A., Liu, J.S. and Wong, W.H. (1994). Sequential imputations and Bayesian missing data problems. *J. Amer. Statist. Assoc.* **89**, 278-288.

Pfanzagl, J. (1979). Conditional distributions as derivatives. *Ann. Probab.* **7**, 1046–1050.

Tierney, L. (1990). *Lisp-Stat: An Object-Oriented Environment for Statistical Computing and Dynamic Graphics.* Wiley, New York.

Turnbull, B.W. (1974). Nonparametric estimation of a survivorship function with doubly censored data. *J. Amer. Statist. Assoc.* **69**, 169–173.

Turnbull, B.W. (1976). The empirical distribution function with arbitrarily grouped, censored, and truncated data. *J. Roy. Statist. Society, Ser. B* **38**, 290–295.

5

Semiparametric Bayesian Methods for Random Effects Models

Joseph G. Ibrahim
Kenneth P. Kleinman

ABSTRACT The linear mixed effects model with normal errors is a popular model for the analysis of repeated measures and longitudinal data. The generalized linear model is useful for data that has non-normal errors but where the errors are uncorrelated. A descendant of these two models generates a model for correlated data with non-normal errors, called the generalized linear mixed model (GLMM). Frequentist attempts to fit these models generally rely on approximate results and inference relies on asymptotic assumptions. Recent methodological and technological advances have made Bayesian approaches to this class of models feasible. Markov chain Monte Carlo methods can be used to obtain 'exact' inference for these models, as demonstrated by Zeger and Karim (1991). In the linear or generalized linear mixed model, the random effects are typically taken to have a fully parametric distribution, such as the normal distribution. In this chapter, we extend the normal random effects model and the GLMM by allowing the random effects to have a nonparametric prior distribution. We do this using a Dirichlet Process prior for the general distribution of the random effects. The approach can be easily extended to more general population models. Computations for the models are accomplished using the Gibbs sampler.

5.1 Introduction

Generalized linear models (McCullagh and Nelder 1989; Nelder and Wedderburn 1972) are a unified approach to regression methods. They can be applied to a wide array of discrete, continuous, and censored outcomes, and are most commonly used when the outcomes are independent. However, in many applications, this independence is not a reasonable assumption. This is particularly obvious in longitudinal studies, where multiple measurements made on the same individual are likely to be correlated.

For continuous outcomes with normal errors, Laird and Ware (1982) present the random effects model, which we refer to throughout as the normal linear random effects model. In this model, a subject-specific covariance structure is generated by assuming that each individual has a unique set of regression coefficients, the random effects, which are distributed around the mean regression coefficients for the population, also known as the fixed effects. There may also be regression coefficients which are equal for all individuals. Conditional on the random effects, repeated observations on a subject are considered independent, while marginalizing over the random effects, a unique covariance structure for the observations within each subject is obtained.

Zeger and Karim (1991) present a generalization of the normal linear random effects model to the class of generalized linear models, generating a generalized linear mixed model (GLMM). They frame the model from the Bayesian perspective and fit it using a Gibbs sampler. They point out that attempts to fit this model using classical (frequentist) techniques are limited by the need for multidimensional numerical integrations, except in special cases. As a result of these analytically intractable integrations, classical approaches to GLMMs have relied on approximations to maximum likelihood techniques (Breslow and Clayton 1993).

Kleinman and Ibrahim (1998a, 1998b) describe a semi-parametric Bayesian version of the normal linear random effects model and the GLMM, where a nonparametric prior distribution is specified for the random effects. We summarize their work in this chapter. We note that attempts to fit models of this class using classical techniques are hindered by the need for additional multidimensional numerical integrations, except in special cases (Aitkin 1996).

The desirability of this approach from a Bayesian perspective is immediately clear, since restricting the model to normally distributed random effects may be contrary to our prior beliefs. In any event, the nonparametric prior will be robust to misspecification at this stage of the model. For the normal linear random effects model, Kleinman and Ibrahim (1998a) present an example where inference about the regression coefficients is sensitive to the assumption of normality about the random effects. This example is discussed in Section 7.1. From the classical perspective, it has been shown that large changes in parameter estimates can be caused by small changes in the distribution of the random effects (Heckman and Singer 1984, Davies 1987). Verbeke and Lesaffre (1996) show that the normal linear random effects model can also perform poorly when the random effects have a mixture distribution,

The nonparametric Bayesian approach for the random effects is to specify a prior distribution on the space of all possible distribution functions. This prior is applied to the general prior distribution for the random effects. This can be accomplished with a Dirichlet Process prior distribution. Thus, for random effects models, this means that the usual Normal prior on the random effects is replaced with a non-parametric prior, followed by a Dirichlet Process prior on that general distribution. Methodological details of this approach are discussed elsewhere in this volume by MacEachern, and Escobar and West.

The rest of this article is organized as follows. In Section 1.2 we provide a more detailed description of the normal linear random effects model. In Section 1.3 we show how to apply the MDP structure to the normal random effects model and in Sections 1.4 and 1.5, we apply the MDP technology to the GLMM. In Section 1.6 we illustrate our methodology with real data and, in addition, compare our approach to the fully parametric Bayesian model. In Section 1.7 we discuss our results and propose directions for future research.

5.2 Normal Linear Random Effects Models

First, we define the normal linear random effects model, then introduce random effects into generalized linear models. For individual i, with n_i repeated measurements, the Normal linear random effects model for outcome vector y_i is given by

$$y_i = X_i\beta + Z_i b_i + e_i, \quad i, \ldots, N$$

where y_i is $n_i \times 1$, X_i is an $n_i \times p$ matrix of fixed covariates, β is a $p \times 1$ parameter vector of regression coefficients, commonly referred to as fixed effects in these models, Z_i is an $n_i \times v$ matrix of covariates for the $v \times 1$ vector of random effects b_i, and e_i is an $n_i \times 1$ vector of errors. It is standard in implementations of this model to assume e_i and b_i are independent and that both are distributed Normal, with $e_i \sim N_{n_i}(0, \sigma^2 I_{n_i})$ and $b_i \sim N_v(0, D)$, where I_s is the $s \times s$ identity matrix and $N_s(\mu, \Sigma)$ denotes the s-dimensional multivariate Normal distribution with mean μ and covariance matrix Σ. Under these assumptions,

$$[y_i|\beta, b_i] \sim N_{n_i}(X_i\beta + Z_i b_i, \sigma^2 I_{n_i}). \tag{1}$$

Throughout, we denote the conditional distribution of A given B by $[A|B]$. For the sake of convenience, we call model (1) the normal random effects model. In this chapter, we relax the normal assumption for the b_i, and allow $b_i \sim G$ where G is a general distribution. To accomplish that goal, we use the Dirichlet process prior, discussed elsewhere in this volume by MacEachern, and Escobar and West.

5.3 DP priors in the Normal Linear Random Effects Model

In this section we describe how the MDP model can be applied to the normal linear random effects model. We show the full conditional distributions and how Gibbs sampling can be implemented for both the conjugate and a non-conjugate case. Section (5.3.1) describes in more detail a special case of a model presented in West, Müller, and Escobar (1994). Section (5.3.2) presents an extension of that work to a non-conjugate MDP model.

5.3.1 Multivariate Normal base measure

If we assume the sampling distribution for y to be Normal, then a Normal base measure for the random effects completes a conjugate MDP model. Let the distribution of the vector of outcomes y_i for subject i be:

$$[y_i|\beta, b_i, \sigma^2] \sim N_{n_i}(X_i\beta + Z_ib_i, \sigma^2 I_{n_i}). \tag{2}$$

Letting $\tau = \sigma^{-2}$, the prior specifications are

$$\begin{aligned} \tau &\sim Gamma(\frac{\alpha_0}{2}, \frac{\lambda_0}{2}) \\ \beta &\sim N_p(\mu_0, \Sigma_0) \\ b_i &\sim G \end{aligned}$$

$$G \sim DP(M \cdot N_v(0, D)). \tag{3}$$

The model (2) implies that there are p 'fixed' effects and v random effects.

When G is a fully parametric prior the joint posterior can be easily found, as in Wilks et al. (1993). Then, following usual algebraic routes, the full conditionals for β and τ are

$$[\beta|b, \tau, y] \sim N_p(Ta, T) \tag{4}$$

where $T = (\tau \sum_{i=1}^n X_i^T X_i + \Sigma_0^{-1})^{-1}$ and $a = \sum_{i=1}^n X_i^T(y_i - Z_ib_i) + \Sigma_0^{-1}\mu_0$ and

$$[\tau|\beta, b, y] \sim Gamma(\frac{n + \alpha_0}{2}, \frac{\sum_i r_i^T r_i + \lambda_0}{2}) \tag{5}$$

where $r_i = y_i - X_i\beta - Z_ib_i$. These full conditionals are unchanged by the MDP model, and can be inserted as dsiplayed above.

Applying the usual MDP results, we find as in West, Müller, and Escobar (1994)

$$p(b_i|\beta, \tau, y, b_{-i}) \propto \sum_{j \neq i} \phi(y_i|X_i\beta + Z_ib_j, \tau I_{n_i}) \cdot \delta_{b_j} +$$

$$\left\{ M \int \phi(y_i|X_i\beta + Z_ib_i, \tau I_{n_i}) \phi(b_i|0, D) db_i \right\} \cdot \phi(b_i|0, D) p(y_i|b_i, \beta, \tau, y_j)$$

where b_{-i} denotes the random effects for the subjects excluding subject i. Also, δ_s is a degenerate distribution with point mass at s. After some algebra, we get

$$p(b_i|\beta,\tau,y,b_{-i}) \propto$$
$$\left(\sum_{j\neq i} \tau^{\frac{n_i}{2}} \exp\left[\frac{-\tau}{2}(y_i - X_i\beta - Z_i b_j)^T(y_i - X_i\beta - Z_i b_j)\right]\cdot \delta_{b_j}\right) +$$
$$M|Q_i|^{\frac{1}{2}}|D|^{\frac{-1}{2}}\tau^{\frac{n_i}{2}} \exp\left\{\frac{\tau}{2}\left[(y_i - X_i\beta)^T U_i (y_i - X_i\beta)\right]\right\}\cdot$$
$$\phi(b_i|0,D)p(y_i|b_i,\beta,\tau) \quad (6)$$

where $Q_i = (D^{-1} + \tau\sum_i Z_i^T Z_i)^{-1}$ and $U_i = (\tau Z_i Q_i Z_i^T - I)$. So with probability proportional to

$$\tau^{\frac{n_i}{2}} \exp(\frac{-\tau}{2}(y_i - X_i\beta - Z_i b_j)^T(y_i - X_i\beta - Z_i b_j))$$

we select from distribution δ_{b_j}, which means that we set $b_i = b_j$. Also, with probability proportional to

$$M|Q_i|^{\frac{1}{2}}|D|^{\frac{-1}{2}}\tau^{\frac{n_i}{2}} \int p(y_i|b_i,\beta,\tau,y_j)\phi(b_i|0,D)db_i,$$

we select from

$$p(b_i|\beta,\tau,y_i) \propto \phi(b_i|0,D)p(y_i|b_i,\beta,\tau,y_j),$$

meaning we sample b_i from its full conditional,

$$[b_i|\beta,\tau,y_i] \sim N_v(\tau Q_i Z_i^T(y_i - X_i\beta), Q_i).$$

This results in a mixture distribution where one piece is a Normal distribution and all of the others are point masses.

There is some plausible intuition behind the above mixture scheme. If subject i has a relatively large residual using subject j's random effect b_j, then b_j is relatively less likely to be chosen as the random effect for subject i. Conversely, if subject i has a relatively small residual using subject j's random effect, then the random effect b_j is relatively more likely to be chosen as the random effect for subject i. On the other hand, the greater the residuals for subject i, assuming its random effect parameters are all 0, the greater the probability that subject i gets a new value as its random effect.

Typically, the covariance matrix D in the base measure of the Dirichlet Process in model (3), is unknown, and therefore a suitable prior distribution must be specified for it. Note that once this has been accomplished, the base measure is no longer marginally Normal.

For convenience, suppose

$$D^{-1} \sim Wishart(d_0, c_0 R_0),$$

where $d_0 \geq v$, $c_0 > 0$ and R_0 is a $v \times v$ positive definite matrix. Then *a priori*

$$p(D^{-1}|d_0, c_0, R_0) \propto |D^{-1}|^{\frac{d_0 - v - 1}{2}} \exp\left\{\frac{-1}{2} tr((c_0 R_0)^{-1} D^{-1})\right\}$$

where $tr(\cdot)$ denotes trace.

After choosing random effects for each subject, the subjects will be grouped into clusters in which the subjects have equal b_is. That is, after selecting a new b_i for each subject i in the sample, there will be some number k, $0 < k \leq n$ of unique values among the b_is. Denote these unique values by $\gamma_l, l = 1 \ldots k$. Additionally, let l represent the set of subjects with common random effect γ_l. Note that knowing the random effects is equivalent to knowing k, all of the γs, and the cluster memberships l. Then for the purposes of calculating the full conditionals of D^{-1}, the γ_l are k independent observations from $N_v(0, D)$. Thus, as in West, Müller, and Escobar (1994),

$$p(D^{-1}|b, \beta, y, \tau) = p(D^{-1}|\gamma, l, \beta, y, \tau)$$

$$\propto |D^{-1}|^{\frac{d_0 + k - v - 1}{2}} \exp\left\{-\frac{1}{2} tr((c_0 R_0)^{-1} D^{-1} - \frac{1}{2}\sum_{l=1}^{k} \gamma_l^T D^{-1} \gamma_l)\right\},$$

so that

$$[D^{-1}|b, \beta, y, \tau] \sim Wishart\left(d_0 + k, ((c_0 R_0)^{-1} + \sum_{1}^{k} \gamma_l \gamma_l^T)^{-1}\right). \quad (7)$$

One additional piece of the model is recommended by Bush and MacEachern (1996) as an aid to convergence for the Gibbs sampler. To speed mixing over the entire parameter space, they suggest moving around the γs after determining how the b_is are grouped. The conditional density of γ_l is

$$p(\gamma_l|\beta, \tau, b, D, y) \propto \phi(\gamma_l|0, D) \prod_{i \in l} p(y_i|\beta, b, D, \tau),$$

which implies that

$$[\gamma_l|\beta, \tau, b, y] \sim N(\tau Q_l \sum_{i \in l} Z_i^T (y_i - X_i \beta), Q_l), \quad (8)$$

where $Q_l = (D^{-1} + \tau \sum_{i \in l} Z_i^T Z_i)^{-1}$.

5.3.2 A Modified MDP Model

Samples of the b_i from a modified distribution can be obtained by specifying the base measure as a multivariate normal scale mixture. The multivariate t distribution can be obtained as a scale mixture of the multivariate Normal distribution as follows. If

$$[x|\eta] \sim N_p(\mu, \eta^{-1}\Sigma), \tag{9}$$

$$\eta \sim Gamma(\frac{s}{2}, \frac{s}{2}) \ , \tag{10}$$

then the marginal distribution of x is $St_p(s, \mu, \Sigma)$, where $St_p(s, \mu, \Sigma)$ is a p-dimensional Student-t distribution with s degrees of freedom, mean μ, and dispersion matrix Σ. From a sampling perspective, sampling from

$$\eta \sim Gamma(\frac{s}{2}, \frac{s}{2})$$

and then from

$$[x|\eta] \sim N_q(\mu, \eta^{-1}\Sigma),$$

is equivalent to sampling from $St_p(s, \mu, \Sigma)$. Thus, marginally, the distribution of x is multivariate t, with η being integrated out. This representation is often used in the Gibbs sampling literature; see Wakefield, Smith, Racine-Poon, and Gelfand (1995) and some of the references therein.

The representation (9) and (10) of the multivariate t allows a closed-form solution to q_0 and yields samples of the b_i from a modified distribution. The specifications for the modified MDP model are as follows.

$$\begin{aligned}
[y_i|\beta, b_i, \sigma^2] &\sim N_{n_i}(X_i\beta + Z_i b_i, \tau I_{n_i}) \\
\tau &\sim Gamma(\frac{\alpha_0}{2}, \frac{\lambda_0}{2}) \\
\beta &\sim N_p(\mu_0, \Sigma_0) \\
b_i &\sim G
\end{aligned}$$

$$G \sim DP(M \cdot N_v(0, \eta^{-1}D)) \tag{11}$$

and

$$\eta \sim Gamma(\frac{\rho}{2}, \frac{\rho}{2}). \tag{12}$$

We note that with a common η for all i, we generate dependent samples for the b_i's. To obtain independent samples for the b_i's, we must specify a separate η_i for each b_i, and take the η_i's to be $i.i.d.$ gamma variates for $i = 1, \ldots, n$.

Once again, it may be helpful to examine the parametric random effects case where $G = N(0, \eta^{-1}D)$. Here

$$p(\beta, \tau, b, \eta|y) \propto e^{\left\{\frac{-\tau}{2}\left[\sum_{i=1}^{n}(y_i - X_i\beta - Z_ib_i)^T(y_i - X_i\beta - Z_ib_i) + \lambda_0\right] - \frac{1}{2}(\beta-\mu_0)^T\Sigma_0^{-1}(\beta-\mu_0)\right\}} \cdot e^{\left\{\frac{-\eta}{2}\left[\sum_{i=1}^{n} b_i^T D^{-1} b_i + \rho\right]\right\}} \eta^{\frac{nv+\rho}{2}-1} \tau^{\left(\frac{\alpha_0 + \sum n_i}{2} - 1\right)} \quad (13)$$

The full conditionals for β and τ in the parametric multivariate t model are the same as in the parametric Normal model, but

$$[b_i|\eta, \beta, \tau, y] \sim N(\tau Q_i^\eta Z_i^T(y_i - X_i\beta), Q_i^\eta)$$

where $Q_i^\eta = (\tau Z_i^T Z_i + \eta D^{-1})^{-1}$, and

$$[\eta|b, \beta, \tau, y] \sim Gamma\left(\frac{\rho + nv}{2}, \frac{\rho + \sum_i b_i^T D^{-1} b_i}{2}\right).$$

For the modified MDP model in (11) and (12), the full conditional distributions for β and τ are the same as in the Normal MDP case.

From the usual MDP results we find

$$p(b_i|\beta, \tau, y, b_{-i}, \eta) \propto \sum_{j \neq i} \phi(y_i|X_i\beta + Z_ib_j, \tau I_{n_i}) \cdot \delta_{b_j} +$$

$$\left\{M \int \phi(y_i|X_i\beta + Z_ib_i, \tau I_{n_i})\phi(b_i|0, \eta^{-1}D)db_i\right\} \cdot \phi(b_i|0, \eta^{-1}D)p(y_i|b_i, \beta, \tau).$$

After some algebraic reduction, we get

$$p(b_i|\beta, \tau, y, b_{-i}) \propto \left(\sum_{j \neq i} \tau^{\frac{n_i}{2}} \exp\left[\frac{-\tau}{2}(y_i - X_i\beta - Z_ib_j)^T(y_i - X_i\beta - Z_ib_j)\right] \cdot \delta_{b_j}\right) + M|Q_i^\eta|^{\frac{1}{2}}|D|^{\frac{-1}{2}}\tau^{\frac{n_i}{2}}\eta^{\frac{\rho}{2}} \exp\left\{\frac{\tau}{2}\left[(y_i - X_i\beta)^T U_i^\eta (y_i - X_i\beta)\right]\right\} \cdot \phi(b_i|0, D)p(y_i|b_i, \beta, \tau, y_j) \quad (14)$$

where $U_i^\eta = (\tau Z_i Q_i^\eta Z_i^T - I)$ and Q_i^η is as defined in the parametric multivariate t model. So with probability proportional to

$$\tau^{\frac{n_i}{2}} \exp\left(\frac{-\tau}{2}(y_i - X_i\beta - Z_ib_j)^T(y_i - X_i\beta - Z_ib_j)\right),$$

$b_i = b_j$. And with probability proportional to

$$M|Q_i|^{\frac{1}{2}}|D|^{\frac{-1}{2}}\tau^{\frac{n_i}{2}}\eta^{\frac{v}{2}} \exp\left\{\frac{\tau}{2}\left[(y_i - X_i\beta)^T U_i^\eta (y_i - X_i\beta)\right]\right\},$$

$$[b_i|\beta, \tau, \eta, y] \sim N(\tau Q_i^\eta Z_i^T(y_i - X_i\beta), Q_i^\eta).$$

Also,

$$p(\eta|b, \beta, \tau, y) \propto \eta^{\frac{kv+\rho}{2}-1} \exp\left\{\frac{-\eta}{2}\left[\sum_{l=1}^{k} \gamma_l^T D^{-1}\gamma_l + \rho\right]\right\},$$

which leads to

$$[\eta|b, \beta, \tau, y] \sim Gamma\left(\frac{\rho + kv}{2}, \frac{\rho + \sum_1^k \gamma_l^T D^{-1}\gamma_l}{2}\right). \quad (15)$$

As in the Normal case, it is desirable to introduce a Wishart prior on D^{-1}. Suppose $D^{-1} \sim Wishart(d_0, c_0 R_0)$. Then

$$p(D^{-1}|b, \beta, \tau, \eta, y) = p(D^{-1}|\gamma, l, \beta, \tau, \eta, y)$$

$$\propto |D^{-1}|^{k/2} \exp(\frac{-1}{2}\sum_{l}^{k} \eta\gamma_l^T D^{-1}\gamma_l) \cdot |D^{-1}|^{\frac{d_0-v-1}{2}} \exp(-\frac{1}{2}tr((c_0R_0)^{-1}D^{-1}))$$

so that

$$[D^{-1}|b, \eta, \beta, y, \tau] \sim Wishart(d_0 + k, ((c_0R_0)^{-1} + \sum_{l}^{k} \eta\gamma_l\gamma_l^T)^{-1}) \quad (16)$$

Also as in the Normal MDP case, we will want to move around the γ_l. For the modified MDP model,

$$p(\gamma_l|\beta, b, \tau, D, \eta, y) \propto \phi(\gamma_l|0, \eta^{-1}D) \prod_{i \in l} p(y_i|\beta, b, \tau, D)$$

so that

$$\gamma_l|\beta, b, \tau, D, y \sim N(\tau Q_l^\eta(\sum_{i \in l} Z_i(y_i - X_i\beta), Q_l^\eta) \quad (17)$$

where $Q_l^\eta = (\eta D^{-1} + \tau \sum_{i \in l} Z_i^T Z_i)^{-1}$.

5.4 Generalized Linear Mixed Models

Suppose the sampling distribution of y_{it}, $t = 1, \ldots, n_i$ is from the exponential family, so that

$$p(y_{it}|\theta_{it}, \tau) = \exp\{\tau [y_{it}\theta_{it} - a(\theta_{it})] + c(y_{it}, \tau)\}$$

where

$$\mu_{it} = E(y_{it}|\theta_{it}, \tau) = \frac{da(\theta_{it})}{d\theta_{it}}$$

and

$$v_{it} = \mathrm{var}(y_{it}|\theta_{it}, \tau) = \tau^{-1} \frac{d^2 a(\theta_{it})}{d\theta_{it}^2}$$

where τ is a scalar dispersion parameter.

In the generalized linear mixed model, the canonical parameter θ_{it} is related to the covariates by

$$h(\theta_{it}) = \eta_{it} = x_{it}^T \beta + z_{it}^T b_i.$$

where x_{it} and z_{it} are rows of the X_i and Z_i matrices, $h(\cdot)$ is a monotonic differentiable function, often referred to as the θ-link, and η_{it} is called the linear predictor. Throughout, we write

$$p(y_{it}|\theta_{it}, \tau) \equiv p(y_{it}|\beta, b_i, \tau),$$

where

$$p(y_{it}|\beta, b_i, \tau) = \exp\left\{\tau\left[y_{it} h^{-1}(\eta_{it}) - a(h^{-1}(\eta_{it}))\right] + c(y_{it}, \tau)\right\}. \qquad (18)$$

When $h(\theta_{it}) = \theta_{it} = \eta_{it}$, then the link is said to be the canonical link.

For example, in GLMM logistic regression, we have

$$p(y_{it}|\beta, b_i, \tau) = \exp\left\{y_{it}(x_{it}^T\beta + z_{it}^T b_i) - \log(1 + e^{x_{it}^T\beta + z_{it}^T b_i})\right\}$$

where $\theta_{it} = \eta_{it} = x_{it}^T\beta + z_{it}^T b_i$ and $\tau \equiv 1$.

Note that the GLMM imitates the normal random effects model in that we assume that conditional on the random effect b_i, the repeated observations on subject i are independent. Thus the likelihood for N subjects in the GLMM is

$$p(y|\beta, b, \tau) \propto \prod_{i=1}^{N} \prod_{t=1}^{n_i} p(y_{it}|\beta, b_i, \tau) \qquad (19)$$

where $b = (b_1, \ldots, b_N)^T$ and $y = (y_{11}, \ldots, y_{Nn_N})^T$.

There are several attractive properties of (19). First, it takes within-subject correlation into account while allowing each individual to have a unique correlation structure and maintaining independence between subjects. Second, the model accommodates unbalanced data, in that response vectors need not be of the same length. Similarly, irregularly timed measurements can be fit with this model without any adjustment. Finally, posterior distributions or estimates of the random effects have interpretive value when the trend of the mean function of individuals is of interest.

Zeger and Karim (1991) assume

$$\beta \sim N_p(\mu_0, \Sigma_0),$$

where (μ_0, Σ_0) are considered known and fixed and there are p elements in β. Typically, it is assumed that

$$b_i \sim N_v(0, D)$$

where there are v random effects. Again, we assume that D^{-1} has a Wishart prior and we relax the normal assumption for the b_i, and allow

$$b_i \sim G$$

where G is a general distribution.

5.5 DP priors in the Generalized Linear Mixed Model

In this section we describe how the MDP model can be applied to the generalized linear mixed model (GLMM). Assume that the base measure for the b_is is Normal. Then any exponential family sampling distribution completes a MDP GLMM. This will be a non-conjugate MDP model except when the data have a normal sampling distribution. Let the distribution of the outcome y_{it} for subject i at time t be $p(y_{it}|\beta, b_i, \tau)$ as given in (18). The prior specifications for the parameters of the MDP GLMM are

$$\begin{aligned} \tau &\sim Gamma(\alpha_0, \lambda_0) \\ \beta &\sim N_p(\mu_0, \Sigma_0) \\ b_i &\sim G \\ G &\sim DP(M \cdot N_v(0, D)). \end{aligned} \tag{20}$$

When G is a fully parametric prior, we can write down the joint posterior density up to a constant of proportionality. Suppose G is a v-dimensional normal distribution with mean 0 and covariance matrix D, as in the standard (fully parametric) GLMM. Given D, the joint posterior for the parameters is

$$p(\beta, \tau, b|y) \propto \exp\{\log p(y|\beta, b, \tau)\}\ \pi(\beta, b, \tau) \propto$$
$$e^{\left\{\sum_{i=1}^N \sum_{t=1}^{n_i} \log p(y_{it}|\beta, b_i, \tau) - \frac{1}{2}(\beta-\mu_0)^T \Sigma_0^{-1}(\beta-\mu_0) - \tau\lambda_0 - \frac{1}{2}\sum_{i=1}^N b_i^T D^{-1} b_i\right\}} \tau^{\alpha_0 - 1}. \tag{21}$$

where $\pi(\cdot)$ denotes the joint prior density and $b = (b_1, \ldots, b_n)$. However if G has the form of (20), it is impossible to write down the joint posterior density of the parameters, because there is not a common dominating measure. The special case where $G = N_v(0, D)$ is included here because the conditional distributions of β and τ can be found through it in the usual way.

From equation (21) and the usual MDP results, we can get the full conditional distributions needed for Gibbs sampling.

Following usual algebraic routes, we get

$$p(\beta|b,\tau,y) \propto \exp\left(\sum_{i=1}^{N}\sum_{t=1}^{n_i}\log p(y_{it}|\beta,b_i,\tau) - \frac{1}{2}(\beta-\mu_0)^T\Sigma_0^{-1}(\beta-\mu_0)\right).$$

Unless y_{it} has the normal distribution, sampling from this full conditional will not be straightforward. But it can still be accomplished, using for example a Metropolis step. (See Metropolis et al. 1953, and Hastings 1970.) The full conditional distribution of τ is

$$p(\tau|\beta,b,y) \propto \tau^{\alpha_0-1}e^{\{\sum_{i=1}^{N}\sum_{t=1}^{n-i}c(y_{it},\tau)\}}e^{\{\sum_{i=1}^{N}\sum_{t=1}^{n_i}[y_{it}\theta_{it}-a(\theta_{it})]-\tau\lambda_0\}}.$$

Sampling from the full conditional distribution of τ can also be accomplished through a Metropolis step, unless $\log(c(y_{it},\tau))$ takes a form proportional to $\tau^{f(y_{it})}$. In such a case,

$$[\tau|\beta,b,y] \sim Gamma(f(y_{it})+\alpha_0, \lambda_0 - [y_{it}\theta_{it} - a(\theta_{ij})]).$$

Using the usual MDP results, we find

$$p(b_i|\beta,\tau,y,b_{-i}) \propto \sum_{j\neq i}^{N}\exp\left\{\sum_{t=1}^{n_i}\log p(y_{it}|\beta,b_j,\tau)\right\}\cdot\delta_{b_j} +$$

$$\left[M\int e^{\{\sum_{t=1}^{n_i}\log p(y_{it}|\beta,b_i,\tau)\}}\phi(b_i|0,D)db_i\right]\cdot\phi(b_i|0,D)\prod_{t=1}^{n_i}p(y_{it}|\beta,b_i,\tau) \tag{22}$$

where b_{-i} denotes the random effects for the subjects excluding subject i. The intuition behind equation (22) is identical to that in the normal rendom effects model.

When the sampling distribution is not normal, q_0 generally will not have a closed form. To avoid numerical integration or approximation, we use the algorithm of MacEachern and Müller (1996). Recall that there are $k \leq N$ unique random effects among the N subjects. These random effects, which we label γ_l, $l = 1,\ldots,k$ are independent draws from $G_0(\cdot)$ which in this case is $N_v(0,D)$. The algorithm requires that we sample an additional $N-k$ values from $N_v(0,D)$; we label these values $\gamma_{k+1},\ldots,\gamma_N$. Then, if $n_l > 1$, $i \in l$ we sample from the following full conditional;

$$p(b_i|y,\gamma,\beta,\tau) = \sum_{l=1}^{k}n_l^{-}\prod_{t=1}^{n_i}p(y_{it}|\beta,\tau,\gamma_l)\delta_{\gamma_l} +$$

$$\frac{M}{k^*+1}\prod_{t=1}^{n_i}p(y_{it}|\beta,\tau,\gamma_{k+1})\delta_{\gamma_{k+1}} \tag{23}$$

where $k^* = k$ and n_l^- is the number of subjects sharing the random effect γ_l excluding subject i. The valuc of $\prod_{t=1}^{n_i} p(y_{it}|\beta, \tau, \gamma_l)$ is the likelihood of subject i's data using the random effect belonging to some group of subjects. Thus the effect of the distribution (23) has a sensible interpretation. The greater subject i's likelihood with random effect γ_l and the greater number of other subjects who share that random effect, the more likely it is that γ_l will be selected as subject i's random effect. On the other hand, the scalar parameter M regulates the probability that subject i gets a new random effect, meaning that they start a new cluster. If $n_l = 1$, $i \in l$, then with probability $\frac{k}{k+1}$ we leave b_i unchanged. Otherwise, we let b_i be distributed according to equation (23) except with $k^* = k - 1$.

We may end up with one fewer clusters after drawing b_i from (23), though this can only happen when $n_l = 1$, $i \in l$. In other words, we may have the case that $n_l = 0$ after drawing from the conditional distribution of b_i. If this occurs we switch the values γ_l for the empty group and γ_k for the last group. The set memberships l and k are also switched. Before drawing from the full conditional in such a case, the number of clusters is k, and afterwards there are $k - 1$ clusters. Before the draw, there were n_k subjects in set k sharing the random effect γ_k. After the draw and the switching, the set these subjects are in simply has a different label l, $l < k$, and the random effect formerly held by subject i is labeled γ_k. In the full conditional of b_{i+1}, there are only $k - 1$ clusters, thus the value γ_k from the full conditional of b_i will be used as γ_{k+1} in drawing b_{i+1} from (23).

As in Section (5.3.1) we place a Wishart prior on D^{-1} with the result that

$$[D^{-1}|b, \beta, y, \tau] \sim Wishart\left(d_0 + k, ((c_0 R_0)^{-1} + \sum_1^k \gamma_l \gamma_l^T)^{-1}\right). \quad (24)$$

Additonally, we again shuffle the γs. The conditional density of γ_l is

$$p(\gamma_l|\beta, \tau, b, D, y) \propto \phi(\gamma_l|0, D) \left\{ \prod_{i \in l} \prod_{t=1}^{n_i} p(y_{it}|\beta, \gamma_l, \tau) \right\}$$

A metropolis sampler or some other technique must be applied to draw a sample from this distribution as well.

5.6 Applications

Here we present examples involving the normal linear random effects model, as well as the logistic and Poisson GLMM's. The logistic and Poisson GLMM's are simpler than the general framework described above, in that $\tau \equiv 1$. The purpose of the analyses is to make comparisons between the MDP and fully parametric analyses of the same data. We also demonstrate the computations and inference for these models.

5.6.1 Continuous Longitudinal Repeated Measures

This section contains an analysis of a longitudinal repeated measurements data set. The data come from a study of AIDS patients performed by the AIDS Clinical Trials Group of the National Institute of Allergy and Infectious Disease. The data are public, and come from the study labeled ACTG 116B/117. For a full description of the study, see Kahn, Lagakos, and Richman, et al. (1992).

The study was intended to test the effectiveness of switching from zidovudine (AZT) to didanosine (ddI), in light of the fact that the effectiveness of zidovudine is known to diminish with time. The design called for subjects who had been taking zidovudine for at least 4 months prior to enrollment in the study. After enrolling in the study, subjects were randomized to receive a continuation of zidovudine, 500 milligrams per day of didanosine, or 750 milligrams per day of didanosine. As part of the data collection, CD4 cell counts were measured at baseline and at 2, 8, 12, 16, and 24 weeks after enrollment in the study.

For the purposes of simplifying the example, only the zidovudine group and the 750 mg didanosine group are included in this analysis. Additionally, subjects were further required to have been taking zidovudine for 6 months in order to be included in the analysis. Finally, at least two CD4 count measurements were required for inclusion. These restrictions left a total of 151 individuals with 659 measurements between them.

We fit a model to the log CD4 cell counts including an intercept (x_0), time in weeks (x_1), a treatment group indicator (x_2), and a time by treatment interaction (x_3) as fixed covariates; an intercept(z_0) and time in weeks (z_1) were included as covariates for the random effects. The time was centered around its mean in both the fixed and random covariate matrix. This was done to reduce correlation among the parameters, and is recommended in Gibbs sampling for linear models to decrease autocorrelation among the samples. See Gilks and Roberts (1996) for a further discussion.

The data from the 500 mg didanosine group, which was not used in the analysis, was used to elicit parameters for the prior distribution of the covariance matrix D in the base measure. This elicitation technique is analogous to using data from a previous experiment to specify prior distributions. For more information about elicitation procedures, see Ibrahim and Laud (1994), Laud and Ibrahim (1995), Ibrahim, Ryan, and Chen (1998), or Chen, Ibrahim, and Yiannoutsos (1998). We emphasize that the data used to generate the priors are not part of the data used in the analysis. In any case, this elicitation is for illustrative purposes.

Prior parameters for the Wishart prior on D^{-1} were obtained in the following manner. First, d_0 was chosen to be 10, and $c_0 = 1$. Then $d_0^{-1}R_0$ was taken to have the same diagonal elements as $\hat{D}^{-1}$, where $\hat{D}$ is the maximum likelihood estimate of D for the 500 mg didanosine group. The off diagonal elements of R_0 were taken to be 0. Thus

$$R_0 = \begin{pmatrix} ...0926 & 0 \\ 0 & 113.9 \end{pmatrix}.$$

Though it might be desirable in some circumstances to have a flatter prior on the covariance matrix in the base measure, this can lead to non-convergence in parametric random effects models. (See Cowles, Carlin, and Connett (1996) for an example.)

Relatively flat priors were chosen for all other parameters. In particular, $\mu_0 = (0\ 0\ 0\ 0)^T$, $\Sigma_0 = 10{,}000 I_4$, $\alpha_0 = 2$, and $\lambda_0 = .0002$. A range of values for the parameter M were chosen to reflect small, moderate, and large departures from normality for the distribution of the random effects. A value of $M = 27$ reflects a small departure from normality, with the average number of clusters $\bar{k} \approx 50$. A value of $M = 5$ reflects a moderate departure from normality, with $\bar{k} \approx 25$. A value of $M = 1.25$ reflects a large departure from normality, with $\bar{k} \approx 12$. Table 1 shows posterior medians and highest posterior density (HPD) regions for parameters in the multivariate Normal parametric and MDP models with the three values of M.

	Normal MDP model			Usual RE
Par.	$M = 1.25$	$M = 5$	$M = 27$	$M = 10^7$
β_0	(2.67,3.52,4.30)	(2.98,3.63,4.50)	(3.22,3.74,4.28)	(3.40,3.67,3.94)
β_1	(-.049,-.028, -.0082)	(-.044,-.028, -.011)	(-.039,-.027, -.015)	(-.037,-.028, -.018)
β_2	(-.29,.45,.62)	(-.29,.37,.50)	(-.38,.08,.53)	(-.33,.03,.39)
β_3	(-.002,.009,.021)	(-.002,.009,.022)	(-.004,.010,. 022)	(-.003,.010,.024)
σ^2	(.16,.20,.21)	(.16,.18,.21)	(.15,.17,.20)	(.15,.17 ,.19)
$b_{0,148}$	(-1.4,-.50,.53)	(-1.53,-.56,.21)	(-1.38,-.71,-.00)	(-1.15,-.63,-.07)
$b_{0,149}$	(.18,.84,.1.9)	(.08,.80,1.57)	(.33,.89,1.42)	(.59, 1.01,1.42)
$b_{1,149}$	(-.03,.01,.04)	(-.03,.00,.04)	(-.03,.00,.0 32)	(-.03,.00,.04)
$D(0,0)$	(.84,1.53,3.22)	(.80,1.37,2.40)	(.80,1.20,1.86)	(.86,1.07,1.34)
$D(0,1)$	(-.02,.00,.02)	(-.01,.00,.02)	(-.01,.00,.01)	(-.01,.00,.01)
$D(1,1)^1$	(.38,.73,1.66)	(.38,.68,1.73)	(.34,.5 89,1.04)	(.36,.58,.90)
$\bar{k}$	12.0	21.2	48.0	151

TABLE 1. 2.5%-, 50%-, and 97.5%-iles for various parameters from Normal MDP models and the parametric Normal model. β_0 is the intercept, β_1 is the slope over time, β_2 is the treatment effect, β_3 is the slope for the treatment-time interaction, and σ^2 is the error variance. $b_{0,i}$ is the intercept for subject i, and $b_{1,i}$ is its slope over time. $D(0,0)$ is the variance of the intercepts in the base measure, $D(1,1)$ is the variance of the slopes, and $D(0,1)$ is their covariance. $\bar{k}$ is the average number of clusters observed in the course of sampling.

Table 2 reproduces these results for multivariate t parametric and Modified base measure MDP models with the same values of M. Gibbs samplers were run for 22,000 iterations, with the first 500 being discarded as a burn-in period. In addition, due to high autocorrelation, every tenth iteration was used, and the rest discarded, making a total sample size of 2150. Convergence of the Gibbs sampler was assessed via the Geweke (1992) method, using the CODA (Best, Cowles, and Vines, 1995) suite of diagnostics in S-plus. Most of the parameters had Geweke statistics within ±1.96, indicating convergence is plausible. A more appropriate measure might be the technique of Raftery and Lewis (1992), which is designed for testing the convergence of estimates of percentiles. However, results from this technique depend on the amount of autocorrelation in the iterates. Thus in this case, it indicated that infeasibly large runs of the Gibbs sampler would be necessary for convergence.

	Modified MDP model			Usual RE
Parameter	$M = 1.25$	$M = 5$	$M = 27$	$M = 10^7$
β_0	(2.75,3.54,4.71)	(2.90,3.61,4.36)	(3.36,3.78,4.27)	(3.43,3.68 ,3.92)
β_1	(-.05,-.03,-.01)	(-.05,-.03,-.01)	(-.04,-.03,-.02)	(-.04,-.03,-.02)
β_2	(-.30,.45,.63)	(-.34,.31,.62)	(-.42,.06,.53)	(-.30,.03,.38)
β_3	(-.003,.009,.02)	(-.003,.010,.02)	(-.003,.010,.02)	(-.003,.010,.023)
σ^2	(.16,.19,.22)	(.16,.18,.21)	(.15,.18,.20)	(.15,.17,.19)
$b_{0,148}$	(-1.79,-.50,.51)	(-1.48,-.58,.30)	(-1.38,-.76,-.08)	(-1.14,-.62,-.09)
$b_{1,148}$	(-.04,-.01,.02)	(-.04,-.01,.03)	(-.04,-.01, .03)	(-.05,-.00,.04)
$b_{0,149}$	(-.01,.82,1.64)	(.25,.86,1.59)	(.34,.86,1.36)	(.56,.99,1.39)
$b_{1,149}$	(-.03,.01,.04)	(-.03,.00,.04)	(-.03,.00,.03)	(-.03,.00,.04)
$D(0,0)$	(.81,1.60,3.42)	(.82,1.50,2.92)	(.84,1.45,2.62)	(.83,1.36,2.18)
$D(0,1)^1$	(-1.95,.16,2.32)	(-1.59,.12,1.73)	(-1.09,.093,1.38)	(-.76,.10,.99)
$D(1,1)^2$	(.38,.74,1.61)	(.37,.70,1.47)	(.37,.65,1.22)	(.38,.66,1.13)
k	11.9	21.5	48.1	151

TABLE 2. 2.5%-, 50%-, and 97.5%-iles for various parameters from the modified MDP models and the parametric multivariate t model. Parameters have the same meaning as in the Normal case.

The values in table 1 and table 2 show an unusual result. Except for β_2, the coefficient for the treatment indicator, the medians of the posterior distributions of the regression coefficients $\beta = (\beta_0\ \beta_1\ \beta_2\ \beta_3)^T$ remain approximately the same. The reason for this deviation is shown in the density estimates presented in figures 1a-d. We see that while there is a piece of the density which is near 0 for all values of M, that as M decreases, the distribution of β_2 takes on a bimodal appearance with a great deal of weight centered around .5. While the HPD region includes 0 for all the values of M used here, it is reasonable to suppose that for a suitably small M, the analysis would indicate that a treatment difference of 0 was unlikely.

This is a clear example of why it is important to correctly model the distribution of the random effects; very different results may be obtained as a result of using the MDP model rather than the fully parametric model. Thus the model we have presented is certainly valuable in this case, and may be useful in others as well. The 750 mg didanosine group has been coded as '1' in this example, and the zidovudine group as '0', so these results suggest that in fact switching to 750 mg of didanosine may be better than staying on zidovudine, a fact not in evidence using a fully parametric analysis. Note that the time covariate has been centered, so that this represents the group difference at the middle of the study.

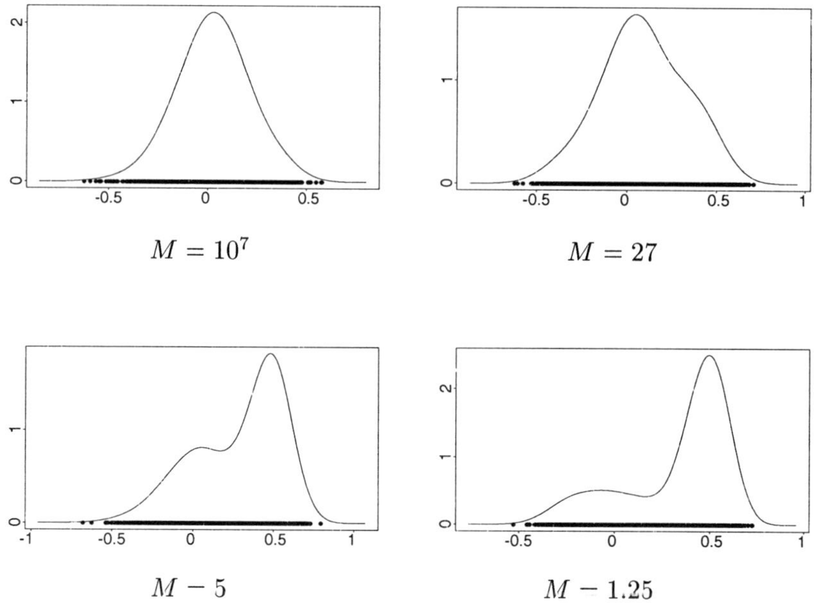

FIGURE 1. Vague prior on β.

From the posterior distribution of the selected random effects, different facts are revealed. Not only are the HPD regions different, but the median values are very different as well. This is to be expected, as the random effects are directly affected by the relaxation of the Normal assumption. Some experimentation with smaller data sets shows that subjects' random effects tend to be shrunken from OLS regression estimates for a subject toward the nearest largest mode in the distribution of the random effects. The shrinkage is more pronounced in subjects where the deviations of the single measurements from a straight line are large. This behavior seems intuitive in light of the mixing distribution shown in equation (6). The elements of the matrix D are not easy to interpret in the MDP model, since they have a very complex role in the marginal posterior distribution of the b_i.

Another conclusion may be drawn from the evidence in table 1 and table 2. The main difference between the two tables is that the HPD regions for the random effects are wider under the multivariate t base measure. The similarity of the results in table 1 and table 2 indicates that for these data, the results are not sensitive to the choice of one of these two base measures.

5.6.2 Correlated Binary Data

This section presents an analysis of longitudinal repeated binary measurements. Zeger and Karim (1991) analyze a subset of data from a study of respiratory infections in Indonesian children. An analysis of the full data set can be found in Sommer, Katz, and Tarwotjo (1983). The children, all preschoolers, were seen quarterly for up to six quarters. At each examination, the presence or absence of respiratory infection was noted and is the outcome in this analysis. The covariates modeled by Zeger and Karim are an intercept, age in months, presence/absence of xeropthalmia, cosine and sine terms for the annual cycle, height for age as a percentage of the National Center for Health Statistics standard, and presence/absence of stunting, defined as being below the 85th percentile in height for age. Age in months was centered at 36 and height for age was centered at 90%. Xeropthalmia is a symptom of chronic vitamin A deficiency and height for age is an indicator of long-term nutritional status. In addition, Zeger and Karim model a random intercept for each child. To facilitate comparison with the results of Zeger and Karim, we use the same model that they use. The original study followed 3000 children. Zeger and Karim use a subset of 250 of these children. To speed computations we take a random subset of 50 from the set of Zeger and Karim.

Since no prior information regarding respiratory infections in this population was available, parameters for the Wishart prior on D^{-1} were chosen in the following fashion. First, d_0 was chosen to be 10. Based on experience with fully parametric logistic GLMM, we wanted $d_0^{-1}R_0$, the prior expected value of D^{-1} to be 0.5. Thus we chose $c_0 = 1$ and $R_0 = 5$. Note that since D is scalar, this is equivalent to a Gamma prior on D^{-1}.

Relatively flat priors were chosen for the other parameters. We let $\mu_0 = (0\ 0\ 0\ 0\ 0\ 0\ 0\ 0)^T$ and $\Sigma_0 = 10,000 I_8$. This is equivalent to saying that the *a priori* probability of respiratory infection is .5 for all subjects across all seasons, but that great uncertainty exists as to the accuracy of this assumption. Without previous experience in this population, this vague prior seems appropriate.

Three values of the parameter M were chosen to reflect large, moderate, and small departures from normality for the distribution of the random effects. A value of $M = .75$ reflects a large departure from normality with the average number of clusters $\bar{k} \approx 5$. A value of $M = 200$ reflects a moderate departure from normality with $\bar{k} \approx 20$. A value of $M = 10^8$ suggests that the distribution of the random effects is very nearly normal with $\bar{k} \approx 45$. Finally, the fully parametric case was also modeled by choosing M large enough that $\bar{k} = 50$.

In table 3, we present results from the MDP GLMM analysis along with the fully parametric GLMM results. Gibbs samplers were run for 25,000 iterations, with the first 3,000 discarded as a burn-in. In addition, due to high autocorrelation, only every tenth iterate was used, with the remainder discarded. This makes for a total sample size of 2,200. Convergence was assessed by the methods of Geweke (1992) and Raftery and Lewis (1992) using the CODA (Best, Cowles, and Vines, 1995) suite of diagnostics in S-plus. Most of the parameters in each of the four samplers had Geweke statistics within ± 1.96, indicating that convergence is plausible. Under the conditions of burn-in and thinning described above, the 2.5 percentile of the posterior is within .01 of the observed 2.5 percentile with probability .8.

	MDP GLMM model			FP GLMM
Par.	$M = .75$	$M = 200$	$M = 10^8$	$M \approx \infty$
β_0	(-7.03,-3.76,-1.97)	(-5.06,-3.47,-2.26)	(-4.56,-3.27,-2.25)	(-4.56,-3.29,-2.24)
β_1	(-.088,-.047,-.014)	(-.094,-.048,-.011)	(-.092,-.048,-.013)	(-.091,-.047,-.01)
β_2	(-1.08,1.15,3.12)	(-.84,1.44,3.43)	(-.65,1.50,3.48)	(-.54,1.32,3.30)
β_3	(-1.58,-.26,.86)	(-1.94,-.52,.77)	(-1.69,-.31,.93)	(-1.65,-.34,.91)
β_4	(-2.18,-1.08,-.11)	(-2.23,-1.14,-.191)	(-2.14,-1.08,-.14)	(-2.17,-1.09,-.13)
β_5	(-2.01,-.97,-.05)	(-2.16,-.99,-.001)	(-2.07,-.94,.01)	(-2.08,-.97,-.08)
β_6	(-.36,-.15,.03)	(-.35,-.14,.04)	(-.36,-.16,.01)	(-.36,-.16,.02)
β_7	(-2.43,-.19,2.1)	(-2.20,.05,2.20)	(-2.49,-.12,1.92)	(-2.24,-.04,1.86)
D	(.36,.79,2.22)	(.50,1.26,3.41)	(.45,.98,2.34)	(.48,.99.2.32)
b_1	(-1.02,1.22,4.58)	(-.87,1.09,3.45)	(-2.16,.31,3.34)	(-2.06,.26,2.46)
b_2	(-2.32,.46,4.24)	(-3.16,-.01,2.66)	(-3.06,.02,2.31)	(-2.36,-.05,2.28)
b_3	(-.78,1.38,4.56)	(-.73,1.26,3.53)	(-1.78,.52,2.91)	(-1.73,.57,3.05)
$\bar{k}$	4.47	19.0	43.2	50

TABLE 3. Posterior 2.5%-, 50%-, and 97.5%-iles for various parameters from MDP logistic GLMM and the fully parametric Poisson GLMM. β_0 is the intercept, β_1 is the effect of age, β_2 is the effect of xeropthalmia, β_3 is the gender effect, β_4 is the effect of the seasonal cosine, β_5 is the effect of the seasonal sine, β_6 is the effect of height for age, and β_7 is the effect of stunting. b_i is the intercept for subject i. For the MDP models, $D(0,0)$ is the variance of the random effects in the base measure. For the fully parametric model, D is the variance of the random effects. $\bar{k}$ is the average number of clusters observed in the course of sampling.

The results in table 3 can be read as follows. In general, many of the medians and 95% highest posterior density (HPD) regions for the population-mean effects are strikingly similar. The exceptions to this rule are the effects for gender, xeropthalmia, stunting, and the intercept. The posterior median of the intercept is similar in all four models, but the 2.5 percentile is progressively more negative as the distribution of the random effects becomes less normal. The same may be said for the effects of xeropthalmia. There is no discernible pattern for the effects of gender or stunting. In no case does the 95% HPD region change from excluding 0 to including it across the models, with the exception of the seasonal cosine, where the 95% HPD region hovers around 0 in all the models. The posterior distributions are most affected by the changing model if the covariates are binary. These parameters are the ones with the highest correlation with the intercept, the continuous covariates being roughly centered at 0. Since it is the distribution of the random intercepts which are directly affected by the changing models, it is unsurprising that the posterior distributions of parameters highly correlated with the population-mean intercept should be most affected.

The largest effect of introducing the non-parametric piece of the model is on the intercept, where it indicates that an even more extreme value may be the population mean. In this model, this means that the probability of respiratory infection when all the other covariates are 0 may plausibly be smaller under the MDP model than under the fully parametric GLMM. This is also demonstrated in the kernel estimates of the posterior distributions presented in figure 2.

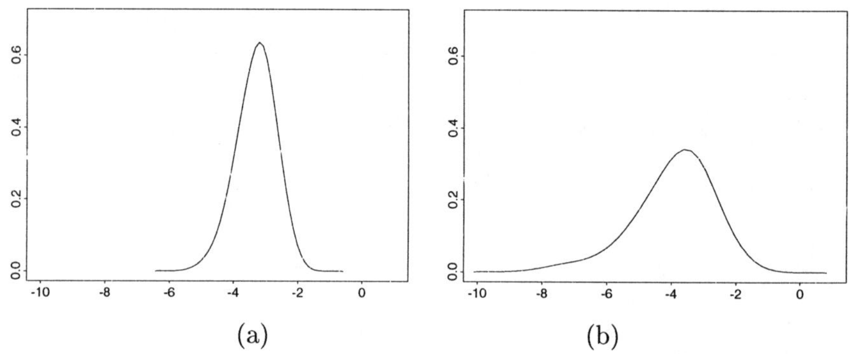

FIGURE 2. Posterior distribution of the intercept: Fully parametric model (panel (a)) and MDP model with $M = .75$ (panel (b)).

In comparing the distributions in figures 2ab, we see that the marginal posterior distribution of β_0 based on the semiparametric model (figure 2b) has much heavier tails than the one based on the fully parametric model (figure 2a). Also, we see that the distribution in figure 2a appears to be symmetric, while the distribution in figure 2b is quite asymmetric. However, the medians and modes of the two distributions are similar.

The introduction of the MDP stage in the model introduces larger changes to the posterior distributions of the random effects. For the three particular random effects tabulated, the medians are markedly different under each of the models. If one were interested in the probability that a particular child would have a respiratory infection, the MDP models would give results very unlike those from the fully parametric model. Finally, the posterior distribution of D^{-1} seems different under the MDP models, but the role of this parameter is not the same under the MDP and fully parametric models and therefore no straightforward comparison can be made.

5.6.3 Correlated Count Data

This section presents an analysis of longitudinal repeated count measurements. Thall and Vail (1990, table 2) present data from a study of seizures in epileptic patients. They use a classical analysis to fit a random intercept to each individual's data and a random effect for each visit across patients. This data was also analyzed by Breslow and Clayton (1993) using an approximate classical method called penalized quasilikelihood (PQL). We will fit a model described by Breslow and Clayton.

The study included 59 epileptic patients who were randomized to either a treatment or a placebo as an adjuvant treatment to standard chemotherapy. Subjects were evaluated every two weeks to determine the number seizures that occurred during the previous two-week period. This process was repeated for 4 measurements, or 8 weeks after randomization. Baseline data available as covariates include the number of seizures in the 8 weeks prior to enrollment in the study and age in years. To be consistent with the previous analysis, we use the log of one fourth of the number of seizures (*Base*) and the the log of age in years (*Age*) as population-mean covariates. The full set of population-mean effects includes an intercept (*Int*), *Base*, a treatment effect (*Trt*), an interaction effect between *Base* and *Trt* (*Base*× *Trt*), and *Age*. Finally, the visit time in weeks is centered and divided by 10 (*Visit*) to give the final population-mean covariate. The random effects include *Int* and a *Visit* effect. Our model thus corresponds to model IV of Breslow and Clayton.

In the absence of similar analyses of this treatment, parameters for the Wishart prior on D^{-1} were obtained in the following manner. First, d_0 was chosen to be 10. Based on previous experience with fully parametric Poisson GLMM, we wanted $d_0^{-1}R_0$, the prior expected value of D^{-1}, to be diagonal with values of 0.5 on the diagonal. Thus we chose $c_0 = 1$ and

$$R_0 = \begin{pmatrix} 5 & 0 \\ 0 & 5 \end{pmatrix}.$$

Relatively vague priors were chosen for the population-mean parameters. In particular, $\mu_0 = (0\ 0\ 0\ 0\ 0\ 0)^T$ and $\Sigma_0 = 10,000 I_6$. Two values of the parameter M were chosen to reflect moderate and large departures from normality for the distribution of the random effects. A value of $M = 100$ reflects a moderate departure from normality, with the average number of clusters $\bar{k} \approx 20$. A value of $M = 1.5$ reflects a large departure from normality, with $\bar{k} \approx 10$.

In table 4, we present results from the MDP GLMM analysis of the model along with the fully parametric Bayesian GLMM analysis and the PQL results of Breslow and Clayton. Recall that the fully parametric model is equivalent to the MDP model when $M \to \infty$. The fully parametric Bayesian results were generated by the BUGS program. Gibbs samplers were run for 25,000 iterations, with the first 3,000 discarded as a burn-in. In addition, due to high auto-correlation among the samples, every tenth iterate was used to make posterior inference, the rest being discarded. This makes for a total sample size of 2,200. Convergence of the Gibbs sampler was assessed by the methods of Geweke (1992), and Raftery and Lewis (1992) using the CODA (Best, Cowles, and Vines, 1995) suite of diagnostics in S-plus. Most of the parameters in each model had Geweke statistics within ± 1.96, indicating that convergence is plausible. Due to the lower autocorrelation in the iterates from this analysis than from the binary outcome example, the Raftery and Lewis diagnostic shows that the burn-in and the use of every tenth iterate suggest that the 2.5 percentile of the posterior is within .01 of the observed 2.5 percentile with probability .9.

	MDP GLMM model		FP GLMM	PQL
Parameter	$M = 1.5$	$M = 100$		
β_0	(-7.0,-3.13,.51)	(-6.84,-2.87,.48)	(-4.57,-.72,2.86)	(-3.6,-1.27,1.08)
β_1	(.54,.97,1.4)	(.62,.98,1.36)	(.59,.90,1.14)	(.60,.87,1.14)
β_2	(-2.9,-1.32,.39)	(-2.85,-1.35,-.01)	(-1.78,-.81,-.09)	(-1.71,-.91,-.11)
β_3	(-.15,.55,1.18)	(.00,.55,1.15)	(-.09,.26,.74)	(-.08,.33,.74)
β_4	(-.16,.88,1.74)	(-.09,.92,2.04)	(-.76,.32,1.34)	(-.25,.46,1.17)
β_5	(-2.54,-.44,1.74)	(-1.49,-.37,.78)	(-.58,-.27,.05)	(-.57,-.26,.05)
$D(0,0)$	(.28,.48,.94)	(.18,.35,.82)	(.23,.35,.52)	(.40,.52,.64)
$D(0,1)$	(-.37,.02,.51)	(-.34,.02,.41)	(-.17,-.00,.17)	(-.07,-.01,.05)
$D(1,1)$	(.41,.84,2.79)	(.42,.98,2.38)	(.31,.55,1.02)	(.43,.74,1.05)
$\bar{k}$	10.6	17.7	-	-

TABLE 4. Posterior 2.5%-, 50%-, and 97.5%-iles for various parameters from MDP Poisson GLMM, fully parametric Poisson GLMM, and the 2.5% confidence limit, the estimated value, and the 97.5% confidence limit for Breslow and Clayton's PQL analysis. β_0 is the intercept, β_1 is the Base effect, β_2 is the Trt effect, β_3 is the Base$\times$Trt effect, β_4 is the Age effect, and β_5 is the slope over Visit. $b_{0,i}$ is the intercept for subject i, where i is the subject identifier from table 2 of Thall and Vail (1990), and $b_{1,i}$ is its slope over Visit. For the MDP models, $D(0,0)$ is the variance of the intercepts in the base measure, $D(1,1)$ is the variance of the slopes, and $D(0,1)$ is their covariance. For the fully parametric model and the PQL model, $D(\cdot,\cdot)$ are the elements of the covariance matrix of the random effects. $\bar{k}$ is the average number of clusters observed in the course of sampling.

The Bayesian results are different from the PQL results, though in all cases the 95% highest posterior density (HPD) regions overlap with the 95% confidence intervals. In general, the posterior medians of the fully parametric GLMM results are very close to the estimates from the PQL analysis. The main differences are that the 95% HPD regions for *Int* and *Age* are much wider than the 95% confidence intervals (CI). In addition, the PQL estimates of the elements of the covariance matrix of the random effects are different from the posterior means: both the variance of the intercept and the variance of the slopes is smaller under the fully parametric Bayesian model. These results are evidence that PQL can be a fairly accurate technique.

Comparing the MDP GLMM models to the fully parametric model, we notice that among the population-mean effects, the 95% HPD region for the *Base* effect excludes 0, just as the 95% CI does. However, an apparent *Trt* effect, seen in the PQL and fully parametric Bayesian GLMM is much more tenuous under the MDP model. When $M = 1.5$, i.e. when the distribution of the random effects is least similar to a normal, 0 is well within the 95% HPD region for the *Trt* effect. Conversely, the effect of *Age*, seems less likely to be 0 in the MDP models, while 0 is well within the CI for the PQL model and the HPD region for the fully parametric Bayesian model. Other parameters seem more or less equivalent across models, although the 95% HPD regions are generally wider under the MDP models than under the fully parametric or PQL fits. Finally, the elements of the D matrix seem different under the MDP models, but it must be noted that the role of this matrix is not the same under the MDP models and the others and therefore no straightforward comparison can be made. Based on the MDP GLMM analysis, we would conclude that there is a slightly larger effect of *Base*, that the treatment may not be effective, and that *Age* may be a useful predictor.

One focus of earlier analyses was outlier detection. This is an area where our semi-parametric method can be very different from fully parametric techniques. In figure 3a, we present a graphical display of the posterior means of the random effects. This display shows markedly different results than those of Breslow and Clayton, (1993), which are reproduced as figure 3b here.

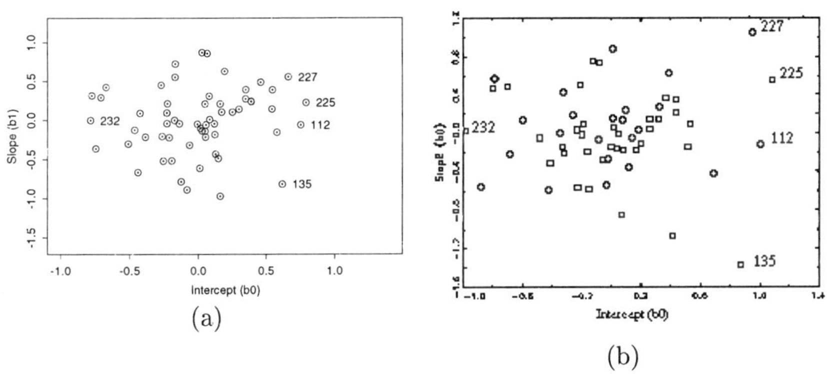

FIGURE 3. Posterior means of random slopes and intercepts from the epilepsy example: MDP GLMM with $M = 1.5$ (panel a); PQL Estimates, reproduced from Breslow and Clayton (1993) (panel (b))

They identified five patients with unusual random effects, who are labeled here with their ID numbers from Thall and Vail. In contrast to the PQL results, we notice that four of the five (subjects 112, 225, 227, and 234) no longer seem to be so unusual. This calls to greater attention the unusual improvement over time of subject 135 in both analyses, given the high initial seizure rate. However, even this subject is drawn toward the central group, and is no longer the subject with the greatest improvement.

5.7 Discussion

In this article we applied a general technique for Bayesian non-parametrics to an important class of models, the normal linear effects model and the generalized linear mixed model (GLMM). We showed that the GLMM can be freed from the parametric assumption for the random effects. Our technique involved specifying a non-parametric prior for the distribution of the random effects, and a Dirichlet Process prior on the space of prior distributions for that non-parametric prior. The resulting model was then fit with the Gibbs sampler. The approach can be quite easily extended to population models more generally. The procedure represents a new application of the MDP model to correlated data. We also demonstrated how the model can be effectively and usefully applied to longitudinal data which come from the normal, Bernoulli, and Poisson families, each of which corresponds to popular study designs in the medical literature. These semi-parametric models may be of some use to applied biostatisticians, who often use random effects models. There has been little direct focus on this model, and little data analysis of semi-parametric random effects models.

For our examples, the Gibbs sampler can take about 1 hour to run for 8,000 iterations on a Sparcstation 4. The time needed increases dramatically with the number of subjects and the number of observations. A potential drawback of the Gibbs sampler for this problem is that more iterations are needed for small values of M. However, reparameterization techniques, such as those suggested in Gilks and Roberts (1996) and Gelfand, Sahu, and Carlin (1995) may speed up convergence.

Future work includes allowing other base measures, in particular a Uniform base measure might be used. In our experience with both the MDP linear random effects model and the MDP GLMM, autocorrelation among the iterates has been a cause for concern. Another area of future interest would be determining to what degree this problem stems from the MDP model, and if there is any way to address it.

References

Aitkin M (1996). A general maximum likelihood analysis of variance components in generalized linear models. Personal communication.

Best N.G., Cowles M.K., Vines S.K. (1995). CODA: Convergence diagnostics and output analysis software for Gibbs sampling output. Version 0.3. MRC Biostatistics Unit, Cambridge.

Breslow N.E., Clayton D.G. (1993). Approximate inference in generalized linear mixed models. *Journal of the American Statistical Association* **88** 9-25

Bush C.A., MacEachern S.N. (1996). A semi-parametric Bayesian model for randomized block designs. *Biometrika* **83**, 275-285.

Chen, M.H., Ibrahim, J.G., Yiannoutsos, C (1998) Prior Elicitation, Variable Selection, and Bayesian Computation for Logistic Regression Models, *Journal of the Royal Statistical Society, Series B*, in press.

Davies R.B. (1987). Mass point methods for dealing with nuisance parameters in longitudinal studies. In *Longitudinal Data Analysis* (ed. R. Crouchley). Avebury, Aldershot, Hants

Geweke J (1992). Evaluating the accuracy of sampling-based approaches to calculating posterior moments. In *Bayesian Statistics 4* (eds: JM Bernardo, JO Berger, AP Dawid, and AFM Smith). Clarendon Press, Oxford.

Gilks W.R., Roberts G.O. (1996). Strategies for improving MCMC. In *Markov Chain Monte Carlo in Practice* (eds: Gilks WR, Richardson S, and Spiegelhalter DJ). Chapman and Hall, London.

Hastings W.K. (1970). Monte Carlo sampling methods using Markov chains and their applications. *Biometrika* **57**, 97-109

Heckman J.J., Singer B (1984). A method for minimizing the impact of distributional assumptions in econometric models of duration. *Econometrica* **52** 271-320

Ibrahim J.G., Laud P.W. (1994). A Predictive Approach to the Analysis of Designed Experiments. *Journal of the American Statistical Association.* **89**, 309-319

Ibrahim, J.G., Ryan, L.M., Chen, M.H. (1998). Use of Historical Controls to Adjust for Covariates in Trend Tests for Binary Data. *Journal of the American Statistical Association*, to appear.

Kahn JO, Lagakos S.W., Richman D.D., et al. (1992). A controlled trial comparing zidovudine with didanosine in human immunodeficiency virus infection. *New England Journal of Medicine* **327** 581-587

Kleinman K.P., Ibrahim, J.G. (1998a). A Semi-parametric Bayesian Approach to the Random Effects Model. *Biometrics*, to appear.

Kleinman K.P., Ibrahim, J.G. (1998b). A Semi-parametric Bayesian Approach to Generalized Linear Mixed Models, *Statistics in Medicine*, to appear.

Laird N.M., Ware J.H. (1982). Random-effects models for longitudinal data. *Biometrics* **38**, 963-974.

Laud P.W., Ibrahim, J.G. (1995). Predictive Model Selection, *Journal of the Royal Statistical Society, Series B*, **57**, 247-262.

MacEachern S.N., Müller P. (1996). Estimating Mixture of Dirichlet process models. Personal communication.

McCullagh P, Nelder J.A. (1989). *Generalized Linear Models* (2nd ed.), London, Chapman & Hall

Metropolis N., Rosenbluth A.W., Rosenbluth M.N., Teller A.H., Teller E (1953) Equations of state calculations by fast computing machine. *Journal of Chemical Physics* **21** 1087-1091

Nelder J.A., Wedderburn R.W.M. (1972). Generalized Linear Models. *Journal of the Royal Statistical Society, Series A* **135** 370-384

Raftery A.L., Lewis S (1992). How many iterations in the Gibbs sampler? In *Bayesian Statistics 4* (eds: JM Bernardo, JO Berger, AP Dawid, and AFM Smith). Clarendon Press, Oxford.

Sommer A, Katz J, Tarwotjo I (1983). Increased mortality in children with vitamin A deficiency. *American Journal of Clinical Nutrition* **40** 1090- 1095

Thall P.F., Vail S.C. (1990). Some covariance models for longitudinal count data with overdispersion. *Biometrics* **46** 657-671

Verbeke G, Lesaffre E (1996). A linear mixed-effects model with heterogeneity in the random-effects population. *Journal of the American Statistical Association* **91** 217-221

Wakefield J.C., Smith A.F.M., Racine-Poon A, Gelfand A.E. (1994). Bayesian analysis of linear and non-linear population models by using the Gibbs sampler. *Applied Statistics* **43** 201-221

West M, Müller P, Escobar M.D. (1994). Hierarchical priors and mixture models, with applications in regression and density estimation. In *Aspects of Uncertainty: A Tribute to D. V. Lindley* (eds. A. F. M. Smith, P. R. Freeman). Wiley, London

Wilks W.R., Wang C.C., Yvonnet B., Coursaget P. (1993). Random effects models for longitudinal data using Gibbs sampling. *Biometrics* **49** 441-453

Zeger S.L., Karim M.R. (1991). Generalized Linear Models with Random Effects: A Gibbs sampling approach. *Journal of the American Statistical Association* **86** 79-86

6

Nonparametric Bayesian Group Sequential Design

Pantelis K. Vlachos
Alan E. Gelfand

ABSTRACT The customary objective of a group sequential design is the comparison of several treatments or populations. The evolutionary nature of such trials encourages the use of the Bayesian paradigm in the design, monitoring and analysis of these trials. Here we focus on the design issue for the case of continuous, possibly multivariate response at each trial. Our approach is descriptive. Given a model specification, a design is characterized by a number of interim evaluations, the group size for each interim look, and a set of stopping criteria which determine our decision at a given look. By simulating replications of the design we can summarize design performance in terms of when the trial was stopped and reason for stopping. Such simulation and evaluation requires a fully Bayesian model specification for each treatment. We take a nonparametric perspective for the likelihood specification by assuming that the data is drawn from a distribution which arises through Dirichlet process mixing. However, we distinguish a sampling or "what if" prior, reflecting illustrative differences between populations, from a fitting or skeptical prior assuming no differences. By drawing trials under the sampling model, while fitting the model under the fitting prior, Bayesian learning moves us from prior indifference to detection of differences. We illustrate with two examples, one having univariate response, the other bivariate response.

6.1 Introduction

The customary objective of a group sequential design is the comparison of several treatments or populations. See, for example, Pocock (1977), O'Brien and Fleming (1979), Geller and Pocock (1987), etc. The evolutionary nature of such trials, which lends itself to the Bayesian paradigm, along with recent computational advances have made Bayesian-based methodology more attractive than ever. Work of Spiegelhalter and Freedman (1988), Freedman and Spiegelhalter (1989), Berry (1993), Spiegelhalter, Freedman and Parmar (1994) as well as Kadane (1995) provides explicit comparison of frequentist to Bayesian approaches, detailing the advantages of the latter. In this paper we focus on the nonparametric Bayesian design of group sequential trials for the case of continuous, possibly multivariate response at each trial of each treatment.

In the existing Bayesian literature, two basic design approaches are discussed. One of them proposes a design and then investigates its sequential performance. For example, Thall, Simon and Estey (1995) present an approach for monitoring multiple categorical outcomes in a single-arm clinical trial, using possibly restrictive multiple simultaneous stopping rules. Gelfand and Vlachos (1995) extend this approach to multiple arms and more general stopping criteria. The second approach allows for the incorporation of losses and costs leading to a formal decision-theoretic evaluation of design performance as well as the possibility of obtaining the optimal Bayesian design under such losses. Berry and Ho (1988) consider the case of a single outcome leading to one-sided sequential boundaries. Lewis and Berry (1994) use a "$0 - K$" loss function to deal with the comparison of two treatments for a disease, with binary outcomes, while Vlachos and Gelfand (1996) assess optimal group sequential designs for trials with multiple treatment arms and multivariate response vectors using backward induction, implemented via Monte Carlo Integration. Carlin, Kadane and Gelfand (1998) take a similar approach but present a forward sampling algorithm as an alternative while Kadane and Vlachos (1998) propose the use of a hybrid algorithm which uses the strengths of both methods while compensating for their weaknesses.

Adopting the first approach, our goal is to simulate a design, i.e., a Bayesian model under a prescribed stopping rule, to learn about its operating characteristics, i.e., anticipated performance with regard to the distribution of stopping time and also, given multiple criteria, the distribution of reason for stopping. We address the continuous case in a nonparametric fashion to free us from the customary, but possibly inappropriate, assumption of normally distributed measurements. Hence design performance is not attached to a Gaussian specification but rather to the assumed treatment differences, the group size and the stopping rule. In fact, we assume that the vector of response at each trial is continuous, having a distribution which is random and arises through Dirichlet process (DP) mixing. For instance, in the context of medical trials, components of the vector might be time in remission, heart or respiratory measurements, weight differences, blood measurements etc. Recording multivariate response vectors suggests multiple study objectives and implies multiple stopping criteria. Again, in the medical context, the criteria usually deal with the trade off between efficacy and adverse outcomes. The trial must be able to adapt not only to the possibility of a new treatment being potentially more effective but also to the possibility of its introducing severe side effects too frequently to be an acceptable therapy.

Until recently, implementation of DP mixing (Antoniak, 1974) was viewed as computationally intractable except for simple cases with few observations (Kuo, 1986). However, Escobar and West (1992, 1995) overcame this hurdle by an iterative sampling based approach which is, in fact, an example of the Gibbs sampler (Gelfand and Smith, 1990). In sections 6.2 and 6.3 we briefly review the DP mixing approach and the sampling based fitting methods to implement DP mixed model fitting. In section 6.4 we show how this modeling framework may be employed in a preposterior fashion to provide nonparametric assessment of group sequential design performance. Illustrative examples are given in section 6.5.

6.2 The DP mixing approach applied to the Group Sequential Framework

Ferguson (1973) introduced the Dirichlet process as a distribution on $\mathcal{P}$, the collection of all probability measures on a measurable space $(\boldsymbol{\Theta}, \mathcal{B})$. For a given probability distribution G_o, defined on $(\boldsymbol{\Theta}, \mathcal{B})$, and for a positive number α, a random probability measure G is said to follow a DP with *precision* α and *base distribution* G_o denoted by $G \sim DP(\alpha G_o)$, if, for any finite measurable partition $B_1, \ldots, B_k$ of $\boldsymbol{\Theta}$, the vector of probabilities $(G(B_1), \ldots, G(B_k)) \sim Dir(\alpha G_o(B_1), \ldots, \alpha G_o(B_k))$. Therefore, for any measurable set B, the random probability $G(B)$ has expectation $G_o(B)$ which we summarize as $E(G) = G_o$, while the variance of $G(B)$ is inversely proportional to α. Since the distribution of G cannot be expressed in terms of a family of cdf's indexed by a finite dimensional parameter, this specification is considered to be *nonparametric.*

We introduce G as follows. Let $F(\cdot; \boldsymbol{\theta})$ be a parametric family of distributions indexed by $\boldsymbol{\theta} \in \boldsymbol{\Theta}$ with corresponding densities $f(\cdot; \boldsymbol{\theta})$. For a proper G on $\boldsymbol{\Theta}$ define the mixture distribution

$$F(\cdot; G) = \int F(\cdot; \boldsymbol{\theta}) G(d\boldsymbol{\theta}). \tag{1}$$

We may think of $G(d\boldsymbol{\theta})$ as the conditional distribution of $\boldsymbol{\theta}$ given G. Differentiating both sides of (1) with respect to $(\cdot)$ yields

$$f(\cdot; G) = \int f(\cdot; \boldsymbol{\theta}) G(d\boldsymbol{\theta}). \tag{2}$$

If $G \sim DP(\alpha G_o)$ then $F(\cdot; G)$ is random and arises according to a distribution obtained from a Dirichlet process, hence via DP mixing. As in Gelfand and Mukhopadhyay (1995), *a priori* $EF(\cdot; G) = F(\cdot; G_o)$ and $Ef(\cdot; G) = f(\cdot; G_o)$. If we collect data $D_n = (Y_1, \ldots, Y_n)$ such that the Y_i's are independent and identically distributed from $F(\cdot; G)$ then inference regarding $F(\cdot; G)$ proceeds from its posterior. Previous work on Dirichlet processes has focused on the mean of this posterior distribution which is the predictive distribution of a future observation given D_n. Gelfand and Mukhopadhyay work more generally with linear functionals of F, say

$H(F(\cdot;G))$. Using the bracket notation of Gelfand and Smith (1990), the associated posterior of such functionals is denoted by $[H(F(\cdot;G))|D_n]$. In practice (and in the examples of section 6.5), we might partition $\boldsymbol{\theta}$ as $(\boldsymbol{\theta}_{(1)}, \boldsymbol{\theta}_{(2)})$ and DP mix over $\boldsymbol{\theta}_{(1)}$, using a parametric prior for $\boldsymbol{\theta}_{(2)}$. Under this *semiparametric* specification (1) yields the random distribution $F(\cdot;G,\boldsymbol{\theta}_{(2)})$.

In the group sequential framework, we assume that data is collected in blocks of trials. While block sizes obtained in data collection will not typically be constant either within or across treatments, with regard to investigating design performance we adopt a constant bock size m. Thus the jth trial of the ith treatment provides response vector $\boldsymbol{Y}_{ij} \sim F(\cdot;G_i)$, $i = 1,2,\ldots,k$ and $j = 1,\ldots,m$. The design specification is completed with the definition of a stopping rule. Such rules are built from stopping criteria which, in the spirit of Gelfand and Mukhopadhyay (1995), we assume involve the posterior expectation of a linear functional of the $F(\cdot;G_i)$, $i = 1,\ldots,k$, i.e.,

$$E\{H(F(\cdot;G_1), F(\cdot;G_2), \ldots F(\cdot,G_k))|\text{ Data }\} \tag{3}$$

where

$$H(F(\cdot;G_1), F(\cdot;G_2), \ldots F(\cdot,G_k)) = \int H(F(\cdot;\boldsymbol{\theta}_1), F(\cdot;\boldsymbol{\theta}_2), \ldots, F(\cdot;\boldsymbol{\theta}_k)) \prod_{i=1}^{k} G_i(d\boldsymbol{\theta}_i). \tag{4}$$

Illustrating in the case of $k = 2$, comparison between $F(\cdot,G_1)$ and $F(\cdot,G_2)$ would arise through $H(F(\cdot;G_1),F(\cdot;G_2)) = H(F(\cdot;G_1)) - H(F(\cdot;G_2))$ where $H(F(\cdot)) = E\{T(\boldsymbol{Y})|\boldsymbol{Y} \sim F(\cdot)\}$ for various choices of $T(\boldsymbol{Y})$. Some choices are:
(*i*) $T(\boldsymbol{Y}) = Y_i$ yielding a marginal stopping criterion based upon the difference in mean functions.
(*ii*) $T(\boldsymbol{Y}) = Y_i^2$ which along with (*i*) enables the variance functional for comparison of reliability between treatments.
(*iii*) $T(\boldsymbol{Y}) = 1_C(\boldsymbol{Y})$ which stops the trial if the response vector falls in some set, perhaps a "tail" region, C.
(*iv*) $T(\boldsymbol{Y}) = 1_{[c_i,\infty)}(Y_i)$ which enables stopping if the chance that the ith component of the response vector falls below some prespecified constant c_i under one treatment differs significantly from the corresponding chance under the other treatment. Using $1 - T(\boldsymbol{Y})$ provides the "cdf-at-a-point" functional for Y_i.
(*v*) $T(\boldsymbol{Y}) = Y_iY_j$, $i,j \in \{1,\ldots,p\}$ which enables along with (*i*) stopping based on differences of association of two components of the response vector.
See Gelfand and Mukhopadhyay (1995) for other choices of $T(\boldsymbol{Y})$. We cannot handle, for instance, the quantile functional, since it is not linear.

The aforementioned choices of $T(\boldsymbol{Y})$ permit consideration of a large class of stopping rules. In our examples in section 6.5, we use (i), (iii) and (iv) to form stopping rules based upon whether absolute differences of posterior expectations are larger (or smaller) than a specified constant, i.e.,

$$E\{H(F(\cdot;G_1))|\text{Data}\} - E\{H(F(\cdot;G_2))|\text{Data}\} > (<)\ c. \tag{5}$$

Other stopping rules of interest may involve relative differences

$$\frac{E\{H(F(\cdot;G_1))|\text{Data}\} - E\{H(F(\cdot;G_2))|\text{Data}\}}{E\{H(F(\cdot;G_1))|\text{Data}\}} > (<)\ c \tag{6}$$

or signal to noise differences

$$\frac{E\{H(F(\cdot;G_1))|\text{Data}\} - E\{H(F(\cdot;G_2))|\text{Data}\}}{\sqrt{Var\{H(F(\cdot;G_1))|\text{Data}\} + Var\{H(F(\cdot;G_2))|\text{Data}\}}} > (<)\ c. \tag{7}$$

Unfortunately, we can not handle posterior probabilities of differences of linear functionals, i.e.,

$$P(H(F(\cdot;G_1)) - H(F(\cdot;G_2)) > (<)\ c|\text{Data}) > (<)\ d \tag{8}$$

where c and d are specified constants.

6.3 Model Fitting Techniques

In practice, working with mixtures of Dirichlet processes requires the introduction of latent variables, in particular, for each $\boldsymbol{Y}_{ij}, i = 1, 2, j = 1, \ldots, m$, a latent $\boldsymbol{\theta}_{ij}$. For each i, the joint density of the $\boldsymbol{Y}_{ij}$ given the $\boldsymbol{\theta}_{ij}$ is $\prod_{j=1}^{m} f(\boldsymbol{y}_{ij}; \boldsymbol{\theta}_{ij})$. The $\boldsymbol{\theta}_{ij}$ are assumed to be independent and identically distributed draws from G_i so that the joint density of the $\boldsymbol{Y}_{ij}$ marginalized over the $\boldsymbol{\theta}_{ij}$ is

$$\prod_{j=1}^{m} f(\boldsymbol{y}_{ij}; G_i) = \prod_{j=1}^{m} \int f(\boldsymbol{y}_{ij}; \boldsymbol{\theta}_{ij}) G_i(d\boldsymbol{\theta}_{ij}).$$

The Bayesian model specification is completed by adding $G_i \sim DP(\alpha_i G_{oi})$, also denoted as $[G_i|G_{oi}, \alpha_i]$. The elegance of the DP mixing approach is that we may instead marginalize over the G_i's to obtain

$$\prod_{j=1}^{m} f(\boldsymbol{y}_{ij}; \boldsymbol{\theta}_{ij})[\boldsymbol{\theta}_{i1}, \ldots, \boldsymbol{\theta}_{im}|G_{oi}, \alpha_i]. \tag{9}$$

Inference proceeds through the posterior of $(\boldsymbol{\theta}_{i1}, \ldots, \boldsymbol{\theta}_{im})$ which is proportional to (9) rather than that of the G_i's. In particular, for each i:

$$[\boldsymbol{\theta}_{i1}, \ldots, \boldsymbol{\theta}_{im}|G_{oi}, \alpha_i, \boldsymbol{Y}_i] \propto$$
$$\prod_{j=1}^{m} f(\boldsymbol{y}_{ij}; \boldsymbol{\theta}_{ij}) \frac{\alpha_i G_{oi}(d\boldsymbol{\theta}_{ij}) + \sum_{k<j} \delta(d\boldsymbol{\theta}_{ij}|\boldsymbol{\theta}_{ik})}{\alpha_i + j - 1} \tag{10}$$

where $\delta(d\boldsymbol{\theta}_{ij}|\boldsymbol{\theta}_{ik})$ is the distribution which has a point mass at $\boldsymbol{\theta}_{ik}$ and $\boldsymbol{Y}_i$ is $\{\boldsymbol{Y}_{ij}\}$. Escobar and West (1995) showed that the Gibbs sampler can be used to obtain samples from (10). The complete conditional distribution of $\boldsymbol{\theta}_{ij}$ is a mixed distribution having a continuous part and $m-1$ atoms and assumes the form,

$$q_{io}[\boldsymbol{\theta}_{ij}|\boldsymbol{\theta}_{i(-j)}, \boldsymbol{Y}_i, G_i = G_{oi}] + \sum_{k\neq j} q_{ik}\delta(d\boldsymbol{\theta}_{ij}|\boldsymbol{\theta}_{ik}) \tag{11}$$

where $[\boldsymbol{\theta}_{ij}|\boldsymbol{\theta}_{i(-j)}, \boldsymbol{Y}_i, G_i = G_{oi}]$ is the continuous complete conditional distribution of $\boldsymbol{\theta}_{ij}$ if G_{oi} is the prior for $\boldsymbol{\theta}_{ij}$, and where $q_{io} \propto \alpha_i f(\boldsymbol{y}_{ij}; G_{oi})$ and $q_{ik} \propto f(\boldsymbol{y}_{ij}; \boldsymbol{\theta}_{ik})$ with $q_{io} + \sum_{k\neq j} q_{ik} = 1$ for every i. $f(\boldsymbol{y}_{ij}; G_{oi})$ is the marginal distribution of $\boldsymbol{Y}_{ij}$ when G_{oi} is the prior for $\boldsymbol{\theta}_{ij}$ and $f(\boldsymbol{y}_{ij}; \boldsymbol{\theta}_{ik})$ is the likelihood that $\boldsymbol{Y}_{ij}$ is from $\boldsymbol{\theta}_{ik}$.

In the implementation of the above sampling scheme, computations can be reduced because the $\boldsymbol{\theta}_{ij}$ need not all be distinct. In fact, if m_i^* is the number of distinct $\boldsymbol{\theta}_{ij}$ with n_{ik} of them taking the value $\boldsymbol{\theta}_{ik}^*$, we can rewrite (11) as

$$q_{io}[\boldsymbol{\theta}_{ij}|\boldsymbol{\theta}_{i(-j)}, \boldsymbol{Y}_i, G_i = G_{oi}] + \sum_{k=1}^{m_i^*} n_{ik} q_{ik}^* \delta(d\boldsymbol{\theta}_{ij}|\boldsymbol{\theta}_{ik}^*)$$

where now $q_{ik}^* \propto f(\boldsymbol{y}_{ij}; \boldsymbol{\theta}_{ik}^*)$ and $q_{ik} + \sum_{k\neq j} n_{ik} q_{ik}^* = 1$ for every i. Antoniak (1974) showed that the conditional distribution of the m_i^* depends on m and α_i but not on G_{oi} and takes the form

$$[m_i^*|m, \alpha_i] = c(m; m_i^*) m! \alpha_i^{m_i^*} \frac{\Gamma(\alpha_i)}{\Gamma(\alpha_i + m)}$$

$m_i^* = 1, 2, \ldots, m$ where $c(m; m_i^*)$ does not involve α_i.

Learning about α_i from the data may be addressed by incorporating α_i into the Gibbs sampler. Escobar and West (1995) assume a Gamma prior on α_i which produces a mixture of two Gamma densities for its posterior by augmenting a variable η_i having a Beta distribution, i.e., $[\eta_i|m_i^*, \alpha_i] = Beta(\alpha_i + 1, m)$.

The output from the Gibbs sampler will produce, for each treatment (suppressing the index i), independent and identically distributed samples $(\boldsymbol{\theta}_{1l}^*, \ldots, \boldsymbol{\theta}_{ml}^*)$, $l = 1, \ldots, B$ from $[\boldsymbol{\theta}_1, \ldots, \boldsymbol{\theta}_m|\boldsymbol{Y}]$. Our interest here is in the posterior distributions of functionals $[H(F(\cdot; G))|\boldsymbol{Y}]$ and their expected value. By direct calculation, following section 6.2,

$$E\{H(F(\cdot; G))|\boldsymbol{Y}\} = \int\int H(F(\cdot; \boldsymbol{\theta}_o))[\boldsymbol{\theta}_o|\boldsymbol{\theta}_1, \ldots, \boldsymbol{\theta}_m][\boldsymbol{\theta}_1, \ldots, \boldsymbol{\theta}_m|\boldsymbol{Y}]. \tag{12}$$

Following Escobar and West we would draw $\boldsymbol{\theta}_{new,l}^*$ from $[\boldsymbol{\theta}_{new}|\boldsymbol{\theta}_{1l}^*, \ldots, \boldsymbol{\theta}_{ml}^*]$, $l = 1, \ldots, B$. Then a Monte Carlo integration for (12) is

$$\hat{E}\{H(F(\cdot; G))|\boldsymbol{Y}\} = B^{-1} \sum_{l=1}^{B} H(F(\cdot; \boldsymbol{\theta}_{new,l}^*)). \tag{13}$$

6.4 Implementation of the design

Consider assessing the performance of a group sequential design, say for the difference in means between two populations. Then, we would have to specify in some fashion a magnitude of difference for the assumed "true" populations and see how the proposed design does in finding this difference. In particular, we would simulate replications of the design. Each replication consists of draws of blocks of size m from each population. As each pair of blocks comes in we would determine whether or not we stop.

In general, under a Bayesian model specification, we have to specify a full Bayesian model for each population. This collection of models is referred to as the sampling model. It generates the data we analyze under the design. By contrast, we must also specify a fitting model. This is the model under which the data is actually analyzed, and the posterior expectations of the functionals of interest are computed. The fitting model would almost surely not be the sampling model since we would never claim to know the "true" model. In fact, we would anticipate investigating the design under a variety of sampling models. In the Bayesian framework, the difference between the sampling and fitting models is captured through the prior. All models share a common likelihood. The sampling prior may be interpreted as a "what if" prior suggesting plausible magnitudes of differences, while the fitting prior would be rather vague and assume "no differences", i.e., "skeptical", following terminology introduced by Spiegelhalter et al. (1994). Our goal is to witness Bayesian learning which moves us from prior indifference (stopping discouraged by the fitting prior) to detection of differences (stopping encouraged by the sampling prior).

Again suppressing the index for treatments, the generation of the data sequence for a block of size m means that we first need to obtain a draw for $(\boldsymbol{\theta}_1, \ldots, \boldsymbol{\theta}_m)$ from G, where $G \sim DP(\alpha, G_o)$, for each block of patients. This can be done by the algorithm described below:

Step 1 Draw $\boldsymbol{\theta}_1$ from G_o.

Step 2 Given $\boldsymbol{\theta}_1$, draw $\boldsymbol{\theta}_2$ from G_o with probability $\alpha/(\alpha+1)$ or let $\boldsymbol{\theta}_2 = \boldsymbol{\theta}_1$ with probability $1/(\alpha+1)$.

Step j Given $\boldsymbol{\theta}_1, \boldsymbol{\theta}_1, \ldots, \boldsymbol{\theta}_{j-1}$ draw $\boldsymbol{\theta}_j$ from G_o with probability $\alpha/(\alpha + j-1)$ or let $\boldsymbol{\theta}_j = \boldsymbol{\theta}_k$, $k = 1, \ldots, j-1$ with respective probabilities $1/(\alpha + j - 1)$.

Then, given $(\boldsymbol{\theta}_1, \boldsymbol{\theta}_1, \ldots, \boldsymbol{\theta}_m)$, a $\boldsymbol{y}_j$ is drawn for each $\boldsymbol{\theta}_j$ according to $f(\boldsymbol{y}|\boldsymbol{\theta}_j)$.

In implementing the simulation we must generate independent and identically distributed replicates. The steps needed for obtaining each replicate are as follows. Given K, the maximum number of patient groups to be considered, we first make a draw from the sampling prior corresponding to each treatment (still suppressing i), say

$(\boldsymbol{\theta}_1, \ldots, \boldsymbol{\theta}_m, \ldots, \boldsymbol{\theta}_{(K-1)m+1}, \ldots, \boldsymbol{\theta}_{Km})$ as we described above and then we generate the entire independent data sequence, $\boldsymbol{y}_j \sim f(\boldsymbol{y}|\boldsymbol{\theta}_j)$, $j = 1, \ldots, Km$. We then emulate the course of the trial by entering patients in each trial arm in blocks of size m (group sequential sampling). Given the data sequence we fit the Bayesian fitting model and obtain posterior samples for
$(\boldsymbol{\theta}_1, \ldots, \boldsymbol{\theta}_m, \boldsymbol{\theta}_{m+1}, \ldots, \boldsymbol{\theta}_{2m}, \ldots, \boldsymbol{\theta}_{(k-1)m+1}, \ldots, \boldsymbol{\theta}_{km})$, where k is the number of patient groups that have entered the trial. We then obtain samples from the posterior distribution of $[\boldsymbol{\theta}_{new}|\boldsymbol{\theta}^*_{1,l}, \ldots, \boldsymbol{\theta}^*_{km,l}]$ for each arm as described in section 6.3 and use (13) with these samples to compute the $E\{H(F(\cdot;G))|\boldsymbol{Y}\}$ and check for stopping. A maximum number of K patient groups are considered and all trial replicates are run to completion to enable the designer to study the posterior distributions of the sample size and reason for stopping for different cut off levels of the stopping criteria. We finally summarize the results of the replicate and aggregate these summaries across replicates in suitable tabular and graphical displays.

6.5 Examples

In the following illustrative examples we consider two treatments applied to subjects that are studied in groups of size $m = 10$, for a total of $K = 3$ groups. In section 6.5.1 a univariate continuous response is monitored and we are interested in the difference of mean response between the two treatments. A bivariate case is illustrated in section 6.5.2, i.e., for each patient two types of response are monitored. In both examples, we run the trials to completion and then examine their performance for different cut-off values of the stopping criteria.

6.5.1 Univariate patient response

Consider a clinical trial with a parallel design for comparing an experimental drug (treatment 1) with a control (treatment 2). A large response is assumed to be beneficial. A maximum number of N patients are to be randomized, half to treatment 1 and half to treatment 2 and we assume that the data are analyzed after each group of $2m$ patients (m per arm) respond. The total number of analysis epochs is K, hence $N = 2mK$. The treatment response for each group of patients is denoted by Y_{ij}, $i = 1, 2$, $j = 1, \ldots, m$ and it is assumed that $Y_{ij} \sim N(\mu_{ij}, \sigma_i^2)$. We place a Dirichlet process prior over the distribution of the μ_{ij}'s and a "parametric" prior over σ_i^2, which results in the following hierarchical model:

$$\begin{aligned} Y_{ij}|\mu_{ij},\sigma_i^2 &\sim N(\mu_{ij},\sigma_i^2) \\ \mu_{ij}|G_i &\sim G_i \\ G_i|G_{oi},\alpha_i &\sim DP(\alpha_i G_{oi}), G_{oi} \equiv N(\nu_i,\tau_i^2) \\ \sigma_i^2 &\sim IGamma(a_i,b_i). \end{aligned}$$

Hence

$$Y_{ij}|G_i,\sigma_i^2 \sim \int N(\mu_{ij},\sigma_i^2)G_i(d\mu_{ij}).$$

For the sampling prior, if historical data are available on the standard treatment, they are used to form the baseline prior G_{o2}. For the experimental treatment we consider a "what if" type of prior reflecting a clinician's *a priori* assessment of the treatment's effectiveness. In our example, since a large response is considered beneficial, this implies that $\nu_1 > \nu_2$. On the other hand, the fitting prior assumes $G_{o1} = G_{o2}$ with a larger τ^2 than under the fitting model. The prior on σ_i^2 is more vague than under the fitting model. The precision parameters α_i are assumed random with $\alpha_i \sim Gamma(\lambda_{1i},\lambda_{2i})$. From section 6.3, to accommodate the α_i in the Gibbs sampler it is convenient to augment the sampler with variables η_i each having a Beta distribution.

The design specification is completed with the stopping rule. For illustration, we consider stopping the trial at an interim stage if:
(i) the response to the experimental treatment is on average a specified amount larger that that of the control, or if
(ii) the chance that the patient response under the new therapy exceeds some prespecified constant is greater than the respective chance under the standard therapy by a specified amount.

Two sampling priors were considered. For the first one we set $\nu_1 = 2$, $\nu_2 = 0$, $\tau_1^2 = \tau_2^2 = 1$, $a_1 = a_2 = 3$, $b_1 = b_2 = 0.5$, while $\lambda_{11} = \lambda_{12} = 0.1$ and $\lambda_{21} = \lambda_{22} = 10$. The second prior was more *enthusiastic* in the sense that $\nu_1 = 3$. For both cases the fitting prior was vague ($\tau_1^2 = 2$, $a_1 = 2$ and $b_1 = 0.2$), centered at the standard therapy ($\nu_1 = 0$). We summarize the simulation results in Tables 1 and 2 for 3 different stopping rules. The first one corresponds to a marginal cut off value for the difference in mean functionals, the second to a marginal cut off value for the "cdf-at-a point" functional while the third one corresponds to joint cut off values. For example, in Table 1 if an improvement in mean patient response of 2 is sought then the clinician should expect termination of the trial after the first group of patients enters the trial 36.6% of the time, 11% of the time after the second group enters the trial and the remaining 52.4% of the time the trial reaches its maximum number of patient groups. Out of that 52.4%, the trial is terminated due to crossing the cut off value of 2, 7.8% of the time (in parentheses). For the same table, if improvements of 1.5 for the mean patient response and 0.08 in the probability the response exceeds 5.5

are sought, then the clinician should expect the trial to terminate 62.6% of the time after the first group of patients enters the trial. Table 2 provides a similar summary for the more enthusiastic prior.

In Table 3, illustrating for the case of considering both criteria, we provide the distribution of decision taken at the completion of the trial. There are 4 possible decisions; 1 = the trial stopped due only to crossing of the cut off value of criterion (*i*), 2 = the trial stopped due only to crossing of the cut off value of criterion (*ii*), 3 = the trial stopped due to violation of both criteria and 0 = the trial reached completion with none of the criteria violated. For instance, if an improvement of 2.0 for the mean patient response and an improvement of 0.09 in the probability the response exceeds 5.5 are sought, under sampling prior 1, the trial terminates because of stopping decision 2 (only due to violation of criterion (*ii*)) 26% of the time. Finally, another tabular display that might be of interest to the clinician is that of the distribution of decision taken per interim stage. For illustration we present such a display in Table 4 using sampling prior 2 and the "stricter" case for both criteria, i.e., for cut off values of 2.5 and 0.10.

6.5.2 Bivariate patient response

In the same spirit as in section 6.5.1 we compare a new treatment to the standard. Correlated bivariate patient responses are monitored, e.g., systolic and diastolic blood pressure. Now the model for $\boldsymbol{Y}_{ij} \sim N_2(\boldsymbol{\theta}_{ij}, \boldsymbol{\Sigma}_i)$, $i = 1, 2$, $j = 1, 2, \ldots, m$. Again, a DP prior is placed over the distribution of the $\boldsymbol{\theta}_{ij}$'s and a "parametric" prior over $\boldsymbol{\Sigma}_i$. In particular:

$$\begin{aligned} \boldsymbol{\theta}_{ij}|G_i &\sim G_i \\ G_i|G_{oi}, \alpha_i &\sim DP(\alpha_i G_{oi}), G_{oi} \equiv N_2(\boldsymbol{\mu}_i, \boldsymbol{\Lambda}_i) \\ \boldsymbol{R}_i \equiv \boldsymbol{\Sigma}_i^{-1} &\sim W_2(\nu_i, \boldsymbol{\Omega}_i) \end{aligned}$$

where the notation $W_p(n, \boldsymbol{M})$ denotes the $p-$variate Wishart distribution with covariance matrix $\boldsymbol{M}$ and n degrees of freedom having mean $n\boldsymbol{M}$.

We assume that the new treatment is expected to increase the first component of the patient response and decrease the second one, i.e., we expect $\mu_{11} > \mu_{21}$ and $\mu_{12} < \mu_{22}$. As with the univariate case the sampling prior is based on such assumptions while the fitting prior assumes no mean differences with a more dispersed covariance matrix. Uncertainty about the precision parameters α_i is reflected in their prior distribution which is a Gamma.

We further assume that the trial can be stopped for any of the following five reasons:

(*i*) If the new treatment increases the first component of the patient response too much compared to the standard treatment.

(*ii*) If the new treatment decreases the second component of the patient response too much compared to the standard treatment.

Stopping Time	Criterion (*i*)			Criterion (*ii*)		
	1.5	2	2.5	0.08	0.09	0.10
1	0.556	0.366	0.210	0.434	0.404	0.366
2	0.174	0.110	0.062	0.180	0.166	0.158
3	0.270 (0.064)	0.524 (0.078)	0.728 (0.020)	0.386 (0.084)	0.430 (0.074)	0.476 (0.068)

Stopping Time	Joint criteria		
	1.5 or 0.08	2.0 or 0.09	2.5 or 0.10
1	0.626	0.508	0.406
2	0.176	0.148	0.154
3	0.198 (0.060)	0.344 (0.072)	0.440 (0.064)

TABLE 1. For the example of section 6.5.1, stopping probabilities under sampling prior 1 for different criteria cut off levels.

Stopping Time	Criterion (*i*)			Criterion (*ii*)		
	1.5	2	2.5	0.08	0.09	0.10
1	0.572	0.354	0.190	0.856	0.830	0.786
2	0.166	0.158	0.070	0.116	0.132	0.168
3	0.262 (0.098)	0.488 (0.102)	0.740 (0.040)	0.028 (0.022)	0.038 (0.028)	0.046 (0.036)

Stopping Time	Joint criteria		
	1.5 or 0.08	2.0 or 0.09	2.5 or 0.10
1	0.924	0.870	0.792
2	0.070	0.106	0.164
3	0.006 (0.004)	0.024 (0.016)	0.044 (0.034)

TABLE 2. For the example of section 6.5.1, stopping probabilities under sampling prior 2 for different criteria cut off levels.

Stopping Decision	Sampling Prior 1		
	1.5 or 0.08	2.0 or 0.09	2.5 or 0.10
0	0.138	0.272	0.376
1	0.288	0.136	0.050
2	0.126	0.260	0.364
3	0.448	0.332	0.210

Stopping Decision	Sampling Prior 2		
	1.5 or 0.08	2.0 or 0.09	2.5 or 0.10
0	0.002	0.008	0.010
1	0.080	0.046	0.006
2	0.060	0.180	0.394
3	0.858	0.766	0.590

TABLE 3. For the example of section 6.5.1, distribution of stopping decisions when both criteria are considered.

(*iii*) If the chance that the first component of the patient response exceeds some prespecified constant c_1 under the new treatment is too much larger than the respective chance under the standard treatment.
(*iv*) If the chance that the second component of the patient response falls below some prespecified constant c_2 under the new treatment is too much larger than the respective chance under the standard treatment and if
(*v*) the chance that the patient response falls in a particular bivariate "tail" region $C = \{(y_1, y_2); y_1 > a_1, y_2 > a_2\}$ under the new treatment is too much larger than the respective chance under the standard treatment.

As in section 6.5.1, two different sampling priors, adapted from an example in Gelfand and Vlachos (1995, §3.2.4), were considered. The first one is characterized by

$$\boldsymbol{\mu}_1 = (-0.714, -0.672), \ \boldsymbol{\mu}_2 = (-0.781, -0.424),$$

$$\boldsymbol{\Lambda}_1 = \begin{pmatrix} 5.33 \times 10^{-3} & -3.29 \times 10^{-18} \\ -3.29 \times 10^{-18} & 5.08 \times 10^{-3} \end{pmatrix},$$

$$\boldsymbol{\Lambda}_2 = \begin{pmatrix} 5.80 \times 10^{-3} & 4.03 \times 10^{-18} \\ 4.03 \times 10^{-18} & 3.96 \times 10^{-3} \end{pmatrix},$$

$\nu_1 = \nu_2 = 7$, $\boldsymbol{\Omega}_1 = \boldsymbol{\Omega}_2 = 10^{-2}I$, $\lambda_{11} = \lambda_{12} = 0.1$, and $\lambda_{21} = \lambda_{22} = 10$. For sampling prior 2 we set $\boldsymbol{\mu}_1 = (-0.612, -0.772)$ reflecting a slight improvement in the first component of the patient response and a bigger improvement in the second component. Finally, the fitting prior was made vague by multiplying the matrices $\boldsymbol{\Lambda}_1$ and $\boldsymbol{\Lambda}_2$ by 10^2.

Several tabular displays, analogous to those used in section 6.5.1 may be used to summarize the design's performance. In Tables 5, 6 and 7 we present the probabilities of stopping the trial at each stage for different criteria cut off levels and different sampling priors. For instance, in Table 7, if an improvement of 0.05 in the average response for the first component, −0.20 for the second component and an improvement of 0.05 in the chance that the patient response falls in the region $\{(x_1, x_2); x_1 > -0.87, x_2 < -0.87\}$ is sought under sampling prior 2, then the probability of terminating the trial after the second group of patients is observed is 0.112. If we consider terminating the trial because of either criterion (*iii*) or (*iv*), then we can accomplish this in four ways say; 0 if we reach the end of the trial without violating any of the two aforementioned criteria, 1 if the trial is terminated only due to violation of criterion (*iii*), 2 if the trial is terminated only due to violation of criterion (*iv*) and 3 if the trial is terminated due to violation of both criteria. In Table 8 we present the probabilities of terminating the trial due to the above 4 reasons, for both sampling priors.

If, on the other hand, we consider stopping the trial after checking for violations on either criterion (*i*), (*ii*) or (*v*) then we can envision the following 8 stopping reasons

Stopping Time	Stopping Decision 0	1	2	3	Total
1	0.000	0.006	0.266	0.520	0.792
2	0.000	0.000	0.102	0.062	0.164
3	0.010	0.000	0.026	0.008	0.044
Total	0.010	0.006	0.394	0.590	1.000

TABLE 4. For the example of section 6.5.1, distribution of stopping decisions per interim stage for sampling prior 2 using joint criteria.

Stopping Time	Criterion (*i*)			Criterion (*iii*)		
	0.03	0.05	0.08	0.01	0.03	0.05
1	0.524	0.362	0.172	0.422	0.204	0.064
2	0.254	0.194	0.080	0.266	0.214	0.086
3	0.222 (0.100)	0.444 (0.142)	0.748 (0.090)	0.312 (0.134)	0.582 (0.116)	0.850 (0.072)

Stopping Time	Joint criteria (*i*) and (*ii*)		
	0.03 or −0.16	0.05 or −0.20	0.08 or −0.24
1	0.898	0.578	0.256
2	0.078	0.188	0.092
3	0.024 (0.022)	0.234 (0.110)	0.652 (0.088)

Stopping Time	Joint criteria (*iii*) and (*iv*)		
	0.01 or 0.01	0.03 or 0.03	0.05 or 0.05
1	0.830	0.302	0.544
2	0.142	0.222	0.260
3	0.028 (0.024)	0.476 (0.158)	0.196 (0.126)

TABLE 5. For the example of section 6.5.2, stopping probabilities under sampling prior 1 for different criteria cut off levels.

Stopping Time	Criterion (*i*)			Criterion (*iii*)		
	0.03	0.05	0.08	0.01	0.03	0.05
1	0.812	0.698	0.480	0.538	0.270	0.108
2	0.152	0.214	0.216	0.258	0.270	0.152
3	0.030	0.088	0.304	0.204	0.460	0.740
	(0.024)	(0.070)	(0.146)	(0.018)	(0.150)	(0.112)

Stopping Time	Joint criteria (*i*) or (*ii*)		
	0.03 or −0.16	0.05 or −0.20	0.08 or −0.24
1	0.968	0.840	0.540
2	0.032	0.126	0.206
3	0.000 (0.000)	0.034 (0.024)	0.254 (0.128)

Stopping Time	Joint criteria (*iii*) or (*iv*)		
	0.01 or 0.01	0.03 or 0.03	0.05 or 0.05
1	0.950	0.868	0.730
2	0.048	0.116	0.196
3	0.002 (0.002)	0.016 (0.014)	0.074 (0.056)

TABLE 6. For the example of section 6.5.2, stopping probabilities under sampling prior 2 for different criteria cut off levels.

Stopping Time	Sampling Prior 1		
	0.03 or −0.16 or 0.03	0.05 or −0.20 or 0.05	0.08 or −0.24 or 0.08
1	0.916	0.620	0.258
2	0.072	0.196	0.096
3	0.012 (0.000)	0.184 (0.110)	0.646 (0.094)

Stopping Time	Sampling Prior 2		
	0.03 or −0.16 or 0.03	0.05 or −0.20 or 0.05	0.08 or −0.24 or 0.08
1	0.972	0.854	0.554
2	0.028	0.112	0.200
3	0.000 (0.000)	0.034 (0.024)	0.246 (0.126)

TABLE 7. For the example of section 6.5.2, stopping probabilities when criteria (*i*), (*ii*) or (*v*) are considered.

Criterion Violated	Stopping Decision							
	0	1	2	3	4	5	6	7
(i)	–	x	–	–	x	x	–	x
(ii)	–	–	x	–	x	–	x	x
(v)	–	–	–	x	–	x	x	x

Table 9 presents the probabilities of terminating the trial due to the above 8 reasons under both sampling priors. Finally, as in section 6.5.1, we can present the probabilities of terminating the trial for each stopping reason per interim stage. In Table 10 such probabilities are given under sampling prior 2 in the case of monitoring for violation of criterion (i), (ii) or (v) for cut-off values 0.05, −0.20 and 0.05 respectively. The same case can also be summarized graphically in the 3 dimensional plot seen in Figure 1 while the marginal distributions of stopping time and reason for stopping are given in Figure 2.

References

Antoniak, C. E. (1974), "Mixtures of Dirichlet Processes with Applications to Nonparametric Problems", *The Annals of Statistics*, **2**, 1152–1174.

Berry, D. A. (1993), "A Case for Bayesianism in Clinical Trials", *Statistics in Medicine*, **12**, 1377–1393.

Berry, D. A. and Ho, C. H. (1988), "One-Sided Sequential Boundaries for Clinical Trials: A Decision-Theoretic Approach", *Biometrics*, **44**, 219–227.

Carlin, B. P., Kadane, J. B., and Gelfand, A. E. (1998), "Approaches for Optimal Sequential Decision Analysis in Clinical Trials", *Biometrics (to appear).*

Escobar, M. D. and West, M. (1992), "Computing Bayesian Nonparametric Hierarchical Models", Technical report, ISDS, Duke University.

Escobar, M. D. and West, M. (1995), "Bayesian Density Estimation and Inference Using Mixtures", *Journal of the American Statistical Association*, **90**, 577–588.

Ferguson, T. S. (1973), "A Bayesian Analysis of Some Nonparametric Problems", *The Annals of Statistics*, **1**, 209–230.

Freedman, L. S. and Spiegelhalter, D. J. (1989), "Comparison of Bayesian with Group Sequential Methods for Monitoring Clinical Trials", *Controlled Clin. Trials*, **10**, 357–367.

Gelfand, A. E. and Mukhopadhyay, S. (1995), "On Nonparametric Bayesian Inference for the Distribution of a Random Sample", *Canandian Journal of Statistics*, **23**, 411–420.

Stopping Decision	Sampling Prior 1		
	0.01 or 0.01	0.03 or 0.03	0.05 or 0.05
0	0.004	0.070	0.318
1	0.504	0.346	0.148
2	0.000	0.000	0.000
3	0.492	0.584	0.534
Stopping Decision	**Sampling Prior 2**		
	0.01 or 0.01	0.03 or 0.03	0.05 or 0.05
0	0.000	0.002	0.018
1	0.944	0.934	0.942
2	0.000	0.000	0.000
3	0.056	0.064	0.040

TABLE 8. For the example of section 6.5.2, distribution of stopping decisions when joint criteria (*iii*) or (*iv*) are considered.

Stopping Decision	Sampling Prior 1		
	0.03 or −0.16 or 0.03	0.05 or −0.20 or 0.05	0.08 or −0.24 or 0.08
0	0.000	0.074	0.552
1	0.290	0.478	0.316
2	0.398	0.292	0.116
3	0.030	0.072	0.014
4	0.176	0.066	0.022
5	0.000	0.000	0.000
6	0.000	0.000	0.000
7	0.106	0.018	0.000
Stopping Decision	**Sampling Prior 2**		
	0.03 or −0.16 or 0.03	0.05 or −0.20 or 0.05	0.08 or −0.24 or 0.08
0	0.000	0.010	0.120
1	0.338	0.642	0.776
2	0.156	0.152	0.076
3	0.004	0.014	0.016
4	0.360	0.166	0.022
5	0.000	0.000	0.000
6	0.000	0.000	0.000
7	0.142	0.016	0.000

TABLE 9. For the example of section 6.5.2, distribution of stopping decisions when joint criteria (*i*), (*ii*), or (*v*) are considered.

Stopping Time	Stopping Decision								Total
	0	1	2	3	4	5	6	7	
1	0.000	0.526	0.142	0.014	0.160	0.000	0.000	0.012	0.854
2	0.000	0.092	0.010	0.000	0.006	0.000	0.000	0.004	0.112
3	0.010	0.024	0.000	0.000	0.000	0.000	0.000	0.000	0.034
Total	0.010	0.642	0.152	0.014	0.166	0.000	0.000	0.016	1.000

TABLE 10. For the example of section 6.5.2, distribution of stopping decisions per interim stage for sampling prior 2 using joint criteria (*i*), (*ii*) or (*v*).

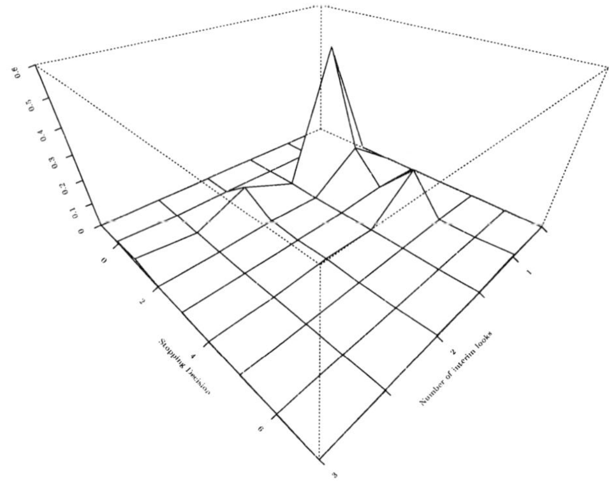

FIGURE 1. For the example of section 6.5.2, joint distribution of stopping time and reason for stopping, under sampling prior 2 when criteria (i), (ii) and (v) are considered.

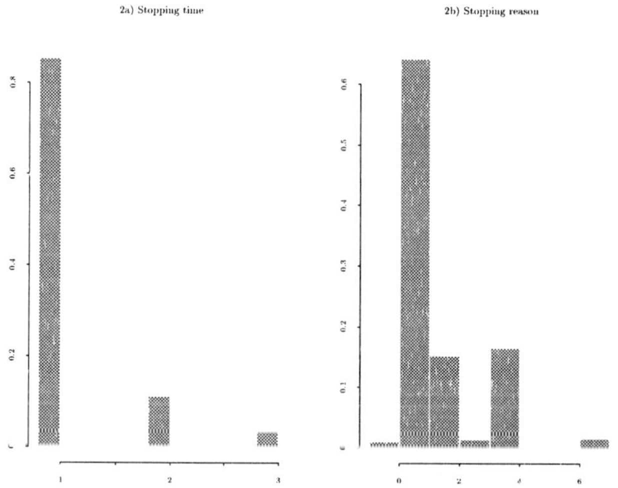

FIGURE 2. For the example of section 6.5.2, distribution of stopping time (2a) and reason for stopping (2b) under sampling prior 2 when criteria (i), (ii) and (v) are considered.

Gelfand, A. E. and Smith, A. F. M. (1990), "Sampling Based Approaches to Calculating Marginal Densities", *Journal of the American Statistical Association*, **85**, 398–409.

Gelfand, A. E. and Vlachos, P. K. (1995), "Bayesian Clinical Trial Design for Multivariate Categorical Response Models", Technical Report 95-01, Department of Statistics, The University of Connecticut.

Geller, N.L. and Pocock, S.J. (1987), "Interim Analyses in Randomized Clinical Trials: Ramifications and Guidelines for Practitioners", *Biometrics*, **43**, 213–223.

Kadane, J. B. (1995), "Prime Time for Bayes", *Controlled Clin. Trials*, **16**, 313–318.

Kadane, J. B. and Vlachos, P. K. (1998), "Hybrid Methods for Calculating Optimal Sequential Strategies: Data Monitoring for a Clinical Trial", Technical report, Carnegie Mellon University.

Kuo, L. (1986), "Computations of Mixtures of Dirichlet Processes", *SIAM J. Sci. Statist. Comput.* **7**, 60–71.

Lewis, R. J. and Berry, D. A. (1994), "Group Sequential Clinical Trials: A Classical Evaluation of Bayesian Decision-Theoretic Design", *Journal of the American Statistical Association.*

O'Brien, P. C. and Fleming, T. R. (1979), "A Multiple Testing Procedure for Clinical Trials", *Biometrics* **35**, 549–556.

Pocock, S.J. (1977), "Group Sequential Methods in the Design and Analysis of Clinical Trials", *Biometrika*, **64**, 191–199.

Spiegelhalter, D. J. and Freedman, L. S. (1988), "Bayesian Approaches to Clinical Trials (with discussion)", in *Bayesian Statistics 3* (eds: Bernardo, J.M., Berger, J.O., DeGroot, M., and Smith, A.F.M.), Oxford University Press, 453–477.

Spiegelhalter, D. J., Freedman, L. S., and Parmar, M. K. B. (1994), "Bayesian Approaches to Randomized Trials (with discussion)", *Journal of the Royal Statistical Society, Series B*, **157**, 357–416.

Thall, P. F., Simon, R., and Estey, E. H. (1995), "Bayesian Sequential Monitoring Designs for Single-arm Clinical Trials with Multiple Outcomes", *Statist. in Medicine*, **14**, 357–379.

Vlachos, P. K. and Gelfand, A. E. (1996), "Bayesian Decision Theoretic Design for Group Sequential Medical Trials Having Multivariate Patient Response", Technical Report 96-03, Department of Statistics, The University of Connecticut.

7

Wavelet-Based Nonparametric Bayes Methods

Brani Vidakovic

ABSTRACT In this chapter, we will provide an overview of the current status of research involving Bayesian inference in wavelet nonparametric problems. In many statistical applications, there is a need for procedures to (i) adapt to data and (ii) use prior information. The interface of wavelets and the Bayesian paradigm provide a natural terrain for both of these goals.

7.1 Introduction

Wavelets are the building blocks of wavelet transformations the same way that the functions e^{inx} are the building blocks of the ordinary Fourier transformation. But in contrast to sines and cosines, wavelets can be (or almost can be) supported on an arbitrarily small closed interval. This feature makes wavelets a very powerful tool in dealing with phenomena that change rapidly in time.

In recent years statistical wavelet modeling and computational research has become a burgeoning area in both theoretical and applied statistics. Also, it is beginning to impact developments in statistical methodology and in various applied scientific fields. Wavelet ideas are developing in many statistical areas: regression; density and function estimation; factor analysis; modeling and forecasting in time series analysis; and spatial statistics; with ranges of application areas in science and engineering. The emerging interests in Bayesian statistical modeling and wavelets is generating exciting new directions for the interface of these two research areas, with significant potential for future impact on applied work.

Wavelet shrinkage is a simple and yet powerful tool in nonparametric statistical modeling. It can be described as a three step procedure. Data are transformed into a set of wavelet coefficients; a shrinkage of the coefficients is performed; and then the shrunken wavelet coefficients are transformed back in the domain of the original data.

The wavelet domain is a good modeling environment; several supporting arguments are listed below.

Discrete wavelet transformations tend to "disbalance" the data. Even though the transformations preserve the ℓ_2 norm of the data, the "energy" (an engineering term for the ℓ_2 norm) of the transformed data concentrates in only a few wavelet coefficients. That concentration narrows the class of plausible models and facilitates the thresholding. Mallat (1989) gives an interesting discussion on modeling from the signal-processing point of view. The disbalancing property also yields a variety of criteria for the best basis selection.

Wavelets, as building blocks in modeling, are localized well in both time and scale (frequency). Signals with rapid local changes (signals with discontinuities, cusps, sharp spikes, etc.) can be well represented with only a few wavelet coefficients. This result is, in general, not true for other standard orthonormal bases which may need many "compensating" coefficients to describe discontinuity artifacts and to suppress Gibbs' effects.

Heisenberg's principle states that in modeling time-frequency phenomena one can not be precise in the time domain and in the frequency domain simultaneously. Wavelets automatically trade-off the time-frequency precision by their innate nature. The parsimony of wavelet transformations can be attributed to the ability of wavelets to handle limitations in the Heisenberg principle in a data-dependent manner.

Also, there is theoretical and empirical evidence that wavelet transformations tend to simplify the dependence structure in the original signal. It is even possible, for any given stationary dependence in the input signal, to construct a biorthogonal wavelet basis such that the corresponding wavelet coefficients become uncorrelated (a wavelet counterpart of Karhunen-Loève transformation). For a discussion and examples see Walter (1994).

These arguments identify wavelet bases as suitable tools for effective statistical modeling. More favorable arguments can be given: computational speed of the wavelet transformation, simple descriptors of self-similarity and so on.

Many nonparametric procedures are in fact infinitely parametric. An example is the orthogonal series regression or density estimator. In order to estimate such functions, shrinkage, tapering or truncation of parameter estimators from an infinite class is necessary (Chencov's orthogonal series density estimators, Stein-type estimation, and so on.). The benefits of shrinkage estimation in statistics were first explored in the mid-50's by C. Stein. In the 70's and 80's, many statisticians were interested in statistical benefits of classical and Bayesian shrinkage estimators. Since then, shrinkage estimation has been an active field in statistical research. Rapid increase of computing power and the use of Markov Chain Monte Carlo methods make shrinkage estimation a burgeoning research field in Bayesian statistics today.

7.2 Discrete Wavelet Transformations

Prior to describing a formal set-up for Bayesian wavelet shrinkage, we provide a brief review of discrete wavelet transformation and traditional wavelet shrinkage.

Basics on wavelets can be found in many texts, monographs, and papers at many different levels of exposition. The interested reader should consult monographs by Daubechies (1992), Walter (1994), and Ogden (1997), among others. For the sake of self-containedness of this chapter, we provide a brief overview of the discrete wavelet transformation (DWT).

7.2.1 *Discrete Wavelet Transformations*

Let $\underset{\sim}{y}$ be a data vector of dimension (size) n. For simplicity we choose n to be a power of 2, say 2^J. We assume that measurements $\underset{\sim}{y}$ belong to an interval and consider periodized wavelet bases. Generalizations to different sample sizes and general wavelet and wavelet-like transformations are straightforward.

Suppose that the vector $\underset{\sim}{y}$ is wavelet-transformed to a vector $\underset{\sim}{d}$. This linear and orthogonal transformation can be described by an orthogonal matrix W of dimension $n \times n$. In practice, one performs the DWT without exhibiting the matrix W explicitly but by using fast filtering algorithms. The filtering procedure is based on so-called quadrature mirror filters which are uniquely determined by the wavelet of choice. The wavelet decomposition of the vector $\underset{\sim}{y}$ can be written as a vector

$$\underset{\sim}{d} = (H^l \underset{\sim}{y}, GH^{l-1}\underset{\sim}{y}, \ldots, GH^2\underset{\sim}{y}, GH\underset{\sim}{y}, G\underset{\sim}{y}). \tag{1}$$

In (1, $\underset{\sim}{d}$ has the same length as $\underset{\sim}{y}$ and l is any fixed number between 1 and $J = \log_2 n$. The operators G and H are defined coordinate-wise via

$$(Ha)_k = \Sigma_m h_{m-2k} a_m, \text{ and } (Ga)_k = \Sigma_m g_{m-2k} a_m,$$

where g and h are high- and low-pass filters corresponding to the decomposing wavelet. Components of g and h are connected via the *quadrature mirror* relation $g_n = (-1)^n h_{1-n}$. For all commonly used wavelet bases, the filters g and h are readily available in the literature or in standard wavelet software packages.

The elements of $\underset{\sim}{d}$ are called "wavelet coefficients." The sub-vectors given in (1) correspond to different levels in a tree-like decomposition where the wavelet coefficients are organized level-wise. For instance, the vector $G\underset{\sim}{y}$ contains $n/2$ coefficients representing the level of the finest detail. When $l = J$, the vectors $GH^{J-1} = \underset{\sim}{y} = \{d_{00}\}$ and $H^J\underset{\sim}{y} = \{c_{00}\}$ contain a single coefficient each and represent, respectively, the coarsest level of detail and the smooth part in wavelet decomposition.

In general, the level j in the wavelet decomposition of $\underset{\sim}{y}$ contains 2^j elements, and is represented as

$$GH^{J-j-1}\underset{\sim}{y} = (d_{j,0}, d_{j,1}, \ldots, d_{j,2^j-1}). \tag{2}$$

7.2.2 Wavelet Shrinkage

Wavelet shrinkage methods are now widely recognized as a useful tool in nonparametric function estimation and signal recovery. The simplest non-linear wavelet shrinkage technique is thresholding. The coordinates of $\underset{\sim}{d}$ are replaced by 0 if they are smaller in absolute value than a fixed threshold λ.

The two most common thresholding policies are **hard** and **soft** thresholding with corresponding transformations given by:

$$\begin{aligned} \theta^h(d,\lambda) &= d\ \mathbf{1}(|d| > \lambda), \\ \theta^s(d,\lambda) &= (d - \mathrm{sign}(d)\lambda)\ \mathbf{1}(|d| > \lambda), \end{aligned}$$

where $\mathbf{1}(A)$ is the indicator of A.

Graphs of hard- and soft-thresholding rules are given in Figure 1.

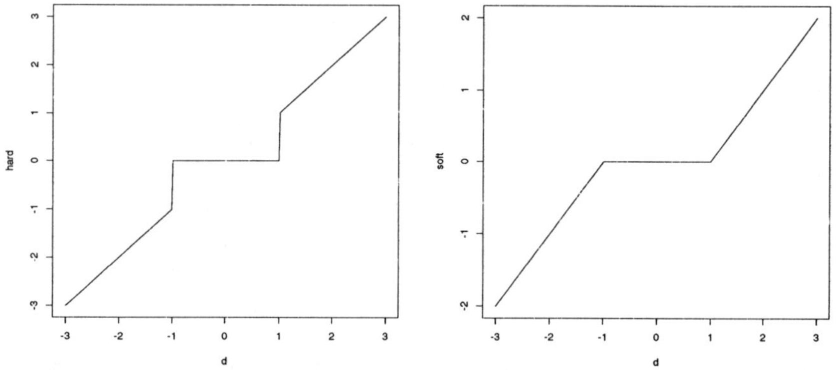

FIGURE 1. Hard and soft thresholding rules with $\lambda = 1$.

7.3 Bayes and Wavelets

Bayesian approaches to choosing the shrinkage method are less ad-hoc than earlier proposals, and have been shown to be effective. It is known that, in general, Bayes rules are *shrinkers* and that their shape in many cases has a desirable property for wavelet shrinkage: it can heavily shrink small arguments and only slightly shrink large arguments. If we use Bayes models for the wavelet coefficients, the resulting optimal actions can be very close to thresholding.

Suppose that y_i are noisy measurements, i.e. the sums of an unknown signal f_i and noise ϵ_i:

$$y_i = f_i + \epsilon_i, \quad i = 1, \ldots, n, \tag{3}$$

or in vector notation, $\underset{\sim}{y} = \underset{\sim}{f} + \underset{\sim}{\epsilon}$. After applying the wavelet transformation W, the resulting vector $\underset{\sim}{d} = W(\underset{\sim}{y})$ is again the sum of the transformation of the signal $\underset{\sim}{\theta} = W(\underset{\sim}{f})$ and the transformation of the noise $\underset{\sim}{\eta} = W(\underset{\sim}{\epsilon})$. This outcome is a consequence of the linearity of W. If the components of the noise vector $\underset{\sim}{\epsilon}$, ϵ_i, are modeled by independent normals with mean 0 and variance σ^2, then the orthogonality of W implies that the components of $\underset{\sim}{\eta}$ have the same distribution. The case of correlated normal noise can be addressed as well, see Johnstone and Silverman (1996).

Thus, instead of estimating the function directly, we estimate its wavelet transformation. Equivalently, we estimate the means θ_i in the model $[d_i] \sim \mathcal{N}(\theta_i, \sigma^2)$. The goal of Bayesian wavelet shrinkage is to exhibit such models so that resulting optimal actions, resulting from the Bayesian inference, mimic thresholding rules.

To illustrate Bayesian shrinkage in the wavelet domain, we start with an example from Vidakovic (1998).

Example 1 Let

$$[d|\theta, \sigma^2] \sim \mathcal{N}(\theta, \sigma^2), \quad \sigma^2 \text{ unknown} \tag{4}$$

be a model for a wavelet coefficient. Because of practical (computational) reasons and noninformative properties,[1] the prior distribution on σ^2 is chosen to be exponential,

$$[\sigma^2] \sim \mathcal{E}(\mu) \quad (f(\sigma^2|\mu) = \mu e^{-\mu\sigma^2}). \tag{5}$$

The marginal model (marginal likelihood) is double exponential,

$$[d|\theta] \sim \mathcal{DE}(\theta, \frac{1}{\sqrt{2\mu}}), \tag{6}$$

with $f(d|\theta) = \frac{1}{2}\sqrt{2\mu}e^{-\sqrt{2\mu}|d-\theta|}$. This follows from the fact that the double exponential distribution is a scale mixture of normals.

The prior on θ is t with location 0, scale τ, and n degrees of freedom,

$$[\theta] \sim t_n(0, \tau). \tag{7}$$

Several graphs of Bayes rules, with respect to this model, are given in Figure 2. The rules are odd functions, and only positive values are plotted. The

[1] Exponential distribution minimizes the Fisher information in the class of all distributions supported on $[0, \infty)$ with a fixed first moment.

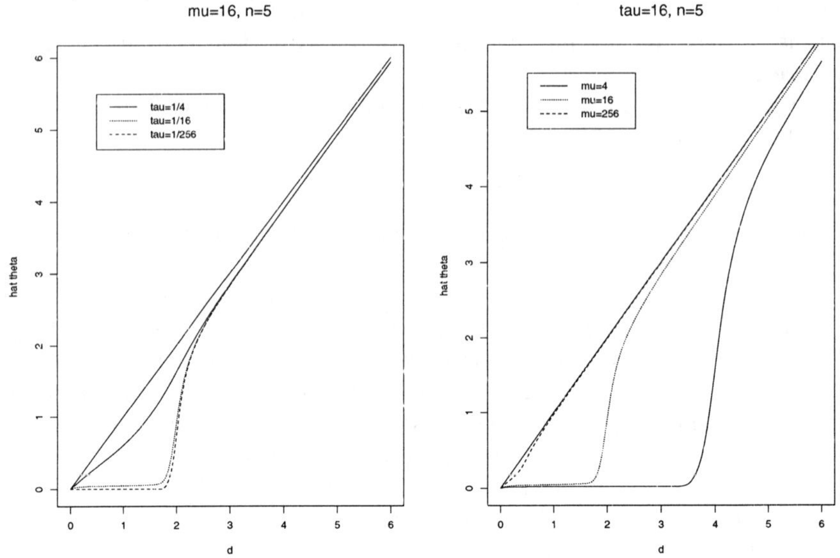

FIGURE 2. Bayes rules for selected values of τ and μ. The rules are plotted only for d positive.

hyperparameters μ, τ, and n can be specified by empirical Bayes arguments.

It is possible to get an analytic expression for the Bayes rule under the model described, when the prior on θ is general but symmetric, i.e. $\pi(\theta) = \pi(-\theta)$. The Bayes rule with respect to the squared error loss is:

$$\delta(d) = d - \frac{\Pi_1'(c) - \Pi_2'(c)}{\Pi_1(c) + \Pi_2(c)}, \tag{8}$$

where Π_1 and Π_2 are the one-sided Laplace transforms of functions $\pi(\theta + d)$ and $\pi(\theta - d)$, $\theta \in (0, \infty)$, and $c = \sqrt{2\mu}$.

7.3.1 *Regression Problems*

The somewhat simplistic model in Example 1 can be replaced with more powerful and coherent Bayesian models that exploit a hierarchical structure.

In the context of wavelet regression, we will discuss two approaches in more detail. The first one is Adaptive Bayesian Wavelet Shrinkage (ABWS) proposed by Chipman, Kolaczyk, and McCulloch (1997). Their approach is based on the stochastic search variable selection (SSVS) model proposed by George and McCulloch (1997), with the assumption that σ is known.

Chipman, Kolaczyk, and McCulloch (1997) start with the model

$$[d|\theta, \sigma^2] \sim \mathcal{N}(\theta, \sigma^2). \tag{9}$$

The prior on θ is defined as a mixture of two normals (Figure 3, Left)

$$[\theta|\gamma_j] \sim \gamma_j \mathcal{N}(0, (c_j\tau_j)^2) + (1-\gamma_j)\mathcal{N}(0, \tau_j^2), \tag{10}$$

where

$$[\gamma_j] \sim Bernoulli(p_j). \tag{11}$$

Because the hyperparameters p_j, c_j, and τ_j depend on the level j to which the corresponding θ (or d) belongs, and can be level-wise different, the method is adaptive.

The Bayes rule under squared error loss for θ (from the level j) has an explicit form,

$$\delta(d) = [P(\gamma_j = 1|d)\frac{(c_j\tau_j)^2}{\sigma^2 + (c_j\tau_j)^2} + P(\gamma_j = 0|d)\frac{\tau_j^2}{\sigma^2 + \tau_j^2}]\, d, \tag{12}$$

where

$$P(\gamma_j = 1|d) = \frac{p_j\pi(d|\gamma_j = 1)}{(1-p_j)\pi(d|\gamma_j = 0)}$$

and

$$\pi(d|\gamma_j = 1) \sim \mathcal{N}(0, \sigma^2 + (c_j\tau_j)^2) \text{ and } \pi(d|\gamma_j = 0) \sim \mathcal{N}(0, \sigma^2 + \tau_j^2).$$

The shrinkage rule (12, Figure 3, Right) can be viewed as smooth interpolation between two lines through the origin with slopes $\tau_j^2/(\sigma^2 + \tau_j^2)$ and $(c_j\tau_j)^2/(\sigma^2 + (c_j\tau_j)^2)$. The authors give very sophisticated empirical Bayes argument for tuning the hyperparameters level-wise. They provide a full posterior analysis on the coefficients and function values. The simulations performed demonstrate that the ABWS method is superior to the VisuShrink and SureShrink over the standard Donoho-Johnstone test functions in terms of MSE.

The approach used by Clyde, Parmigiani, and Vidakovic (1998) is based on a limiting form of the conjugate SSVS prior in George and McCulloch (1997). A similar model was used before in Müller and Vidakovic (1994) but in the context of density estimation.

Clyde, DeSimone, and Parmigiani (1996) consider a prior for θ which is a mixture of a point mass at 0 if the variable is excluded from the wavelet regression and a normal distribution if it is included,

$$[\theta|\gamma_j, \sigma^2] \sim \mathcal{N}(0, (1-\gamma_j) + \gamma_j c_j \sigma^2) \tag{13}$$

The γ_j are indicator variables that specify which basis element or column of W should be selected. As before, the subscript j points to the level to

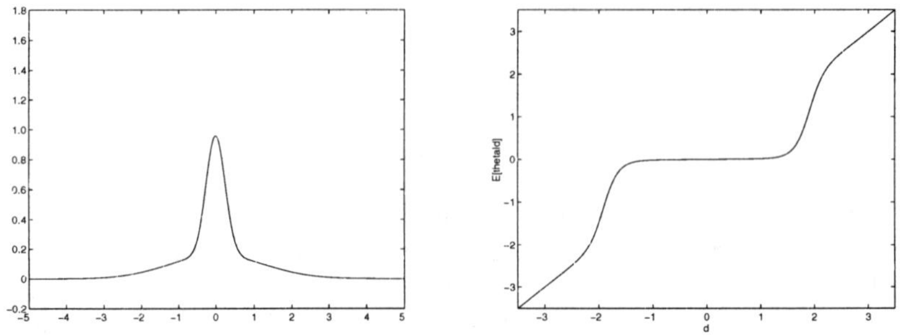

FIGURE 3. Left: Prior on θ as a mixture of two normal distributions with different variances; Right: Shrinkage rule (12) in Chipman, Kolaczyk, and McCulloch (1997).

which θ belongs. The set of all possible vectors $\underset{\sim}{\gamma}$'s will be referred to as the subset space. The prior distribution for σ^2 is inverse χ^2, i.e.,

$$[\lambda\nu/\sigma^2] \sim \chi^2_\nu,$$

where λ and ν are fixed hyperparameters and the γ_j's are independently distributed as *Bernoulli* (p_j) random variables.

The posterior mean of $\underset{\sim}{\theta}|\underset{\sim}{\gamma}$ is

$$E(\underset{\sim}{\theta}|\underset{\sim}{d},\underset{\sim}{\gamma}) = \Gamma(I_n + C^{-1})^{-1}\underset{\sim}{d}, \tag{14}$$

where Γ and C are diagonal matrices with γ_{jk} and c_{jk}, respectively, on the diagonal and 0 elsewhere. For a particular subset determined by the ones in γ, (14) corresponds to thresholding with linear shrinkage.

The posterior mean is obtained by averaging over all models. Model averaging leads to a multiple shrinkage estimator of $\underset{\sim}{\theta}$:

$$E(\underset{\sim}{\theta}|\underset{\sim}{d}) = \Sigma_{\underset{\sim}{\gamma}}\pi(\underset{\sim}{\gamma}|\underset{\sim}{d})\Gamma\left(I_n + C^{-1}\right)^{-1}\underset{\sim}{d}, \tag{15}$$

where $\pi(\gamma|\underset{\sim}{d})$ is the posterior probability of a particular subset γ.

An additional nonlinear shrinkage of the coefficients to 0 results from the uncertainty in which subsets should be selected.

Calculating the posterior probabilities of $\underset{\sim}{\gamma}$ and the mixture estimates for the posterior mean of $\underset{\sim}{\theta}$ above involve sums over all 2^n values of $\underset{\sim}{\gamma}$. The calculational complexity of the mixing is prohibitive even for problems of moderate size, and either approximations or stochastic methods for selecting subsets $\underset{\sim}{\gamma}$ possessing high posterior probability must be used.

In the orthogonal case, Clyde, DeSimone, and Parmigiani (1996) obtain an approximation to the posterior probability of $\underset{\sim}{\gamma}$ which is adapted to the wavelet setting in Clyde, Parmigiani, and Vidakovic (1998). The approximation can be achieved by either conditioning on σ (plug-in approach) or by assuming independence of the elements in $\underset{\sim}{\gamma}$.

The approximate model probabilities, for the conditional case, are functions of the data through the regression sum of squares and are given by:

$$\pi(\underset{\sim}{\gamma}|\underset{\sim}{d}) \approx \tilde{\pi}(\underset{\sim}{\gamma}|\underset{\sim}{y}) = \prod_{j,k} \rho_{jk}^{\gamma_{jk}} (1-\rho_{jk})^{1-\gamma_{jk}} \tag{16}$$

$$\rho_{jk}(\underset{\sim}{d},\sigma) = \frac{a_{jk}(\underset{\sim}{d},\sigma)}{1+a_{jk}(\underset{\sim}{d},\sigma)},$$

where

$$a_{jk}(\underset{\sim}{d},\sigma) = \frac{p_{jk}}{1-p_{jk}}(1+c_{jk})^{-1/2}\cdot \exp\left\{\frac{1}{2}\frac{S_{jk}^2}{\sigma^2}\right\}$$

$$S_{jk}^2 = d_{jk}^2/(1+c_{jk}^{-1}).$$

The p_{jk} can be used to obtain a direct approximation to the multiple shrinkage Bayes rule. The independence assumption leads to more involved formulas. Thus, the posterior mean for θ_{jk} is approximately

$$\rho_{jk}(1+c_{jk}^{-1})^{-1}d_{jk}. \tag{17}$$

Equation (17) can be viewed as a level dependent wavelet shrinkage rule, generating a variety of nonlinear rules. Depending on the choice of prior hyperparameters, shrinkage may be monotonic, if there are no level dependent hyperparameters, or non-monotonic; see Figure 4, Left. Authors report good MSE performance of approximation rules.

Clyde and George (1998) propose a model in which the distributions for the error ϵ and θ are scale mixtures of normals, thus justifying Mallat's paradigm (Mallat, 1989). They use an empirical Bayes approach to estimate the prior hyperparamters, and provide analytic expressions for the shrinkage estimator based on Bayesian model averaging. They report an excellent denoising performance of their shrinkage method for a range of noise distributions.

7.3.2 Bayesian Thresholding Rules

Bayes rules under the squared error loss and regular models are never thresholding rules. We discuss two possible approaches for obtaining *bona fide* thresholding rules in a Bayesian manner. The first one is via hypothesis testing, while the second one uses weighted absolute error loss.

Donoho and Johnstone (1994, 1995) gave a heuristic for the selection of the universal threshold via rejection regions of suitable hypotheses tests. Testing a precise hypothesis in Bayesian fashion requires a prior which has

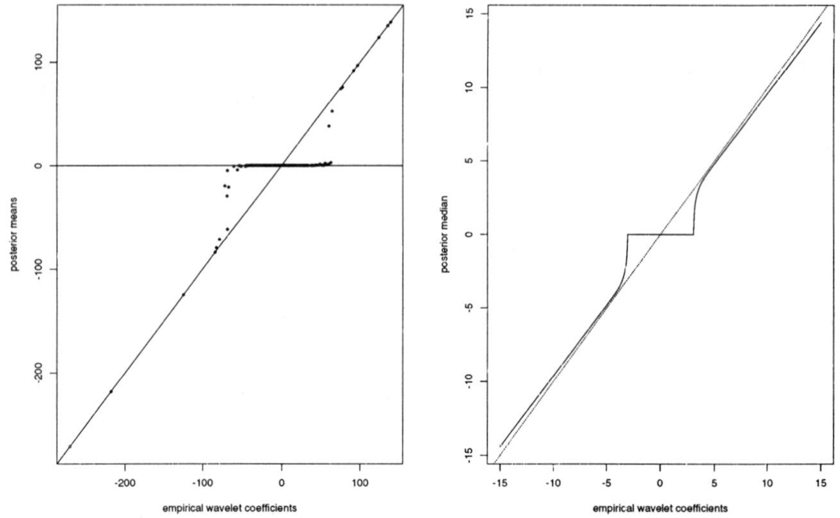

FIGURE 4. Left: Shrinkage rule from Clyde, Parmigiani, and Vidakovic (1998) based on independence approximation (17); Right: Posterior median thresholding rule (23) from Abramovich, Sapatinas, and Silverman (1998).

a point mass component. A method based on Bayes factors is discussed first. For details, see Vidakovic (1994).

Let

$$[d|\theta] \sim f(d|\theta).$$

After observing the coefficient d, the hypothesis $H_0 : \theta = 0$, versus $H_1 : \theta \neq 0$ is tested. If the hypothesis H_0 is rejected, θ is estimated by d. Let

$$[\theta] \sim \pi(\theta) = \pi_0 \delta_0 + \pi_1 \xi(\theta), \tag{18}$$

where $\pi_0 + \pi_1 = 1$, δ_0 is a point mass at 0, and $\xi(\theta)$ is a prior that describes distribution of θ when H_0 is false.

The resulting Bayesian procedure is:

$$\hat{\theta} = d \; \mathbf{1}(P(H_0|d) < \frac{1}{2}), \tag{19}$$

where

$$P(H_0|d) = (1 + \frac{\pi_1}{\pi_0} \frac{1}{B})^{-1}, \tag{20}$$

is the posterior probability of the H_0 hypothesis, and $B = \frac{f(d|0)}{\int_{\theta \neq 0} f(d|\theta)\xi(\theta)d\theta}$ is the Bayes factor in favor of H_0.

For instance, let, as in Example 1,

$$d|\theta \sim \mathcal{DE}(\theta, \frac{1}{\sqrt{2\mu}})$$

and

$$\pi(\theta) = \pi_0\delta_0 + \pi_1\xi(\theta),$$

then d will be thresholded if

$$\frac{\pi_0 e^{-c|d|}}{\pi_0 e^{-c|d|} + \pi_1(\Pi_1(c) + \Pi_2(c))} \geq \frac{1}{2}, \tag{21}$$

where Π_1 and Π_2 are one-sided Laplace transformations of $\xi(\theta - d)$ and $\xi(\theta + d)$.

Abramovich, Sapatinas, and Silverman (1998) use weighted absolute error loss and show that for a prior on θ

$$[\theta] \sim \pi_j \mathcal{N}(0, \tau_j^2) + (1 - \pi_j)\delta(0) \tag{22}$$

and normal $\mathcal{N}(\theta, \sigma^2)$ likelihood, the posterior median is

$$Med(\theta|d) = \text{sign}(d)\max(0, \zeta). \tag{23}$$

Here

$$\begin{aligned}
\zeta &= \frac{\tau_j^2}{\sigma^2 + \tau_j^2}|d| - \frac{\tau_j\sigma}{\sqrt{\sigma^2 + \tau_j^2}}\Phi^{-1}\left(\frac{1 + \min(\omega, 1)}{2}\right), \text{ and} \\
\omega &= \frac{1 - \pi_j}{\pi_j}\frac{\sqrt{\tau_j^2 + \sigma^2}}{\sigma}\exp\{-\frac{\tau_j^2 d^2}{2\sigma^2(\tau_j^2 + \sigma^2)}\}.
\end{aligned}$$

The index j, as before, points to the level containing θ (or d) facilitating adaptivity. The plot of the thresholding function (23) is given in Figure 4, Right.

Abramovich, Sapatinas, and Silverman (1998) assume

$$\tau_j^2 = C_1 2^{-\alpha j} \quad \text{and} \quad \pi_j = \min(1, C_2 2^{-\beta j}), \tag{24}$$

where C_1, C_2, α, and β are non-negative hyperparameters. The hyperparameters α and β are determined from the assumption that the function to be estimated belongs to a particular Besov space, while C_1 and C_2 are determined in an empirical Bayes fashion.

The authors compare their BayesThresh rule (23) with several existing methods (Cross-Validation, False Discovery Rate, VisuShrink and Global Sure) and report very good MSE performance.

7.3.3 The Density Estimation Problem

Delyon and Juditsky (1993), Donoho *et al.* (1996), Hall and Patil (1995), Vannucci (1996), and Walter (1994), among others, applied wavelets in density estimation from a classical and data analytic perspective.

Chencov (1962) proposed projection type density estimators in terms of an arbitrary orthogonal basis. In the case of a wavelet basis, Chencov's estimator has the form

$$\hat{f}(x) \quad = \quad \Sigma_k c_{j_0 k} \phi_{j_0 k}(x) + \Sigma_{j_0 \le j \le j_1} \Sigma_k d_{jk} \psi_{jk}(x), \tag{25}$$

where the coefficients c_{jk} and d_{jk} constituting the vector $\underset{\sim}{d}$ are defined via the standard empirical counterparts of $\langle f, \phi_{jk} \rangle$ and $\langle f, \psi_{jk} \rangle$. Let $X_1, \ldots, X_n$ be a random sample from f. Then

$$c_{jk} = \frac{1}{n} \Sigma_{i=1}^n \phi_{jk}(X_i), \quad \text{and} \quad d_{jk} = \frac{1}{n} \Sigma_{i=1}^n \psi_{jk}(X_i). \tag{26}$$

The local nature of wavelet functions makes the wavelet estimator superior to projection estimators that use classical orthonormal bases (Fourier, Hermite, etc.).

Brunk (1978) first proposed the shrinkage of coefficients in the projection density estimate based on linear Bayesian estimation. When an $\mathbb{L}_2$ density f has the Fourier expansion $f(x) = \Sigma_n a_n \xi(x)$ in some orthonormal basis $\xi_n(x), n \in \mathbb{N}$, Brunk (see also Wahba, 1981) assumed normal $\mathcal{N}(0, b_n)$ and independent priors on the coefficients a_n. The suggested choice for b_n was C/m^n, where C is a constant and $m > 1$.

Brunk's linear Bayes shrinkage paradigm, in the wavelet setup, was explored by Vidakovic and Müller (1995). They assumed

$$[\underset{\sim}{d} | \underset{\sim}{\theta}, \sigma^2] \sim \mathcal{MVN}(\underset{\sim}{\theta}, \sigma^2 I), \tag{27}$$

where the components of $\underset{\sim}{d}$ are empirical wavelet coefficients from (26).

The rationale behind the model in (27) is that both c_{jk} and d_{jk} are averages of n iid random variables, see (26). Given $\underset{\sim}{\theta}$, the components of $\underset{\sim}{d}$ are independent. The model in (27) was completed by specifying the prior distribution on the location $\underset{\sim}{\theta}$ and scale σ.

Let

$$[\underset{\sim}{\theta}, \sigma^2] \sim \mathcal{NIG}(\alpha, \delta, 0, \Sigma), \tag{28}$$

where $\mathcal{NIG}(\alpha, \delta, m, \Sigma)$ is the normal-inverse-gamma distribution with density function $f(\underset{\sim}{\theta}, \sigma^2 | \alpha, \delta, m, \Sigma) = C \cdot (\sigma^2)^{-\frac{\delta+p+2}{2}} \exp[-\{(\underset{\sim}{\theta} - \underset{\sim}{m})' \Sigma^{-1} (\underset{\sim}{\theta} - \underset{\sim}{m}) + \alpha\}/(2\sigma^2)]$.

The posterior is straightforward because of the conjugate structure. The choice of a prior covariance matrix Σ corresponds to the choice of variances b_n in the traditional (Fourier) problems discussed in Brunk (1978) and

Wahba (1981). The authors consider a more general case in which the wavelet coefficients are not independent. More precisely, the coefficients in different levels are assumed independent, but the coefficients in the same level are assumed correlated. If the coefficients between levels j_0 and j_1 are considered in the estimator (25), covariance structure can be described by the following covariance matrix:

$$\Sigma = \lambda_{j_0,\phi}\Sigma_{j_0,\phi} \oplus \bigoplus_{j=j_0}^{j_1} \lambda_{j,\psi}\Sigma_{j,\psi}, \tag{29}$$

where $\oplus$ is direct sum operation. Σ in (29), is a block diagonal matrix in which the block sub-matrices $\Sigma_{j_0,\phi}$ and $\Sigma_{j,\psi}, j = j_0, \ldots, j_1$ describe correlations within their corresponding levels.

The posterior for $[\underset{\sim}{\theta}, \sigma^2]$ is again the normal-inverse-gamma distribution

$$[\underset{\sim}{\theta}, \sigma^2 | \underset{\sim}{d}] \sim \mathcal{NIG}(\alpha^\star, \delta^\star, m^\star, \Sigma^\star),$$

with

$$\begin{aligned} \Sigma^\star &= (I + \Sigma^{-1})^{-1} \\ \hat{\underset{\sim}{\theta}} &= \Sigma^\star \underset{\sim}{d} \quad (= E(\underset{\sim}{\theta} \mid \underset{\sim}{d})) \\ \alpha^\star &= \alpha + ||\underset{\sim}{d}|| + (\theta^\star)'(\Sigma^\star)^{-1}\theta^\star \\ \delta^\star &= \delta + n. \end{aligned}$$

The Bayes estimator of $\underset{\sim}{\theta}$ is $\hat{\underset{\sim}{\theta}} = \Sigma^\star \underset{\sim}{d}$. Naturally, we estimate the density in (25) by replacing c_{jk} and d_{jk} from $\underset{\sim}{d}$ by their affine Bayes estimators.

Example 2 In practical implementations of the method, the matrix Σ was a block diagonal matrix consisting of Laurent submatrices $\Sigma_{j_0,\phi}$, $\Sigma_{j_0,\psi}, \ldots,$ $\Sigma_{j_1,\psi}$ of appropriate dimensions with entries $\sigma_{i,j} = \rho^{|i-j|}$, $|\rho| < 1$.

Figure 5 shows shrinkage of empirical wavelet coefficients $\underset{\sim}{d}$ for the *galaxy velocities* data set (Roeder, 1990). There were 176 empirical coefficients between levels $j_0 = 0$ and $j_1 = 6$. The following values for λ's and ρ were adopted: $\lambda_{0,\phi} = 1000, \lambda_{0,\psi} = 1000, \lambda_{1,\psi} = 100, \lambda_{2,\psi} = 10, \lambda_{3,\psi} = 1, \lambda_{4,\psi} = 0.1, \lambda_{5,\psi} = 0.01, \lambda_{6,\psi} = 0.001$, and $\rho = 0.8$.

Building on (27)-(29), Huerta (1997) proposed the following model

$$\begin{aligned} [\underset{\sim}{d}|\underset{\sim}{\theta}, \sigma^2] &\sim \mathcal{MVN}(\underset{\sim}{\theta}, \sigma^2 I), \quad [\sigma^2] \sim \mathcal{IG}(\alpha_1, \delta_1), \\ [\underset{\sim}{\theta}|\tau^2] &\sim \mathcal{MVN}(\underset{\sim}{0}, \tau^2\Sigma), \quad [\tau^2] \sim \mathcal{IG}(\alpha_2, \delta_2). \end{aligned}$$

The closed form for $E(\underset{\sim}{\theta}|\underset{\sim}{d})$ in the above model is not possible, however the full conditionals can be specified. The shrinkage is performed by Gibbs sampling. The prior covariance matrix Σ was defined as in (29). Obtained simulation results are promising.

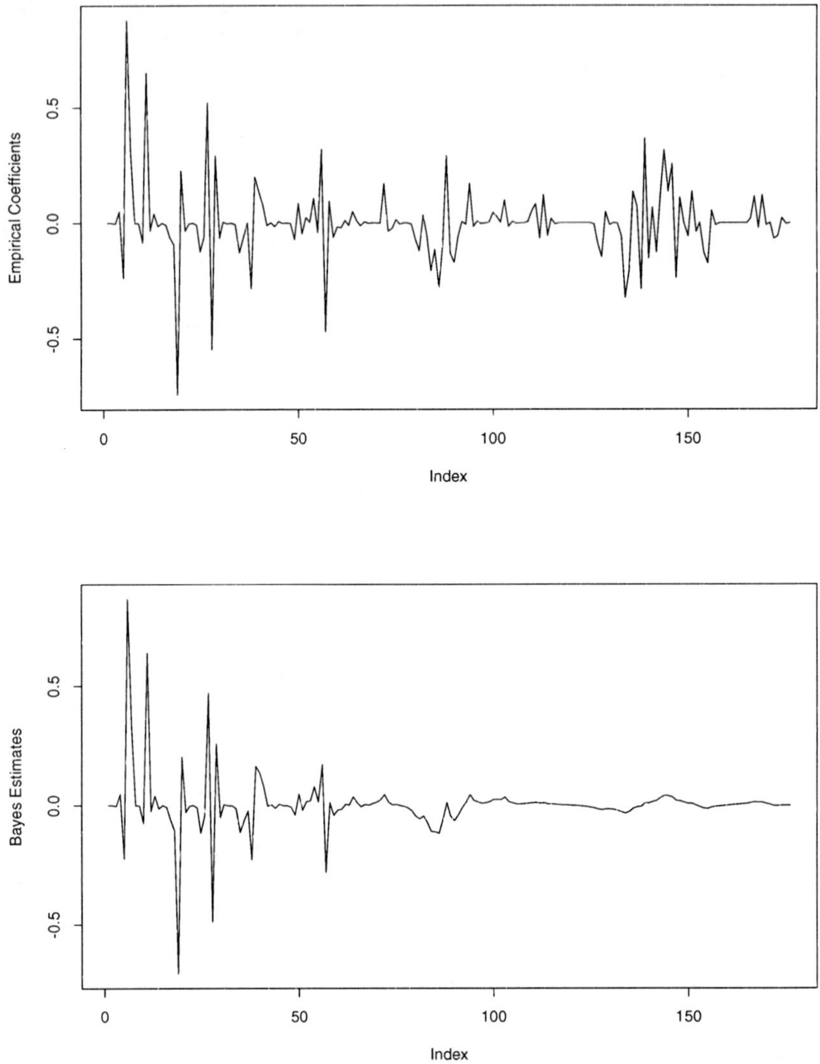

FIGURE 5. Empirical coefficients (26) for the `galaxy` data and their affine Bayes shrinkage. The coefficients are arranged as a concatenation of levels in the decomposition, starting with scaling coefficients and ending with coefficients of fine detail. Notice that the Bayes estimate preserves coarse structure and heavily shrinks detail coefficients that cause overfitting.

Reducing the number of hyperparameters in (27)-(29) and imposing a prior structure on the remaining hyperparameters in a hierarchical fashion was suggested by Vannucci and Corradi (1996). They applied the shrinkage on a regression problem and called their method BayesShrink. Assuming that $\underset{\sim}{\theta}$ corresponds to an autoregressive process in the time domain, the authors demonstrate that the matrix Σ depends on only two hyperparameters, λ and ρ. The parameter ρ is the "autocovariance index," and λ is the precision parameter. The covariance matrix $\Sigma(\lambda, \rho) = \lambda\Sigma(\rho)$ has an interesting "finger-like" structure.

Vannucci and Corradi (1996) suggest

$$[\lambda] \sim \mathcal{IG}(p/2, q/2), \text{ and}$$

$$[\rho] \propto (C-\rho)^{r_1-1}(C+\rho)^{r_2-1}, \ |\rho| < C,$$

and simulate from the posterior distribution of $\underset{\sim}{\theta}$ using MCMC method.

7.3.4 Full Bayesian Model

Müller and Vidakovic (1994) parameterize an unknown density $f(x)$ by a wavelet series on its logarithm, and propose a prior model which explicitly defines geometrically decreasing prior probabilities for non-zero wavelet coefficients at higher levels of detail.

The unknown probability density function $f(\cdot)$ is modeled by:

$$\log f(x) = \Sigma_{k\in Z}\xi_{j_0 k}\phi_{j_0,k}(x) + \Sigma_{j\geq j_0, k\in Z} s_{jk}\theta_{jk}\psi_{jk}(x) - \log K, \tag{30}$$

where $K = \int f(x)dx$ is the normalization constant and $s_{jk} \in \{0,1\}$ is an indicator variable that performs model induced thresholding.

The dependence of $f(x)$ on the vector $\theta = (\xi_{j_0,k}, s_{jk}, \theta_{jk},\ j = j_0, \ldots, j_1,\ k \in Z)$ of wavelet coefficients and indicators is expressed by $f(x) = p(x|\theta)$. The sample $X = \{X_1, \ldots, X_n\}$ defines a likelihood function $p(X|\theta) = \prod_{i=1}^{n} p(X_i|\theta)$.

The model is completed by a prior probability distribution for θ. Without loss of generality, $j_0 = 0$ can be assumed. Also, any particular application will determine a maximum level of detail j_1.

$$\begin{aligned}
[\xi_{0k}] &\sim \mathcal{N}(0, \tau r_0), \\
[\theta_{jk}|s_{jk} = 1] &\sim \mathcal{N}(0, \tau r_j), \quad r_j = 2^{-j}, \\
[s_{jk}] &\sim \textit{Bernoulli } (\alpha^j), \\
[\alpha] &\sim \textit{Beta } (a, b), \\
[1/\tau] &\sim \textit{Gamma } (a_\tau, b_\tau).
\end{aligned} \tag{31}$$

The wavelet coefficients θ_{jk} are non-zero with geometrically decreasing probabilities. Given that a coefficient is non-zero, it is generated from a normal distribution. The parameter vector θ is augmented in order to include all model parameters, i.e. $\theta = (\theta_{jk}, \xi_{jk}, s_{jk}, \alpha, \tau)$.

The scale factor r_j contributes to the adaptivity of the method. Wavelet shrinkage is controlled by both: the factor r_j and geometrically decreasing prior probabilities for non-zero coefficient, α^j.

The conditional prior $p(\theta_{jk}|s_{jk} = 0) = h(\theta_{jk})$ is a pseudo-prior as discussed in Carlin and Chib (1995). The choice of $h(\cdot)$ has no bearing on the inference about $f(x)$. In fact, the model could be alternatively formulated by dropping θ_{jk} under $s_{jk} = 0$. However, this model would lead to a parameter space of varying dimension. Carlin and Chib (1995) argue that the pseudo-prior $h(\theta_{jk})$ should be chosen to produce values for θ_{jk} which are consistent with the data.

The particular MCMC simulation scheme used to estimate the model (30, 31) is described. Starting with some initial values for θ_{jk}, $j = 0, \ldots, j_1$, ξ_{00}, α, and τ, the following Markov chain was implemented.

1. For each $j = 0, \ldots, j_1$ and $k = 1, \ldots, 2^j - 1$ go over the steps **2** and **3.**

2. Update s_{jk}. Let θ_0 and θ_1 indicate the current parameter vector θ with s_{jk} replaced by 0 and 1, respectively. Compute $p_0 = p(y|\theta_0) \cdot (1 - \alpha^j)h(\theta_{jk})$ and $p_1 = p(y|\theta_0) \cdot \alpha^j p(\theta_{jk}|s_{jk} = 1)$. With probability $p_1/(p_0 + p_1)$ set $s_{jk} = 1$, else $s_{jk} = 0$.

3a. Update θ_{jk}. If $s_{jk} = 1$, generate $\tilde{\theta}_{jk} \sim g(\tilde{\theta}_{jk}|\theta_{jk})$. Use, for example, $g(\tilde{\theta}_{jk}|\theta_{jk}) = \mathcal{N}(\theta_{jk}, 0.25\sigma_{jk})$, where σ_{jk} is some rough estimate of the posterior standard deviation of θ_{jk}. We will discuss alternative choices for the probing distribution $g(\cdot)$ below.

Compute

$$a(\theta_{jk}, \tilde{\theta}_{jk}) = \min\left[1, \frac{p(y|\tilde{\theta})p(\tilde{\theta}_{jk})}{p(y|\theta)p(\theta_{jk})}\right],$$

where $\tilde{\theta}$ is the parameter vector θ with θ_{jk} replaced by $\tilde{\theta}_{jk}$, and $p(\theta_{jk})$ is the p.d.f. of the normal prior distribution given in (31).

With probability $a(\theta_{jk}, \tilde{\theta}_{jk})$ replace θ_{jk} by $\tilde{\theta}_{jk}$; else keep θ_{jk} unchanged.

3b. If $s_{jk} = 0$, generate θ_{jk} from the full conditional posterior $p(\theta_{jk}|\ldots, X) = p(\theta_{jk}|s_{jk} = 0) = h(\theta_{jk})$.

4. Update ξ_{00}. Generate $\tilde{\xi}_{00} \sim g(\tilde{\xi}_{00}|\xi_{00})$. Use, for example, $g(\tilde{\xi}_{00}|\xi_{00}) = \mathcal{N}(\xi_{00}, 0.25\rho_{00})$, where ρ_{00} is some rough estimate of the posterior standard deviation of ξ_{00}. Analogously to step **3a,** compute an acceptance probability a and replace ξ_{00} with probability a.

5. Update α. Generate $\tilde{\alpha} \sim g_\alpha(\tilde{\alpha}|\alpha)$ and compute

$$a(\alpha, \tilde{\alpha}) = \min\left[1, \frac{\prod_{jk} \tilde{\alpha}^{js_{jk}}(1 - \tilde{\alpha}^j)^{s_{jk}}}{\prod_{jk} \alpha^{js_{jk}}(1 - \alpha^j)^{s_{jk}}}\right].$$

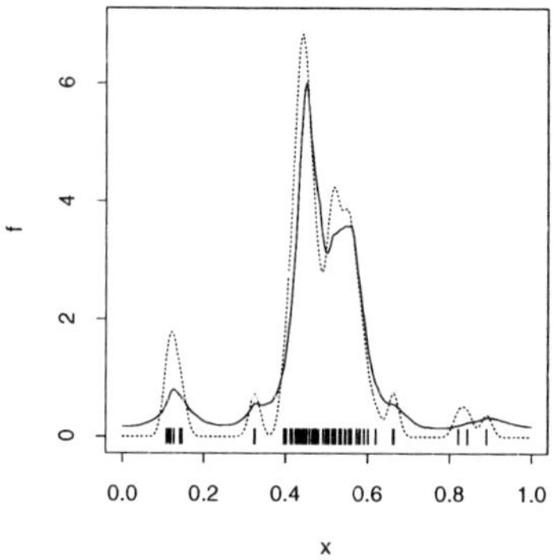

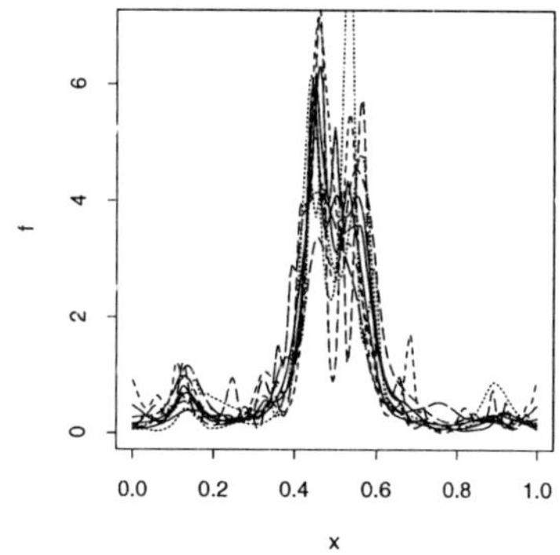

FIGURE 6. Left: The estimated p.d.f. $\hat{f}(x) = \int p(x|\theta)dp(\theta|X)$. The dotted line plots a conventional kernel density estimate for the same data. Right: The posterior distribution of the unknown density $f(x) = p(x|\theta)$ induced by the posterior distribution $p(\theta|X)$. The lines plot $p(x|\theta_i)$ for ten simulated draws posterior $\theta_i \sim p(\theta|X), i = 1, \ldots, 10$.

With probability $a(\alpha, \tilde{\alpha})$ replace α by $\tilde{\alpha}$, else keep α unchanged.

6. Update τ. Resample τ from the complete inverse Gamma conditional posterior.

7. Iterate over steps **1** through **6** until the chain is judged to have practically converged.

The algorithm implements a Metropolis chain changing one parameter at a time in the parameter vector. See, for example, Tierney (1994) for a description and discussion of Metropolis chains for posterior exploration. For a practical implementation, g should be chosen such that the acceptance probabilities a are neither close to zero, nor close to one. In the implementations, $g(\tilde{\theta}_{jk}|\theta_{jk}) = \mathcal{N}(\theta_{jk}, 0.25\sigma_{jk})$ with $\sigma_{jk} = 2^{-j}$, was used.

Example 3 Described wavelet based density estimation model is illustrated on the galaxy data set (Roeder, 1990). The data is rescaled to the interval $[0, 1]$. The hyperparameters were fixed as $a = 10$, $b = 10$, and $a_\tau = b_\tau = 1$. The $Beta(10, 10)$ prior distribution on α is reasonably non-informative compared to the likelihood based on $n = 82$ observations.

Initially, all s_{jk} are set to one, and α to its prior mean $\alpha = 0.5$. The first 10 iterations as burn-in period were discarded, then 1000 iterations of steps 1 through 6 were simulated. For each j, k, Step 3 was repeated three times. The maximum level of detail selected was $j_1 = 5$

Figures 6 and 7 describe some aspects of the analysis.

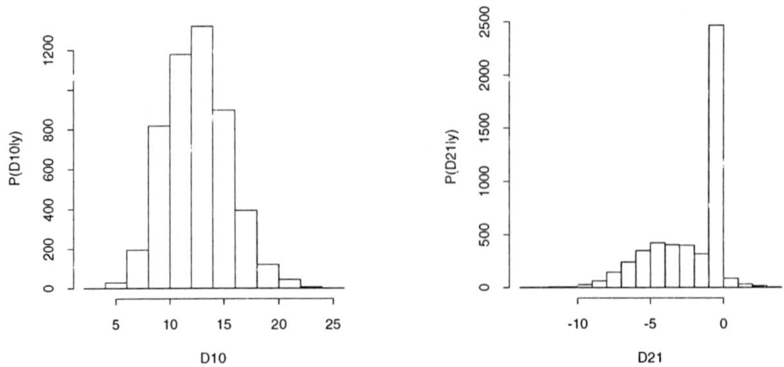

FIGURE 7. Posterior distributions $p((s_{10}\theta_{10})|X)$ and $p((s_{21}\theta_{21})|X)$. While $s_{10}\theta_{10}$ is non-zero with posterior probability close to one, the posterior distribution $p((s_{21}\theta_{21})|X)$ is a mixture of a point mass at zero and a continuous part.

7.4 Other Problems

Lina and MacGibbon (1997) apply a Bayesian approach to wavelet regression with complex valued Daubechies wavelets. To some extent, they exploit redundancy in the representation of real signals by the complex wavelet coefficients. Their shrinkage technique is based on the observation that the modulus and the phase of wavelet coefficients encompass very different information about the signal. A Bayesian shrinkage model is constructed for the modulus, taking into account the corresponding phase.

Simoncelli and Adelson (1996) discuss Bayes "coring" procedure in the context of image processing. The prior on the signal is Mallat's model, see Mallat (1989), while the noise is assumed normal. They implement their noise reduction scheme on an oriented multiresolution representation - known as the *steerable pyramid*. They report that Bayesian coring outperforms classical Wiener filtering.

Malfait and Roose (1995) and Malfait, Jansen, and Roose (1996) employed the theory of Markov random fields in wavelet denoising of images. The idea is that the thresholding of a coefficient in a 2-D array depends not only on its own magnitude but also on the magnitudes of its neighbors. This dependence is modeled by using masks consisting of 0 and 1 and exhibiting properties of a Markov field. Crouse, Nowak, and Baraniuk (1997) also consider hidden Markov fields in a similar setup. They develop the Efficient Expectation Maximization (EEM) algorithm to fit their model.

Green, Sapatinas, and Vidakovic (1998) consider a fully Bayesian adaptive wavelet shrinkage. Their model is in fact hierarchical in two respects: the wavelet coefficients θ_{jk} are modeled conditionally on hyperparameters η, that are in turn drawn from hyperpriors $p(\eta)$, and, further, $p(\underset{\sim}{\theta} \mid \eta)$ is specified hierarchically in j. Shrinkage induced by Bayesian models in which the hyperparameters of the prior are made time dependent in an empirical Bayes fashion is considered in Bielza and Vidakovic (1996).

Pesquet *et al.* (1996) develop a Bayesian-based approach to the best basis problem, while preserving the classical tree search efficiency in wavelet packets and local trigonometric bases. Kohn and Marron (1997) use a model similar to one in Chipman, Kolaczyk, and McCulloch (1997) but in the context of the best basis selection. Gendron (1997) builds on the Bayesian hypothesis testing idea in the wavelet domain. He uses the Bayesian paradigm to construct a cost function which points to the best wavelet packet. Using the best basis he develops a filtering procedure for seismic signals by applying posterior expectations when the prior is a mixture of a uniform distribution and a point mass at 0.

Timmermann and Nowak (1997) introduce multiscale Bayesian estimation of Poisson intensities. To preserve Poisson characteristics in the wavelet domain, they utilize the non-normalized Haar transformation. The proposed mixture Beta priors on the intensity are realistic and lead to an efficient shrinkage estimation.

Ogden (1996) develops a Bayesian wavelet method for estimating the location of a change-point. Initial treatment is for the standard change-point model (that is, constant mean before the change and constant mean after it) but extends to the case of detecting discontinuity points in an otherwise smooth curve. The conjugate prior distribution on the change point τ is given in the wavelet domain, and it is updated by observed empirical wavelet coefficients.

Ruggeri and Vidakovic (1998) discuss Bayesian decision theoretic thresholding. In the set of all hard thresholding rules, they find restricted Bayes rules under a variety of models, priors, and loss functions. They identify model-prior pairs that work well (in sense of exhibiting a hard thresholding rule that minimizes Bayes risk) and show that in the presence of prior information on noise and signal their procedure outperforms SureShrink and VisuShrink methods. Rios Insua and Vidakovic (1998) propose algorithms for construction of wavelet-based random densities and discuss fast Bayesian updating of random priors.

Lu, Huang, and Tung (1997) performed linear Bayesian wavelet shrinkage in a nonparametric mixed-effect model. Their formulation is conceptually inspired by the duality between reproducing kernel Hilbert spaces and random processes as well as by connections between smoothing splines and Bayesian regressions. The unknown function f in the standard nonparametric regression formulation ($y = f(x_i) + \sigma\epsilon_i$, $i = 1, \dots, n$; $0 \le x \le 1$; $\sigma > 0$; $Cov(\epsilon_1, \dots, \epsilon_n) = R$) is given a prior of the form $f(x) = \Sigma_k \alpha_{Jk}\phi_{Jk}(x) + \delta Z(x)$; $Z(x) \sim \Sigma_{j>J}\Sigma_k \theta_{jk}\psi_{jk}(x)$, where θ_{jk} are uncorrelated random variables such that $E\theta_{jk} = 0$ and $E\theta_{jk}^2 = \lambda_j$. The authors propose a linear, empirical Bayes estimator $\mathring{f}$ of f that enjoys Gauss-

Markov type of optimality. Several non-linear versions of the estimator are proposed, as well. Independently, and by using different techniques, Huang and Cressie (1997) consider the same problem and derive a Bayesian estimate.

Acknowledgments: I would like to thank the editors, especially Peter Müller, for the invitation to compile a chapter on Bayesian wavelet nonparametrics. Peter's contributions to this chapter have been numerous.

Several colleagues and two anonymous referees also commented on an early version of this chapter. I am grateful to Fabrizio Ruggeri, Gabriel Huerta, Fanis Sapatinas, Marina Vannucci, and Todd Ogden for their suggestions.

This work was supported in part by NSF Grant DMS-9626159 at Duke University.

References

Abramovich, F., Sapatinas, T. and Silverman, B. W. (1998), "Wavelet thresholding via Bayesian approach," *Journal of the Royal Statistical Society,* 60, No.3.

Bielza, C., and Vidakovic, B. (1996), "Time-dependent wavelet shrinkage," Discussion Paper, 96-24, ISDS, Duke University.

Brunk, H. (1978), "Univariate density estimation by orthogonal series," *Biometrika,* 65, 3, 521-528.

Carlin, B. and Chib, S. (1995), "Bayesian model choice via Markov chain Monte Carlo," *Journal of the Royal Statistical Society B*, 57, 473-484.

Chencov, N. N. (1962), "Evaluation of an unknown distribution density from observations," *Doklady*, 3, 1559-1562.

Chipman, H., McCulloch, R., and Kolaczyk, E. (1997), "Adaptive Bayesian Wavelet Shrinkage," *Journal of the American Statistical Association*, 92, 1413-1421.

Clyde, M., DeSimone, H., and Parmigiani, G. (1996), "Prediction via orthogonalized model mixing," *Journal of the American Statistical Association*, **91**, 1197-1208.

Clyde, M., and George, E. (1998), "Empirical wavelet estimation in wavelets," Discussion Paper, ISDS, Duke University

Clyde, M., Parmigiani, G., and Vidakovic, B. (1998), "Multiple shrinkage and subset selection in wavelets," *Biometrika*, to appear.

Crouse, M., Nowak, R., and Baraniuk, R. (1997), "Statistical signal processing using wavelet-domain hidden Markov models," Proceedings of

SPIE, Wavelet Applications in Signal and Image Processing V, vol. 3169, 248-259.

Daubechies, I. (1992), *Ten Lectures on Wavelets.* S.I.A.M., Philadelphia.

Delyon, B., and Juditsky, A. (1993), "Wavelet estimators, global error measures: Revisited," Publication interne, 782, IRISA-INRIA, France.

Donoho, D., and Johnstone, I. (1994), "Ideal spatial adaptation by wavelet shrinkage," *Biometrika,* 81, 425-455.

Donoho, D., and Johnstone, I. (1995), "Adapting to unknown smoothness via wavelet shrinkage," *Journal of the American Statistical Association,* 90, 1200-1224.

Donoho, D., Johnstone, I., Kerkyacharian, G., and Pickard, D. (1996), "Density Estimation by Wavelet Thresholding," *The Annals of Statistics,* 24, 508-539.

George, E.I., and McCulloch, R. (1997), "Approaches to Bayesian variable selection," *Statistica Sinica,* 7, 339-373.

Green, P.J., Sapatinas, T., and Vidakovic, B. (1998), "Fully Bayesian adaptive wavelet shrinkage," Manuscript in Preparation.

Hall, P., and Patil, P. (1995), "Formulae for the mean integrated square error of non-linear wavelet based density estimators," *The Annals of Statistics,* 23, 905-928.

Hernández, E., and Weiss, G. (1996), *A First Course on Wavelets*, Boca Raton: CRC Press Inc.

Huerta, G. (1997), "Bayes wavelet shrinkage and applications to data denoising," In electronic proceedings of International Workshop on Wavelets in Statistics, Duke University, 12-13 October 1997.

Huang, H-C., and Cressie, N. (1997), "Deterministic/stochastic wavelet decomposition for recovery of signal from noisy data," Technical Report, 97-23, Department of Statistics, Iowa State University.

Johnstone, I., and Silverman B. W. (1996), "Wavelet threshold estimators for data with correlated noise," *Journal of the Royal Statistical Society B,* 59, 319-351.

Kohn R, and Marron J. S. (1997), "Bayesian wavelet shrinkage," International Workshop on Wavelets in Statistics, Duke University, 12-13 October 1997.

Lina, J-M., and MacGibbon, B. (1997), "Non-Linear shrinkage estimation with complex Daubechies wavelets," Proceedings of SPIE, Wavelet Applications in Signal and Image Processing V, vol. 3169, 67-79.

Lu, H. H-S., Huang, S-Y., and Tung Y-C. (1997), "Wavelet shrinkage for nonparametric mixed-effects models," Technical Report, Institute of

Statistics, National Chiao Tung University. Also in: Electronic proceedings of International Workshop on Wavelets in Statistics, Duke University, 12-13 October 1997.

Malfait, M., and Roose, D. (1995), "Wavelets and Markov random fields in a Bayesian framework," In *Wavelets and Statistics* (eds: A. Antoniadis and G. Oppenheim), pp. 225-238, Lecture Notes in Statistics, New York: Springer-Verlag,

Malfait, M., Jansen, M., and Roose, D. (1996), "Bayesian approach to wavelet-based image processing," In Proceedings of The Joint Statistical Meetings, Chicago, August 1996.

Mallat, S. (1989), "A theory for multiresolution signal decomposition: The wavelet representation," *IEEE Trans. Pattern Anal. Machine Intell.*, 11, 674-693.

Meyer, Y. (1992), *Wavelets and Operators,* Cambridge.

Müller, P., and Vidakovic, B. (1995), "Bayesian inference with wavelets: Density estimation," Discussion Paper, 95-33, ISDS, Duke University.

Ogden, T. (1996), "Wavelets in Bayesian change-point analysis," Technical Report, Department of Statistics, University of South Carolina.

Ogden, T. (1997), *Essential Wavelets for Statistical Applications and Data Analysis,* Birkhäuser, Boston.

Pesquet, J., Krim, H., Leporini, D., and Hamman, E. (1996), "Bayesian approach to best basis selection," In: *IEEE International Conference on Acoustics, Speech, and Signal Processing,* 5, 2634-2637.

Rios Insua, D., and Vidakovic, B. (1998), "Wavelet-based random densities," *Computational Statistics,* to appear.

Roeder, K. (1990), "Density estimation with confidence sets exemplified by superclusters and voids in the galaxies," *Journal of the American Statistical Association,* 85 617–624.

Ruggeri, F., and Vidakovic, B. (1998), "A Bayesian decision theoretic approach to wavelet thresholding," *Statistica Sinica*, to appear.

Simoncelli, E., and Adelson, E. (1996), "Noise removal via Bayesian wavelet coring," Presented at: 3rd IEEE International Conference on Image Processing, Lausanne, Switzerland.

Tierney, L. (1994), "Markov chains for exploring posterior distributions," *The Annals of Statistics*, 22, 1701–1728.

Timmermann K. E., and Nowak, R. D.,(1997), "Multiscale Bayesian Estimation of Poisson Intensities," *Conference Record of the Thirty-First Asilomar Conference on Signals, Systems, and Computers*, IEEE Computer Society PRESS, Los Alamitos CA.

Vannucci, M. (1996), *Sull'Applicazione delle "Wavelets" in Statitstica,* Tesi di dottorato, Dipartimento di Statistica G. Parenti, Università degli Studi di Firenze.

Vannucci, M., and Corradi, F. (1997), "Some findings on the covariance structure of wavelet coefficients: Theory and models in a Bayesian perspective," Report UKC/IMS, 97-05, Institute of Maths and Stats, University of Kent at Canterbury.

Vannucci, M., and Vidakovic, B. (1995), "Preventing the Dirac disaster: Wavelet based density estimation," Discussion Paper, 95-27, ISDS, Duke University.

Vidakovic, B. (1998), "Nonlinear wavelet shrinkage with Bayes rules and Bayes factors," *Journal of the American Statistical Association*, 93 173-179.

Vidakovic, B., and Müller, P. (1995), "Wavelet shrinkage with affine Bayes rules with applications," Discussion Paper, 95-34, ISDS, Duke University.

Wahba, G. (1981), "Data-based optimal smoothing of orthogonal series density estimates," *The Annals of Statistics,* 9, 146-156.

Walter, G.G. (1994), *Wavelets and Others Orthogonal Systems with Applications,* CRC Press, Boca Raton, FL.

8

Nonparametric Estimation of Irregular Functions with Independent or Autocorrelated Errors

Michael Smith
Robert Kohn

ABSTRACT This paper shows how to estimate non-smooth univariate functions that are either discontinuous or have discontinuities in the first or higher order derivatives. The estimation is carried out by modelling the function as a linear combination of terms from one of a variety of non-orthonormal bases, such that each basis has elements that are themselves non-smooth. The advantage of such an approach over wavelet based alternatives is that the sample size need not be a power of two, nor do the observations of the independent variable have to be equally spaced. To estimate the nonparametric regression we use a hierarchical Bayesian framework, with the function estimate calculated using its posterior mean. To efficiently capture any irregularities in the functions we use bases with approximately as many elements as there are observations. The computation is made tractable by using a 'focused' Markov chain Monte Carlo sampling scheme. Both the hierarchical Bayesian model and the focused sampling scheme are extended to the case where the errors in a regression are potentially autocorrelated. The efficiency of the resulting estimators for both the independent error case and the autocorrelated case is demonstrated using simulation experiments based on several functions commonly examined in the literature.

8.1 Introduction

In most of the literature on nonparametric regression the underlying regression function (or signal) is assumed to be smooth, which usually means continuous or continuously differentiable up to some order. However, in this paper we follow authors such as Donoho and Johnstone (1994) and Luo and Wahba (1997) who consider the possibility that the function is 'irregular'. By this, we mean functions that are discontinuous, or have discontinuities in the first or higher order derivatives.

A popular approach for estimating such irregular regression functions is to use shrinkage estimators with a wavelet basis, as in Donoho and Johnstone (1994, 1995). Such wavelet bases are highly locally adaptive and are orthogonal, resulting in function estimators that are very fast to compute. However, the wavelet approach usually requires equally spaced data with the number of observations being a power of two. Another approach was suggested by Luo and Wahba (1997) who use a two stage estimation method involving stepwise selection of basis terms, the coefficients of which are then estimated using penalised regression. The method has the advantage that a wide variety of bases can be used.

This paper applies a hierarchical Bayesian framework similar to that of Smith and Kohn (1996) to model irregular regression functions. These functions are approximated as linear combinations of terms from a given basis. In particular, we use and compare four different bases, including a reproducing kernel basis, a linear natural spline, a cubic regression spline and a mixed radial basis. The hierarchical mode is estimated using the 'focused sampler' introduced by Wong, Hansen, Kohn and Smith (1997). This provides an efficient means of traversing the space of all possible subsets of the basis terms and the resulting mixture estimates of the function based on the Monte Carlo sample are of high quality. It is demonstrated that, in examples which employ higher noise sizes and lower sample sizes than traditionally used in the literature, cubic regression splines are not as effective a basis for modelling irregular functions as the other three bases considered. The reproducing kernel basis, in particular, is highly effective at estimating both the irregularities and smooth sections of several test functions. It is also shown that, with the settings that are outlined, the wavelet shrinkage estimator 'SUREshrink' provides poor estimates compared to the Bayesian estimator.

The focused sampling scheme is extended to the case where the errors of a nonparametric regression are possibly autocorrelated and are modelled as an autoregressive process. To enforce stationarity, we re parameterise the autoregressive parameters in terms of the partial autocorrelations. To obtain a parsimonious autoregressive model we explicitly allow each of the partial autocorrelations to be zero. This also allows the estimation of the order of the autoregression which is the order of the highest non-zero partial autocorrelation. The resulting Bayesian estimator provides simultaneous estimates of the irregular regression function, the autoregressive parameters and the order of the autoregression (given a user prescribed maximal order).

We apply this estimator to two distinct cases. In the first the independent variable is time, while in the second the independent variable is more general and can follow any distribution. In both cases we show that the autoregression and function can be separated effectively, but that it is more difficult in the case where the independent variable is time.

8.2 Nonparametric regression for independent errors

8.2.1 *Introduction*

Consider the univariate regression model

$$y_i = f(x_i) + u_i \tag{1}$$

where $\mathbf{y} = (y_1, \ldots, y_n)'$ and $\boldsymbol{x} = (x_1, \ldots, x_n)'$ are vectors of observations on a dependent and independent variable, respectively, while $\boldsymbol{u} = (u_1, \ldots, u_n)' \sim N(0, \sigma^2 I_n)$ is a vector of independently distributed disturbances. In this model we aim to estimate f, which is an unknown function that may be discontinuous or have discontinuous first or higher order derivatives. To undertake this, throughout this article we approximate the function f as a linear combination of terms from an appropriate basis $\mathcal{B} = \{b_1(x), \ldots, b_m(x)\}$, so that

$$f(x) = \sum_{j=1}^{m} \beta_j b_j(x) \tag{2}$$

For most smooth functions m, the number of basis terms in $\mathcal{B}$, does not need to be too large, with around forty usually being sufficient to ensure the regression function is estimated without bias (Smith and Kohn, 1996). However, in the case of irregular functions, larger numbers of basis terms are required to estimate potential discontinuities or spikes. Therefore, we consider the following 'saturated' bases where, in the absence of replicated design points, there are more basis terms than observations.

1. **Reproducing kernel basis:** Here, $\mathcal{B}_1 = \{b_1(x), \ldots, b_n(x), 1, x\}$, $m = n + 2$, and $b_i(x) = (d_i^2 - 1/12)/2$, where $d_i = |x - x_i| - \frac{1}{2}$. This quadratic basis was discussed by Luo and Wahba (1997).

2. **Cubic regression spline:** Here, $\mathcal{B}_2 = \{1, x, x^2, x^3, (x-x_1)_+^3, \ldots, (x-x_n)_+^3\}$, where $(\cdot)_+ = \max(0, \cdot)$ and $m = n + 4$. While subsets of this basis have been used extensively in the literature to estimate smooth functions (Silverman and Friedman, 1989; Friedman, 1991; Smith and Kohn, 1996), its performance on irregular functions is largely untested.

3. **Linear natural spline:** Here, $\mathcal{B}_3 = \{b_1(x), \ldots, b_n(x), 1, x\}$, $m = n + 2$ and $b_i(x) = \min(x, x_i)$. This is a linear basis intended to be suitable for functions with discontinuities in the first and higher order derivatives and is discussed in Wahba (1990).

4. **Mixed Radial Basis:** Here, $\mathcal{B}_4 = \{b_1(x), \ldots, b_n(x), 1, x\}$, $m = n+2$ and $b_i(x) = |x - x_i| + (x - x_i)^2$.

Once a decision on the basis has been made, the regression at (1) can be rewritten as the linear model

$$\mathbf{y} = X\boldsymbol{\beta} + \boldsymbol{u} \qquad \text{where } X = [\mathbf{b}_1|\mathbf{b}_2|\cdots|\mathbf{b}_m] \tag{3}$$

and $\mathbf{b}_i = (b_i(x_1), \ldots, b_i(x_n))'$ is the ith basis term evaluated at each of the n observations of the independent variable. Estimating the coefficients $\boldsymbol{\beta} = (\beta_1, \ldots, \beta_m)'$ using least squares is not possible if $m > n$ or the design matrix is not of full rank. Even in the case when $m < n$, any sizable number of basis terms will make the resulting function estimate highly variable.

Proposed solutions to this problem include the work of Friedman and Silverman (1989), Friedman (1991), Stone (1994) and Luo and Wahba (1997). However, all the above approaches use a deterministic algorithm to search the space of potential subsets and the coefficients estimated conditional upon the distinct 'best' subset of basis terms found.

There are two fundamental problems which such approaches. First, the number of possible subsets is 2^m, which quickly becomes enormous for the bases above as the sample size increases. In such cases the deterministic algorithms only visit a very small percentage of these subsets and can easily get stuck in the complex space they search. Second, finding a single 'best subset' is not necessarily a good idea because even if the 'best' (as measured by a likelihood, posterior probability or other criterion) subset could be found, in a space with support of size 2^m it is exceedingly unlikely to have high absolute probability. Moreover, even if the space could be fully explored it is possible that the most likely subsets are very different in profile, so that focusing on just one of these could be misleading.

However, both these problems can be addressed using a stochastic Bayesian sampling scheme. First, the iterates from a Markov chain Monte Carlo (MCMC) approach form a Markov chain that converges to the exact posterior distribution of the subsets, given the data. A carefully constructed sampling scheme can efficiently traverse the space of likely subsets and visit many more of the relatively more likely models than a deterministic algorithm. Second, it provides a comprehensive framework within which Monte Carlo estimates can be constructed. Using the iterates inference can be undertaken while smoothing over the posterior distribution of the subsets; something Raftery, Madigan and Hoeting (1997) call model averaging.

8.2.2 Hierarchical Linear Model

We model the uncertainty associated with the significance (or otherwise) of each of the basis terms using a vector of binary indicator variables $\boldsymbol{\gamma} = (\gamma_1, \ldots, \gamma_m)'$, where $\gamma_i = 1$ iff $\beta_i \neq 0$ and $\gamma_i = 0$ iff $\beta_i = 1$. We can therefore rewrite the linear model at (3) conditional on $\boldsymbol{\gamma}$ as

$$\mathbf{y} = X_{\boldsymbol{\gamma}}\boldsymbol{\beta}_{\boldsymbol{\gamma}} + \boldsymbol{u}. \tag{4}$$

Here, $\boldsymbol{\beta}_{\boldsymbol{\gamma}}$ is the vector of regression coefficients of the 'active' basis terms (that is, the β_i such that $\gamma_i = 0$), while $X_{\boldsymbol{\gamma}}$ is a matrix of the corresponding columns of X.

To construct the hierarchical model we require a conditional prior for $\boldsymbol{\beta}_{\boldsymbol{\gamma}}|\boldsymbol{\gamma}, \sigma$. Suggestions include that by Mitchell and Beauchamp (1988), who use a spike and slab prior, and George and McCulloch (1993) who use a non-conjugate non-point normal prior. Smith and Kohn (1996) refine these with a conjugate point normal prior, while Wong, *et. al.* (1997) take a fractional prior of the form suggested by O'Hagan (1995). We will follow this latter approach and take $p(\boldsymbol{\beta}_{\boldsymbol{\gamma}}|\boldsymbol{\gamma}, \sigma) \propto p(\mathbf{y}|\boldsymbol{\gamma}, \boldsymbol{\beta}_{\boldsymbol{\gamma}}, \sigma)^{1/n}$, which results in the prior

$$\begin{aligned} \boldsymbol{\beta}_{\boldsymbol{\gamma}}|\boldsymbol{\gamma}, \sigma &\sim N(\hat{\boldsymbol{\beta}}_{\boldsymbol{\gamma}}, n\sigma^2(X_{\boldsymbol{\gamma}}'X_{\boldsymbol{\gamma}})^{-1}) \\ \boldsymbol{\beta}_{-\boldsymbol{\gamma}}|\boldsymbol{\gamma}, \sigma &= 0 \end{aligned}$$

Note that $\boldsymbol{\beta}_{-\boldsymbol{\gamma}}$ are the regression coefficients of the 'inactive' basis terms (that is, where $\gamma_i = 0$). This is a point prior because $\boldsymbol{\beta}_{-\boldsymbol{\gamma}}$ is set exactly equal to zero. It is also based on the data $\mathbf{y}$, with $\hat{\boldsymbol{\beta}}_{\boldsymbol{\gamma}} = (X_{\boldsymbol{\gamma}}'X_{\boldsymbol{\gamma}})^{-1}X_{\boldsymbol{\gamma}}'\mathbf{y}$.

We set the prior $p(\sigma^2) \propto 1/\sigma^2$, which is a commonly used uninformative prior for σ^2 (Box and Tiao, 1973). We make all subsets equally likely *a priori*, by taking $p(\boldsymbol{\gamma}) = 2^{-m}$. This is different than putting a prior on the number of terms in the model m, because the latter puts less weight on individual $\boldsymbol{\gamma}$'s that have more active terms. However, this does not necessarily mean the priors, when considered as a whole, favour $\boldsymbol{\gamma}$'s with more active terms. Using these priors, the posterior probability of any model can be calculated as

$$p(\boldsymbol{\gamma}|\mathbf{y}) \propto S(\boldsymbol{\gamma})^{-n/2}(1+n)^{-m_{\boldsymbol{\gamma}}/2} \tag{5}$$

where $S(\boldsymbol{\gamma}) = \mathbf{y}'\mathbf{y} - \mathbf{y}'X_{\boldsymbol{\gamma}}(X_{\boldsymbol{\gamma}}'X_{\boldsymbol{\gamma}})^{-1}X_{\boldsymbol{\gamma}}'\mathbf{y}$ is the sum of squares and $m_{\boldsymbol{\gamma}} = \sum_{i=1}^{m} \gamma_i$ is the number of active terms. This is almost identical to the Bayesian information criterion (BIC) given by Schwarz (1978) for subset selection in a linear model. The posterior mean of the regression coefficients is

$$E(\boldsymbol{\beta}|\mathbf{y}) = \sum_{\boldsymbol{\gamma}} E(\boldsymbol{\beta}|\mathbf{y}, \boldsymbol{\gamma})p(\boldsymbol{\gamma}|\mathbf{y}) \tag{6}$$

For m larger than around fifty, this posterior cannot be calculated exactly as it would involve 2^m evaluations of the terms in the sum at (6). Consequently, we use the MCMC scheme outlined in section 8.2.3 to obtain iterates $\boldsymbol{\gamma}^{[1]}, \ldots, \boldsymbol{\gamma}^{[J]}$ from the posterior distribution $p(\boldsymbol{\gamma}|\mathbf{y})$ and approximate the expected value using the mixture estimate

$$\hat{\boldsymbol{\beta}} = \frac{1}{J}\sum_{j=1}^{J} E(\boldsymbol{\beta}|\mathbf{y}, \boldsymbol{\gamma}^{[j]}) \tag{7}$$

Here, the conditional expectation is easily calculated as $E(\boldsymbol{\beta}_{-\boldsymbol{\gamma}^{[j]}}|\mathbf{y},\boldsymbol{\gamma}^{[j]}) = \mathbf{0}$, while $E(\boldsymbol{\beta}_{\boldsymbol{\gamma}^{[j]}}|\mathbf{y},\boldsymbol{\gamma}^{[j]}) = (X'_{\boldsymbol{\gamma}^{[j]}}X_{\boldsymbol{\gamma}^{[j]}})^{-1}X'_{\boldsymbol{\gamma}^{[j]}}\mathbf{y}$. The function estimate at a point x is therefore

$$\hat{f}(x) = (b_i(x), \ldots, b_m(x))'\hat{\boldsymbol{\beta}}$$

which is the mixture estimate of $E(f|\mathbf{y})$.

8.2.3 Focused Sampling Scheme

Wong, *et. al.* (1997) introduced the focused sampler and compared it to the Gibbs sampler of Smith and Kohn (1996) and reversible jump sampler (Green, 1995; Denison, Mallick and Smith, 1996) when applied to nonparametric regression. The focused sampler is both faster and more efficient than the Gibbs sampling alternative. It is essential when the problem involves bases with more than around two hundred terms as the Gibbs sampler can get stuck. Wong *et. al.*, (1997) also discuss how the focused sampler generalises the movement steps that reversible jump sampling uses to move around the support of $\boldsymbol{\gamma}$ and how some reversible jump sampling implementations are more ad-hoc algorithms than true MCMC schemes because they do not generate the regression coefficients $\boldsymbol{\beta}_{\boldsymbol{\gamma}}$, but plug in an estimate instead. The latter means that the variance of $\boldsymbol{\beta}$ is not properly accounted for, one result being that correct confidence intervals for the regression function are not available. The authors also demonstrate how these faults can lead to poor performance when reversible jump sampling is applied to univariate and bivariate nonparametric smoothing and then compared to focused sampling. Another related procedure is the importance sampler discussed in Clyde, DeSimone and Parmigiani (1996) who orthogonalise the design matrix X and then draw independent samples with probability that approximates $p(\boldsymbol{\gamma}|\mathbf{y})$. The orthogonal property allows for very fast computation and therefore it is computationally possible to apply the sampler to nonparametric regression with independent errors where m is large, but is hard to extend to the autocorrelated error case.

In essence, the focused sampler runs Gibbs sampling sub-chains on subsets of the γ_i binary indicators. The subsets are made up of terms that, given the data, are likely to be active. The focused sampler is given below.

Focused Sampling Scheme.
`Step (0): Select initial state` $\boldsymbol{\gamma} = \boldsymbol{\gamma}^{[0]}$
`Sequentially iterate the following steps:`
`Step (1): Choose` $\mathcal{C} \subset \{1, 2, \ldots, m\}$ `in a random manner`
`Step (2): Repeat the following` K `times:`
`Sequentially generate` $\gamma_i|\gamma_{j\neq i}, \mathbf{y}$ `for` $i \in \mathcal{C}$

Here, $\mathcal{C}$ is the set of indices of indicator variables on which we focus. If

$\mathcal{C} = \{1, 2, \ldots, m\}$ and $K = 1$, then the above scheme reduces to the Gibbs sampler of Smith and Kohn (1996). However, we chose the subset $\mathcal{C}$ in step (1) of an iteration j, say, to adaptively focus on the variables that are more likely to be active. If $m_\gamma = \sum_{i=1}^{m} \gamma_i^{[j-1]}$ is the number active variables in the previous $(j-1)$th iteration, then we use the following rule at step (1).

Step (1) at iteration j

(1a) $\mathcal{C}_1 = \{i | \gamma_i^{[j-1]} = 1\}$; $\mathcal{C}_2 = \emptyset$

(1b) $P = \min\left(1, \max(m_\gamma, 20)/(m - m_\gamma)\right)$

(1c) Choose each of the indices in $\{i | \gamma_i^{[j-1]} = 0\}$ to be an element in $\mathcal{C}_2$ with probability P

(1d) $\mathcal{C} = \mathcal{C}_1 \cup \mathcal{C}_2$

This random focusing step allows generation of all the active variables in the previous iteration (indexed with $\mathcal{C}_1$), along with an additional randomly selected subset of those that were inactive in the previous iteration (indexed with $\mathcal{C}_2$). These elements are selected with probability P, which is chosen so that on average a minimum of 20 inactive terms are selected for generation. This ensures the focused sampler has the opportunity to move around the space. If the regression function we are trying to estimate requires a lot of basis terms, then an average of m_γ additional terms are included in $\mathcal{C}$, so that there is a larger selection of basis terms to generate from.

Using such a focusing rule makes the the selection of $\mathcal{C} = \mathcal{C}_1 \cup \mathcal{C}_2$ dynamic, in that over several iterations of the sampling scheme any state in the support of $\gamma | \mathbf{y}$ can be easily reached. The scheme is irreducible, as well as being aperiodic and will converge to its invariant distribution $\gamma | \mathbf{y}$ (Tierney, 1994).

By repeating the generations K times in step (2) we have a Gibbs sampling sub-chain which converges to the conditional distribution of the block $\gamma_{i \in \mathcal{C}} | \gamma_{i \notin \mathcal{C}}, \mathbf{y}$ from which an iterate is obtained. Values we consider in this paper for the number of iterations in the sub-chain are $K = 1$ or 15, though we have found that taking $K = 1$ works very well in most cases and produces rapid convergence for the sampling scheme.

8.2.4 Empirical Investigation

In this section we compare the four bases $\mathcal{B}_1, \ldots, \mathcal{B}_4$ defined previously on a number of functions already examined in the literature. These are the four irregular functions used by Donoho and Johnstone (1994) and a smooth function examined in Luo and Wahba (1997), plots of which are provided at panel (b) of figures 2-6. These functions are defined in those papers, though note that we do not rescale the functions as in Donoho and Johnstone (1994). To enable a comparison with the work found in these papers, we restrict the independent variable to be an evenly spaced grid

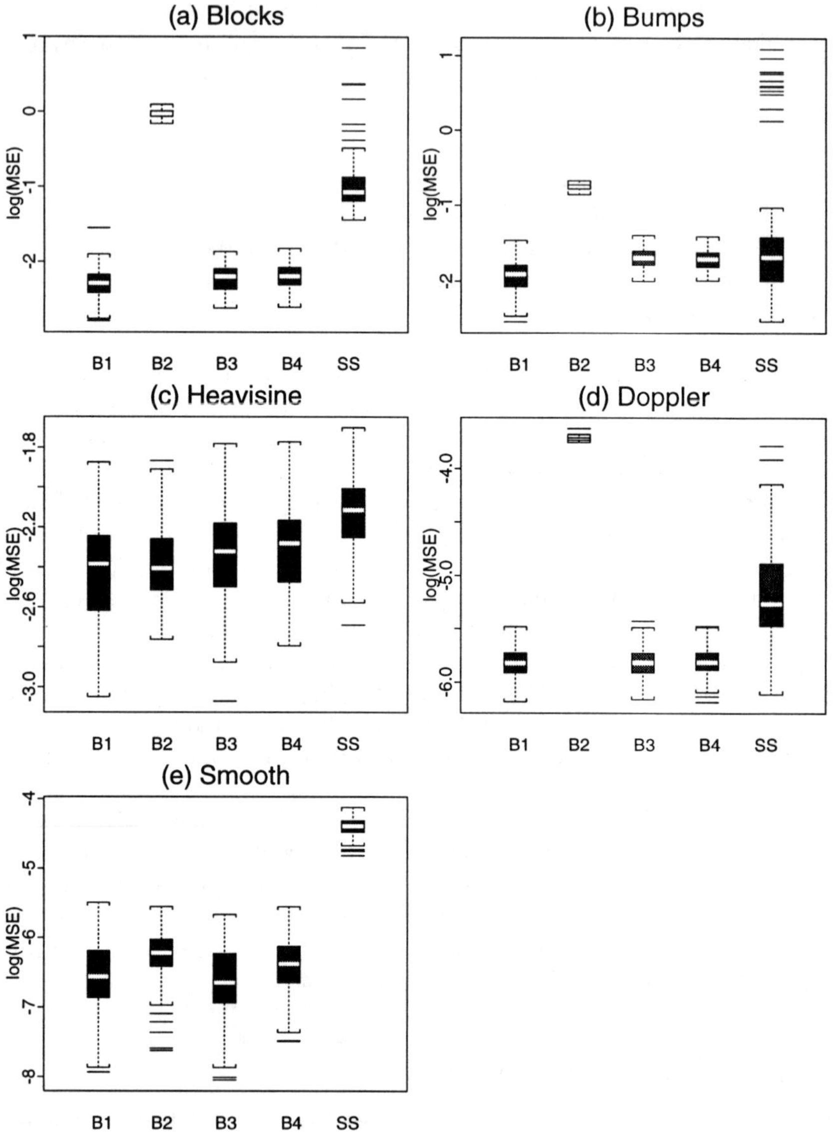

FIGURE 1. Boxplots of the log(MSE) results from the simulation examples using the Bayesian estimator with each of the four bases $\mathcal{B}_1, \mathcal{B}_2, \mathcal{B}_3$ and $\mathcal{B}_4$, and the SUREshrink estimator (SS). Panels (a) through to (e) are for the functions f_1 to f_5, respectively.

on $[0, 1]$, though we note that using our approach the independent variable x_i could follow any distribution. In our examples we used a substantially lower sample size ($n = 2^9 = 512$) and higher noise levels ($\frac{1}{8}\times$function range) than those used in Donoho and Johnstone (1994), with the details given in table 1.

Name	Range(f)	Reference	σ	$SD(f)/\sigma$
Blocks (f_1)	7.2	D&J (1994)	0.9	2.122
Bumps (f_2)	4.18	D&J (1994)	0.523	1.256
Heavisine (f_3)	10	D&J (1994)	1.25	2.376
Doppler (f_4)	0.991	D&J (1994)	0.124	2.334
Smooth (f_5)	2.331	L&W (1997)	0.291	2.411

TABLE 1. Details of simulated examples. The noise, σ, is chosen to be one eighth of the function range over $[0, 1]$. Here 'SD' is 'standard deviation' and the measure $SD(f)/\sigma$ is provided to enable comparison with the results found in Donoho and Johnstone (1994).

One hundred and one replications of the dataset were generated for each of the five functions and the focused sampler applied to estimate the underlying function using the four bases $\mathcal{B}_1, \ldots, \mathcal{B}_4$. We also used the wavelet shrinkage estimator 'SUREshrink' proposed by Donoho and Johnstone (1995) and implemented by Nason and Silverman in their S-PLUS 'wavethresh' package. For this, we use a 'primary resolution level' of 5, the 'DaubLeAsymm' family of wavelets with filter number 8.

To measure the quality of the function estimates we use the mean square error criterion, defined over $[0, 1]$ as

$$MSE = \frac{1}{400}\sum_{i=1}^{400}(f(z_i) - \hat{f}(z_i))^2$$

Here, the series $z_1, z_2, \ldots, z_{400}$ is an evenly spaced grid over the interval $[0, 1]$. Lower quality estimates (both biased and those with high local variability) have higher MSE values.

Figure 1 provides boxplots of the MSE values for each of estimators and five functions considered. For the largely smooth functions f_3 and f_5, the cubic regression spline basis $\mathcal{B}_2$ works well, though it performs very poorly on the three distinctly nonlinear functions f_1, f_2 and f_4. The bases $\mathcal{B}_3$ and $\mathcal{B}_4$ work well on all functions, though are slightly outperformed by the reproducing kernel basis, $\mathcal{B}_1$. Throughout, SUREshrink performs poorly with the given settings, though it should be noted that the results can be very sensitive to these settings, especially the primary resolution level; see Hall and Patil (1996) and Hall and Nason (1997) for some discussion of this issue.

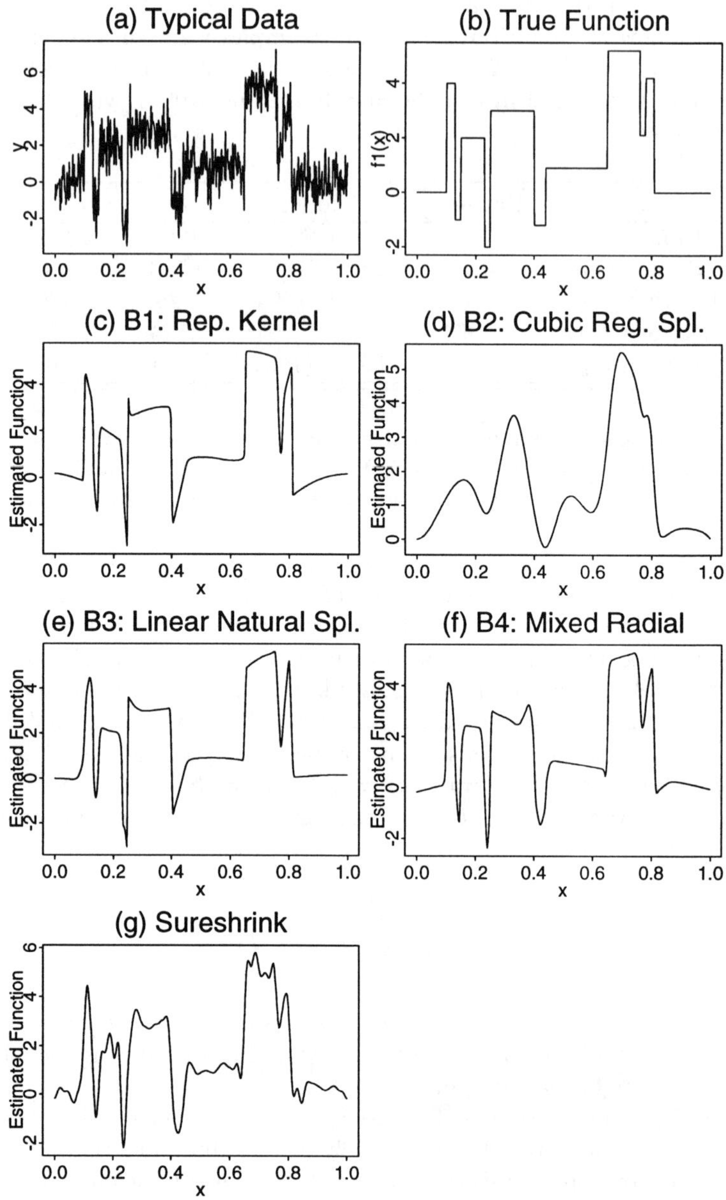

FIGURE 2. BLOCKS: (a) Plot of a typical data set. Panel (b) the true regression function f_1. Panels (d)-(f) plots of the function estimates that correspond to the median MSE value from the simulation for each of the four bases examined. Panel (g) the SUREshrink estimate corresponding to the median MSE value from the simulation.

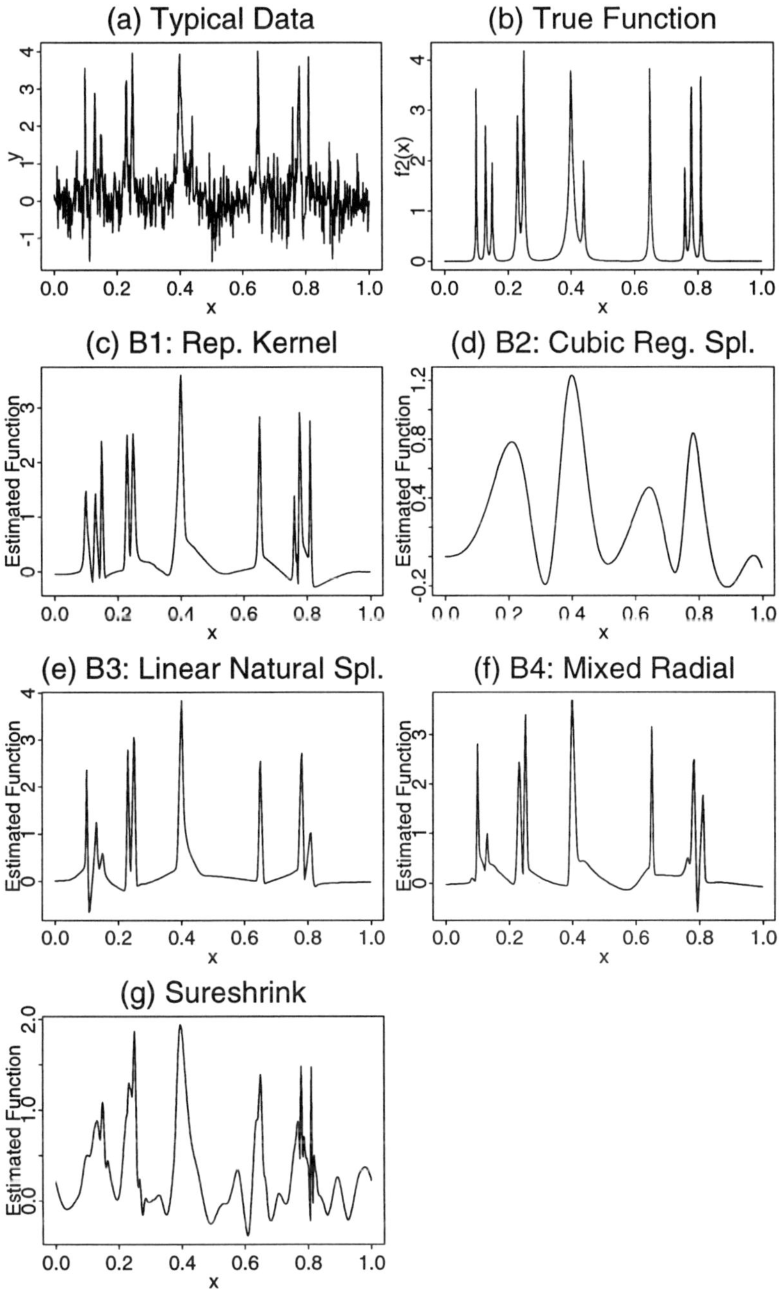

FIGURE 3. BUMPS: The same as the previous plot, except for f_2.

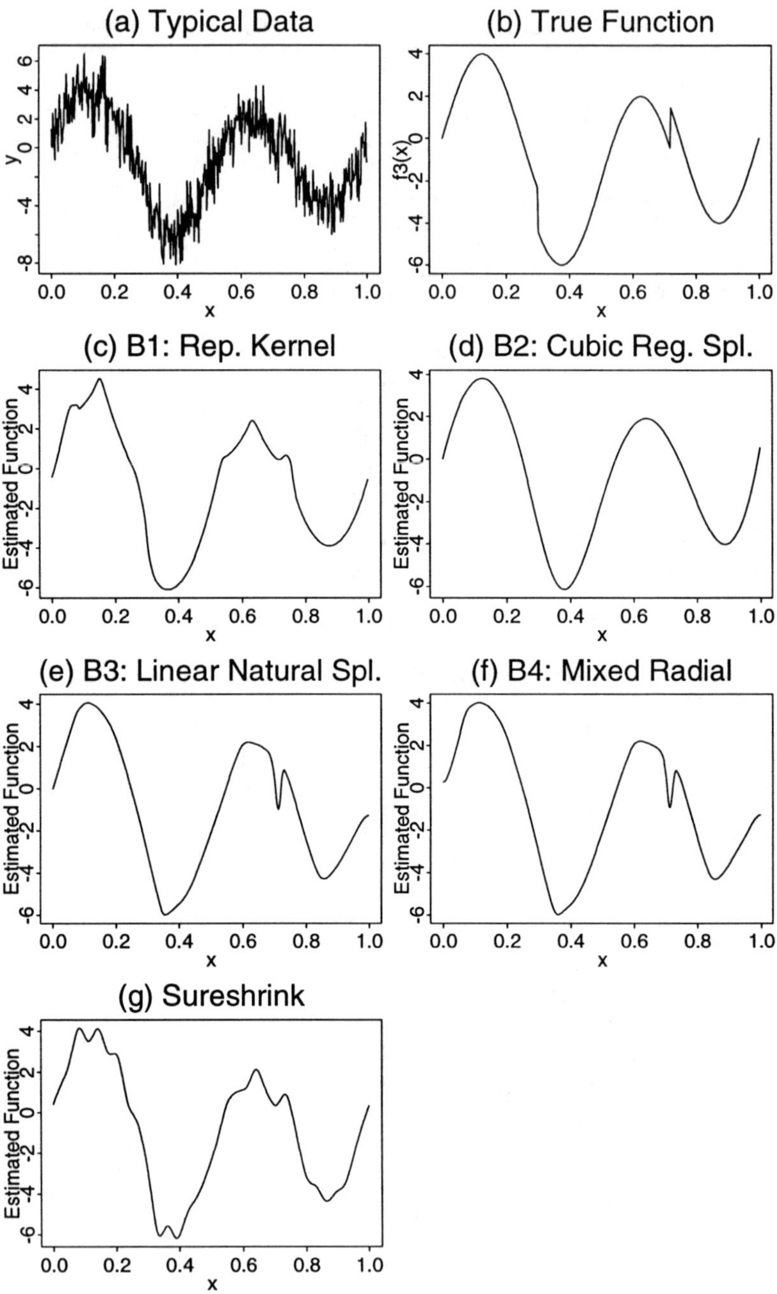

FIGURE 4. HEAVISINE: The same as the previous plot, except for f_3.

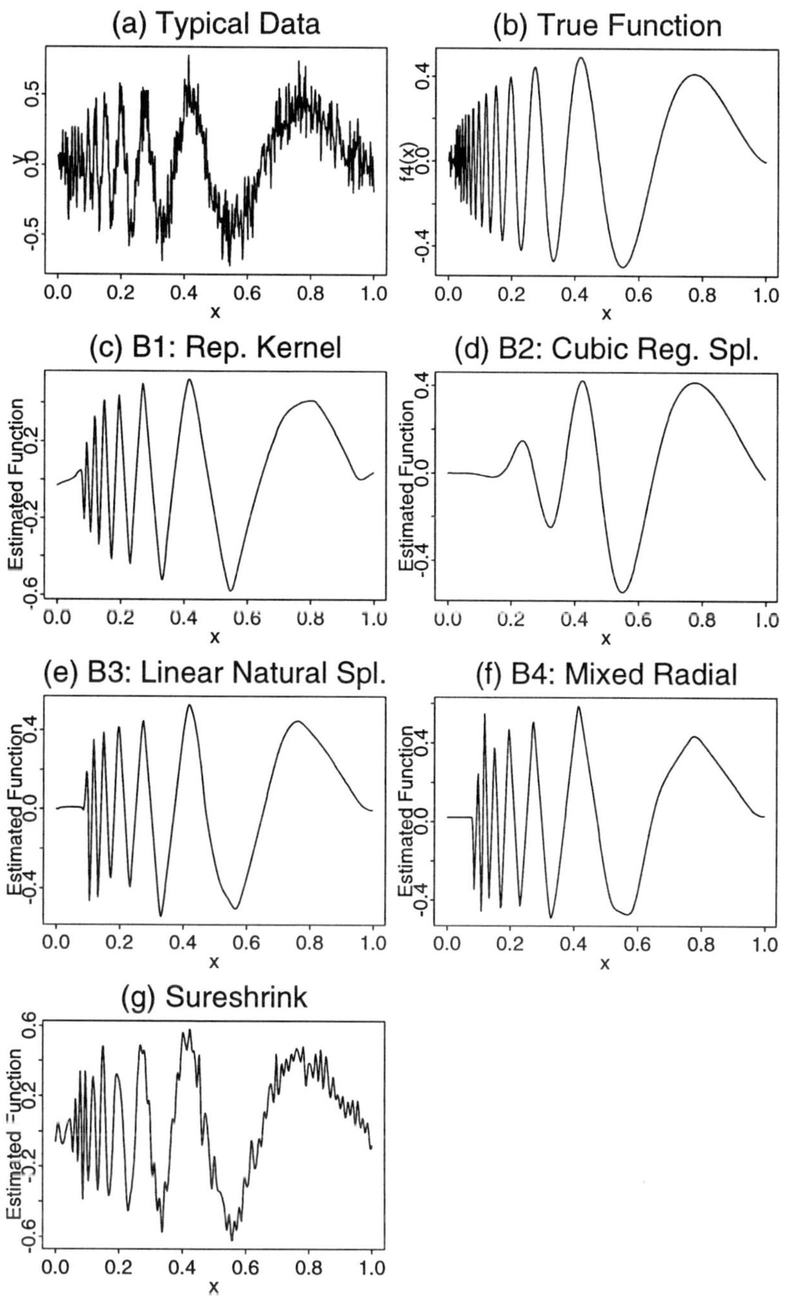

FIGURE 5. DOPPLER. The same as the previous plot, except for f_4.

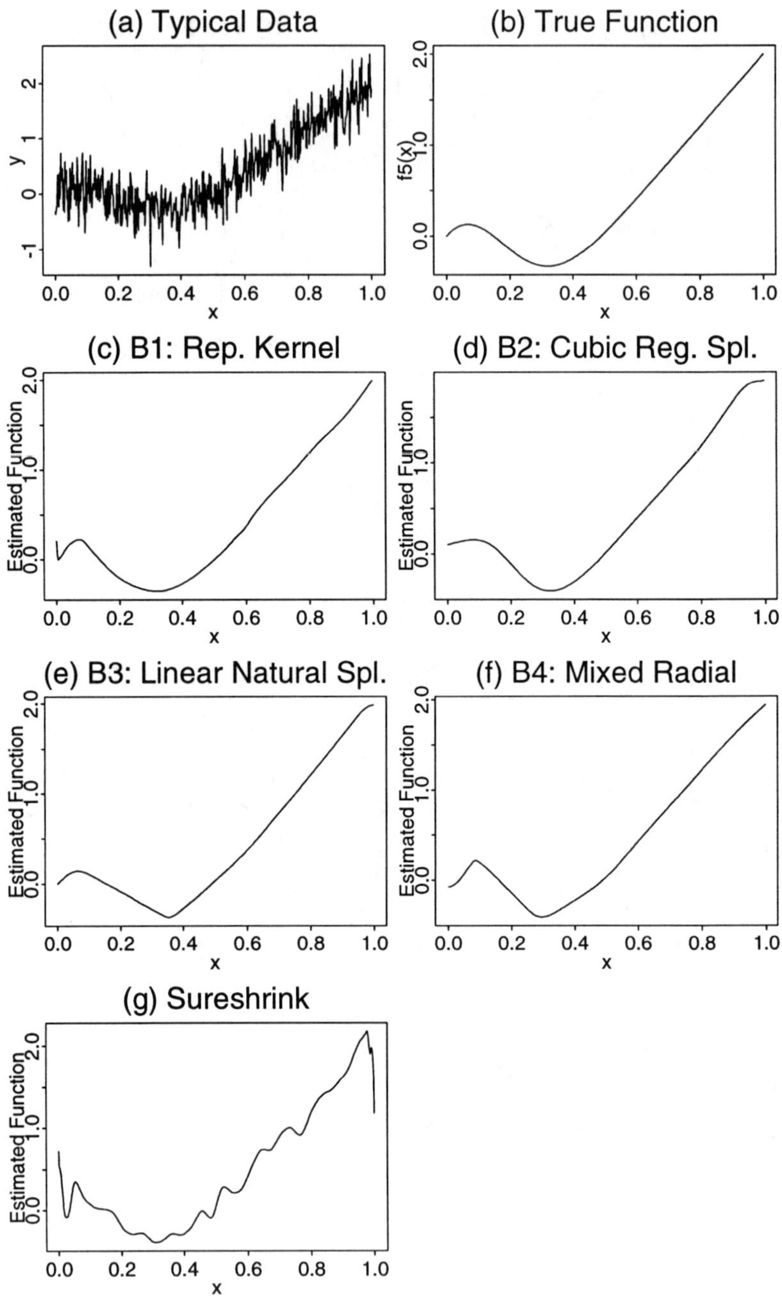

FIGURE 6. SMOOTH: The same as the previous plot, except for f_5.

However, with irregular functions the MSE measure is not necessarily particularly informative on its own. Issues such as whether, or not, 'spikes' (as in the bumps function f_2) or discontinuities (as in the blocks function f_1) were captured correctly are not fairly summarised by this single figure. Consequently, we present a single estimate for each basis which we regard as typical, in that it is the estimate that results in the median MSE value for that basis in our simulation. These are presented for each function in figures 2-6. Also included in these figures are a plot of typical dataset, along with the true underlying function.

Figures 2(d) and 3(d) demonstrate that the poor performance in terms of MSE for the cubic regression spline basis can be seen to be caused by over-smoothing many of the discontinuities in the f_1 (blocks) and spikes in f_2 (bumps). Of particular interest in these figures is the excellent all round performance of the reproducing kernel basis. It is by far the best basis for capturing the profile of the Doppler function (see figure 5(c)), does well on smooth functions, such as that in figure 6(c), and still captures the discontinuities in the blocks and bumps functions (see figures 2(c) and 3(c)). The SUREshrink estimates are found in figures 2(g)-6(g) and display an inability to both reproduce discontinuities and irregularities, and also produce smooth estimates of any continuous parts of the functions.

In our implementation of the Bayesian estimators we set $K = 1$, take a warmup period of 2000 iterations and a sample size of $J = 2000$ iterations, which our experience suggests is conservative.

8.3 Nonparametric regression for data with autocorrelated errors

8.3.1 Introduction

Consider the univariate nonparametric regression model at (1), but where the errors are no longer independent and are modelled as an autoregressive process, so that

$$y_t = f(x_t) + u_t \quad \text{where}$$

$$u_t = \phi_1 u_{t-1} + \cdots + \phi_s u_{t-s} + e_t \tag{8}$$

Here, e_t is independent $N(0, \sigma^2)$ and $\{u_t\}_{t=1}^n$ is a stationary autoregressive error sequence with maximal order s. Again, we approximate f as a linear combination of basis terms from a given basis $\mathcal{B}$, but want to estimate the irregular function f and the autoregressive process simultaneously. Smith, Wong and Kohn (1998) extend the hierarchical model in section 8.2.2 to this situation and used a Gibbs sampler to estimate the unknown parameters.

In this paper, we will apply the focused sampler to estimate a similar hierarchical model to enable the estimation of (i) the irregular function f, (ii) the autoregression coefficients $\boldsymbol{\phi} = (\phi_1, \ldots, \phi_s)'$ and (iii) the order of the autoregression, given a potential maximum s.

8.3.2 Extending the hierarchical model

In our model we re-parameterise the autoregressive parameters $\boldsymbol{\phi}$ in terms of the partial autocorrelations $\boldsymbol{\psi} = (\psi_1, \ldots, \psi_s)'$. This is possible as there is a one-to-one transformation between $\boldsymbol{\psi}$ and $\boldsymbol{\phi}$. As in Monahan (1984) and Barnett, Kohn and Sheather (1996), it is convenient to enforce stationarity by restricting the domain of the partial autocorrelations, so that $-1 < \psi_i < 1$ for $i = 1, \ldots, s$. In the case when $x_t = t$ it is important to enforce stationarity of u_t so as not to confound the model for the errors with the nonparametric estimate of the function.

To facilitate selecting the autoregressive order, indicator variables $\boldsymbol{\kappa} = (\kappa_1, \ldots, \kappa_s)$' are introduced to denote the non-zero partial autocorrelations. In the same spirit as subset selection for regression parameters, a conditional prior $\boldsymbol{\psi}|\boldsymbol{\kappa}, \sigma^2$ can be defined, where $\psi_1, \ldots, \psi_s$ are *a priori* conditionally independent with

$$\begin{aligned} \psi_i|\kappa_i = 0, \sigma^2 &\quad \text{is identically equal to zero} \\ \psi_i|\kappa_i = 1, \sigma^2 &\quad \sim \text{Uniform}(-1, 1) \end{aligned}$$

The maximal order, s, of the autoregression and the prior probabilities $p(\kappa_i = 1), i = 1, \ldots, s$, are prescribed by the user. In all the empirical work presented here a descending prior on the order is used, where $p(\kappa_1 = 1) = 0.5, p(\kappa_2 = 1) = 0.4, p(\kappa_3 = 1) = 0.3, p(\kappa_4 = 1) = 0.2$, and $p(\kappa_i = 1) = 0.1$ for $i = 5, \ldots, s$. This prior favours lower order autoregressions, though, as demonstrated in Smith, Wong and Kohn (1998), the empirical results appear to be insensitive to this prior.

The formulation in terms of $\boldsymbol{\psi}$ allows the error variance to be explicitly calculated as var($\boldsymbol{u}$)=$\sigma^2\Omega_{\boldsymbol{\psi}}$, where $\Omega_{\boldsymbol{\psi}}$ is a matrix depending only upon $\boldsymbol{\psi}$ and the data $\mathbf{y}$. As we are modelling the function f as linear combination of basis terms, we use the hierarchical model introduced at (4) with a fractional prior for $\boldsymbol{\beta}_\gamma$ that conditions on $\boldsymbol{\psi}$. That is, where $p(\boldsymbol{\beta}_\gamma|\gamma, \sigma^2, \boldsymbol{\psi}, \boldsymbol{\kappa}) \propto p(\mathbf{y}|\gamma, \sigma^2, \boldsymbol{\psi}, \boldsymbol{\beta}_\gamma, \boldsymbol{\kappa})^{1/n}$, which results in the prior

$$\boldsymbol{\beta}_\gamma|\gamma, \sigma^2, \boldsymbol{\psi}, \boldsymbol{\kappa} \sim N((X'_\gamma\Omega_{\boldsymbol{\psi}}^{-1}X_\gamma)^{-1}X'_\gamma\Omega_{\boldsymbol{\psi}}^{-1}\mathbf{y}, n\sigma^2(X'_\gamma\Omega_{\boldsymbol{\psi}}^{-1}X_\gamma)^{-1})$$

Finally, the priors for σ^2 and γ are unchanged from those in section 8.2.2 and $\boldsymbol{\kappa}, \gamma, \sigma^2$ are assumed to be *a priori* independent.

8.3.3 Estimation using the focused sampler

We estimate this hierarchical model with autocorrelated errors using a focused sampling scheme with two focusing steps. The first step focuses on the partial autocorrelation indicators, κ_i, that are most likely to be equal to one. The second step is as outlined is section 8.2.3 for focusing on the variables γ_i that are most likely to be equal to one. The sampling scheme is given below.

Focused Sampling Scheme

`Step (0): Choose initial state` $(\boldsymbol{\kappa}^{[0]}, \boldsymbol{\psi}^{[0]}, \boldsymbol{\beta}^{[0]}, \gamma^{[0]})'$
`Sequentially iterate the following steps:`
`Step (1): Generate an iterate from` $\boldsymbol{\beta}|\boldsymbol{\kappa}, \boldsymbol{\psi}, \gamma, \mathbf{y}$
`Step (2): Select` $\mathcal{S} \subset \{1, 2, \ldots, s\}$ `in a random manner`
`Step (3): Repeat the following` K_1 `times:`
    `Generate iterates from` $\kappa_i, \psi_i|\kappa_{j\neq i}, \psi_{j\neq i}, \gamma, \boldsymbol{\beta}, \mathbf{y}$ `for` $i \in \mathcal{S}$
`Step (4): Select` $\mathcal{C} \subset \{1, 2, \ldots, m\}$ `in a random manner`
`Step (5): Repeat the following` K_2 `times:`
    `Generate iterates from` $\gamma_i|\gamma_{j\neq i}, \boldsymbol{\kappa}, \boldsymbol{\psi}, \mathbf{y}$ `for` $i \in \mathcal{C}$

This scheme is a focused sampling version of the Gibbs sampler presented in Smith, Wong and Kohn (1998), who derive the conditional distributions and indicate how to generate from them. Notice that σ^2 is integrated out of the entire sampling scheme and that κ_i and ψ_i are generated as a block. The latter is because $\kappa_i|\psi_i$ is known exactly and a sampler that generated κ_i conditional upon ψ_i would be reducible.

Step (4) is a focusing step exactly as outlined in section 8.2.3. Step (2) is a focusing step for the $\boldsymbol{\kappa}$ variables and if $q_{\boldsymbol{\kappa}} = \sum_{i=1}^{s} \kappa_i$ when $\boldsymbol{\kappa} = \boldsymbol{\kappa}^{[j-1]}$, then this step is undertaken as indicated below.

Step (2) at iteration j

`(2a)` $\mathcal{S}_1 = \{i|\kappa_i^{[j-1]} = 1\}; \quad \mathcal{S}_2 = \emptyset$
`(2b)` $P = \min(1, \max(q_{\boldsymbol{\kappa}}, 3)/(s - q_{\boldsymbol{\kappa}}))$
`(2c) Chose each of the terms in` $\{i|\kappa_i^{[j-1]} = 0\}$ `to be`
    `an element in` $\mathcal{S}_2$ `with probability` P
`(2d)` $\mathcal{S} = \mathcal{S}_1 \cup \mathcal{S}_2$

This is similar to the focused step for the variables in γ. The only difference is that, on average, we generate from subsets of size 3 or $q_{\boldsymbol{\kappa}}$, whichever is larger. This enables easy movement of the Markov chain to areas in the parameter space where a large number of partial autocorrelations are active (that is, where many κ_i's are equal to one). However, where the data provides information that there are only a few active partial autocorrelations, the focus is only on a few κ_i. Note that because $\mathcal{S}$ always encompasses the

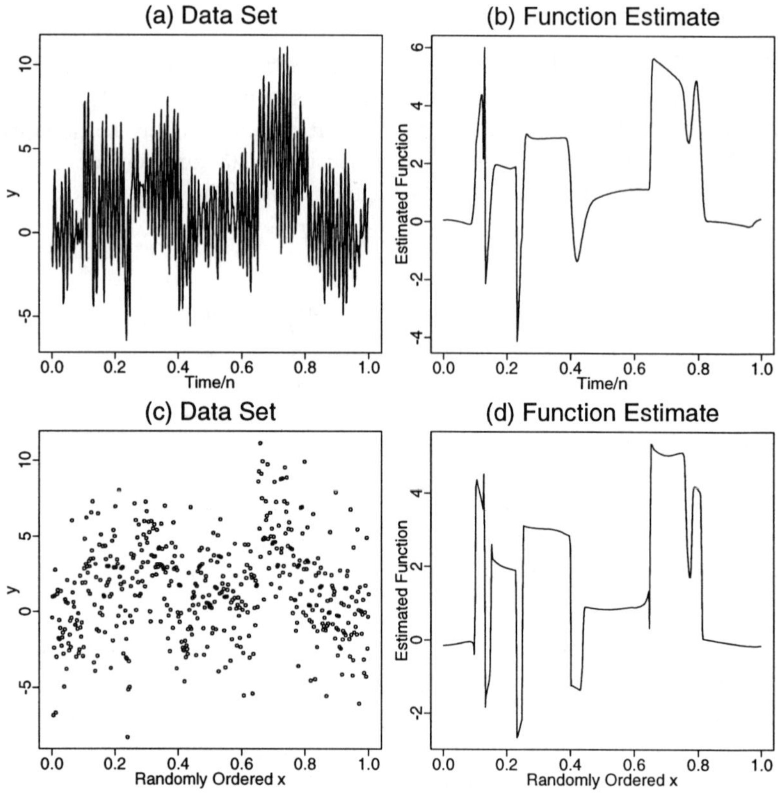

FIGURE 7. (a) Plot of data generated when $x_t = t/n$ with AR(2) errors and the true function is f_1. Panel (b) provides the function estimate based on the dataset in (a) using the Bayesian estimator. Panel (c) is a plot of data generated when x_t is randomly ordered, but equally spaced, while panel (d) presents the Bayesian function estimate using this data.

currently active partial autocorrelations there is no need to generate $\psi_{i \notin S}$ as they are all exactly equal to zero.

The focused sampling steps are Gibbs sampler sub-chains with invariant conditional distributions $\gamma_{i \in C} | \gamma_{i \notin C}, \boldsymbol{\kappa}, \boldsymbol{\psi}, \mathbf{y}$ and $\kappa_{i \in S}, \psi_{i \in S} | \kappa_{i \notin S}, \psi_{i \notin S}, \boldsymbol{\gamma}, \boldsymbol{\beta}, \mathbf{y}$ and iteration lengths K_2 and K_1, respectively. In our empirical work we set $K_2 = 15$ and $K_1 = 1$, though we have found that lower values for K_2 appear to work well. After an initial warmup period the focused sampler produces iterates $\{\boldsymbol{\beta}^{[j]}, \boldsymbol{\kappa}^{[j]}, \boldsymbol{\psi}^{[j]}, \gamma^{[j]}\}_{j=1}^{J}$ upon which inference can be based. We estimate $E(\beta|y), E(\psi|y)$, and $E(\kappa_i|y) = p(\kappa_i = 1|y)$ using the mixture and histogram estimates

$$\hat{\boldsymbol{\beta}} = \frac{1}{J} \sum_{j=1}^{J} E(\boldsymbol{\beta} | \gamma^{[j]}, \boldsymbol{\psi}^{[j]}, \boldsymbol{\kappa}^{[j]}, \mathbf{y}) \quad \hat{\boldsymbol{\psi}} = \frac{1}{J} \sum_{j=1}^{J} \boldsymbol{\psi}^{[j]}$$

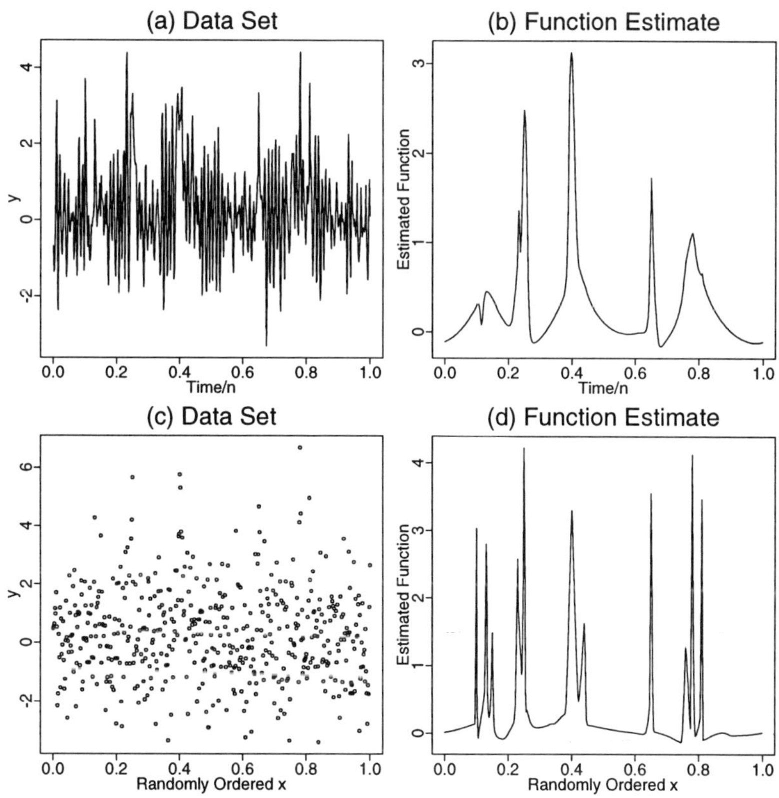

FIGURE 8. Plots as in previous figure, but with f_2 as underlying function

$$\hat{\kappa}_i = \frac{1}{J}\sum_{j=1}^{J} p(\kappa_i = 1|\psi^{[j]}_{k\neq i}, \kappa^{[j]}_{k\neq i}, \gamma, \boldsymbol{\beta}, \mathbf{y}) \quad i = 1, \dots, s$$

The estimate $\hat{\boldsymbol{\phi}}$ of the autoregressive parameters is obtained from $\hat{\boldsymbol{\psi}}$ as there is a one-to-one relationship. As in the previous section we use a warmup of 2000 iterates and a sampling period of $J = 2000$ iterations.

8.3.4 Empirical Examples

There are two distinct ways in which the univariate nonparametric regression problem at (8) can arise. The first is where $x_t = t/n$ and then $f(t)$ is effectively a nonparametric estimate of a time trend, over and above the stationary autoregression component u_t. In this case, it may prove to be difficult to separate the time effect into the two components f and the autoregression process parameterised by $\boldsymbol{\phi}$. This problem does not occur in the second case, where the independent variable is not effectively time.

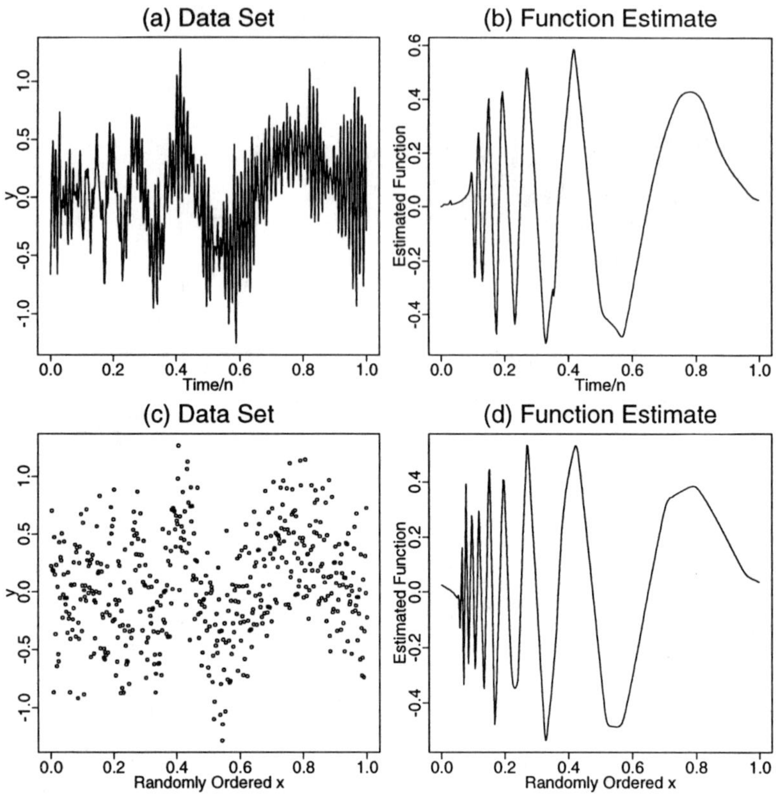

FIGURE 9. Plots as in previous figure, but with f_4 as underlying function

To highlight the performance of the sampling scheme in both cases we simulated some data from the univariate regression model at (8), where the errors were autocorrelated with parameters $\phi_1 = 0.9$ and $\phi_2 = -0.9$. We consider the case when $x_t = t/n$, along with the more general case when x_t is not time. In this second case, to make the results comparable with the previous section we consider equally spaced, but random ordered, values for x_t, though it should be stressed that the independent variable could follow any distribution.

Because Smith, Wong and Kohn (1998) have investigated the empirical performance of the Gibbs sampler for smooth functions from such a univariate regression model, we only consider the more irregular functions f_1, f_2 and f_4 (blocks, bumps and Doppler) used in section 8.2.4. We also only examine the performance for the reproducing kernel basis $\mathcal{B}_1$, which was the best performer for these functions with independent errors. The intrinsic error, σ^2, was set exactly to that outlined in table 1, which was one eighth of the function range. The sample size remains unchanged from the previous section, with $n = 512$, and we selected the order of the autoregression

	For case when $x_t = t/n$					
	f_1		f_2		f_4	
i	$\hat{\phi}_i$	$\hat{\kappa}_i$	$\hat{\phi}_i$	$\hat{\kappa}_i$	$\hat{\phi}_i$	$\hat{\kappa}_i$
1	0.90	1.00	0.86	1.00	0.75	1.00
2	-0.89	1.00	-0.77	1.00	-0.81	1.00
3	0.00	0.05	0.00	0.02	-0.11	0.63
4	0.00	0.01	0.00	0.02	0.00	0.01
5	0.00	0.00	0.00	0.00	0.00	0.01
6	0.00	0.00	0.00	0.01	0.00	0.01
7	0.00	0.01	0.00	0.01	0.00	0.01
8	0.00	0.01	0.00	0.01	0.00	0.01
	For case when x_t is general					
	f_1		f_2		f_4	
i	$\hat{\phi}_i$	$\hat{\kappa}_i$	$\hat{\phi}_i$	$\hat{\kappa}_i$	$\hat{\phi}_i$	$\hat{\kappa}_i$
1	0.93	1.00	0.9	1.00	0.87	1.00
2	-0.90	1.00	-0.9	1.00	-0.89	1.00
3	0.00	0.03	0.0	0.01	-0.01	0.25
4	0.00	0.03	0.0	0.01	-0.02	0.01
5	0.00	0.00	0.0	0.01	0.02	0.15
6	0.00	0.00	0.0	0.00	0.00	0.00
7	0.00	0.00	0.0	0.00	0.00	0.00
8	0.00	0.01	0.0	0.00	0.00	0.00

TABLE 2. The estimates $\hat{\kappa}$ and $\hat{\psi}$ for the functions blocks, bumps and Doppler for both types of independent variables considered.

from a maximum order of $s = 8$.

Figures 7-9 plot the simulated data for both types of independent variables, along with the corresponding function estimates. In the case when the independent variable is not time, but more general, the function estimates are comparable to those obtained from the case where there was no time effect. This remarkably small loss in efficiency is reported upon more fully for smooth functions in Smith, Wong and Kohn (1998). However, the estimates for the case when $x_t = t/n$ are slightly worse, which is due to some of the signal f being captured in the estimated autoregressive process.

Table 2 provides $\hat{\psi}$, the estimates of the autoregressive parameters, and $\hat{\kappa}$ for all six datasets. As the order of an autoregression corresponds to the largest index of the non-zero partial autocorrelations, it can be seen that in five of the examples the correct order is estimated from the data. However, for the Doppler function f_4, when $x_t = t/n$, an AR(3) is just preferred over an AR(2) because a fraction of the signal found in the Doppler function can be captured by an AR(3). Consequently, the estimates of the autoregressive coefficients are slightly off mark, while the function estimate (see figure 9(b)) is slightly inferior to that found when x_t is not time (see figure 9(d)). Similarly, for the bumps function when $x_t = t/n$ some of the structure of f_2 has been confounded with the autoregressive effect, which is reflected in the slight degradation in both function estimates (see figure 8(b)) and autoregressive parameter estimates.

Acknowledgments: : We would like to thank Prof. Guy Nason for making his wavethresh package readily available and for helpful comments with regard to wavelet based curve estimation.

References

Barnett, G., Kohn, R. and Sheather, S. (1996). Bayesian estimation of an autoregressive model using Markov Chain Monte Carlo. *Journal of Econometrics.*, **74**, 237–254.

Box, G. and Tiao, G. (1973). *Bayesian Inference in Statistical Analysis.* New York: John Wiley and Sons.

Clyde, M., Desimone, H. and Parmigiani, G. (1996). Prediction Via Orthogonalized Model Mixing. *Journal of the American Statistical Association*, **91**, 1197-1208

Denison, D., Mallick, B. and Smith, A. (1996). Automatic Bayesian curve fitting. preprint

Donoho, D. and Johnstone, I. (1994). Ideal Spatial Adaptation by Wavelet Shrinkage. *Biometrika*, **81**, 425-455.

Donoho, D. and Johnstone, I. (1995). Adapting to Unknown Smoothness via Wavelet Shrinkage'. *Journal of the Royal Statistical Society*, Ser. B, **57**, 301-370.

Friedman, J. (1991). Multivariate adaptive regression splines. *The Annals of Statistics* (with discussion), **19**, 1-141

Friedman, J. and Silverman, B. (1989). Flexible parsimonious smoothing and additive modelling. *Technometrics*, **31**, 3-39.

George, E., and McCulloch, R. (1993). Variable selection via Gibbs sampling. *Journal of the American Statistical Association*, **88**, 881-889

Green, P. (1995). Reversible jump Markov chain Monte Carlo computation and Bayesian model determination. *Biometrika*, **82**, 711-732

Hall, P. and Patil, P. (1996). On the choice of the smoothing parameter, threshold and truncation in nonparametric regression by nonlinear wavelet methods. *Journal of the Royal Statistical Society*, Series B, **58** 361-377

Hall, P. and Nason, G. (1997). On choosing a non-integer resolution level when using wavelet methods. preprint

Luo, Z. and Wahba, G., (1997). Hybrid Adaptive Splines. *Journal of the American Statistical Association*, **92**, 107-116.

Mitchell, T., and Beauchamp J. (1988). Bayesian variable selection in linear regression. *Journal of the American Statistical Association*, **83**, 1023-1036.

Monahan, J. (1984). Full Bayesian analysis of ARMA time series models. *Journal of Econometrics*, **21**, 307–31.

O'Hagan, A. (1995). Fractional Bayes factors for model comparison (with discussion). *Journal of the Royal Statistical Society*, Series B, **57**, 99-138

Raftery, A., Madigan, D., Hoeting, J. (1997). Bayesian model averaging for linear regression models. *Journal of the American Statistical Association*, **92**, 179-191.

Schwarz, G. (1978). Estimating the dimension of a model. *The Annals of Statistics*, **6**, 461-464.

Smith, M. and Kohn, R. (1996). Nonparametric regression using Bayesian variable selection. *J. Econometrics*, **75**, 317-344.

Smith, M., and Kohn, R. (1997). A Bayesian approach to nonparametric bivariate regression. *Journal of the American Statistical Association*, **92**, 1522-1535.

Smith, M., Wong, C. and Kohn, R. (1998). Additive nonparametric regression with autocorrelated errors. *Journal of the Royal Statistical Society*, Ser. B, in press.

Stone, C. (1994) The use of polynomial splines and their tensor products in multivariate function estimation. (with discussion) *The Annals of Statistics*, **22**, 118-184.

Tierney, L. (1994). Markov chains for exploring posterior distributions. *The Annals of Statistics*, **22**, 1701–1762.

Wahba, G. (1990). *Spline models for observational data*, SIAM: Philadelphia.

Wong, F., Hansen, M., Kohn, R., Smith, M. (1997). 'Focused sampling and its application to nonparametric regression', preprint

9

Feedforward Neural Networks for Nonparametric Regression

David Ríos Insua
Peter Müller

ABSTRACT Feed forward neural networks (FFNN) with an unconstrained random number of hidden neurons define flexible non-parametric regression models. In Müller and Rios Insua (1998) we have argued that variable architecture models with random size hidden layer significantly reduce posterior multimodality typical for posterior distributions in neural network models. In this chapter we review the model proposed in Müller and Rios Insua (1998) and extend it to a non-parametric model by allowing unconstrained size of the hidden layer. This is made possible by introducing a Markov chain Monte Carlo posterior simulation scheme using reversible jump (Green 1995) steps to move between different size architectures.

9.1 Introduction

Outside the mainstream statistical literature there has been enormous interest for *neural networks* (NNs). After a revival due to Rumelhart and McClelland's work (1986), physicists, biologists, philosophers and computer scientists have introduced, developed and expanded various NN models applying them to solve traditional statistical problems like regression or classification. The basic building blocks of NN models are *neurons*, which may be viewed as information processors transforming non-linearly the inputs they receive into one or more outputs. Various ways of linking and relating neurons lead to different NN models, including, among others, *feed-forward neural networks* (FFNNs), Boltzmann machines, or Hopfield nets. Lippman (1987) provides an interesting account from a computational point of view.

In this chapter, we shall focus on FFNNs. In their simplest version, their architecture is defined by three types of nodes (or neurons): *input*, *hidden* and *output* nodes. Input nodes are associated with explanatory variables and feed hidden nodes which nonlinearly transform weighted combinations of inputs to feed outputs, associated with response variables, which, in turn, transform weighted combinations of their inputs. Interest in these models arise from results by Cybenko (1989), Hornik et al (1989) and others, which describe them as *universal approximators*: when nonlinear transformations are of sigmoidal type, as the number of hidden nodes goes to infinity, we may approximate any continuous function in a compact range. If we then introduce an error model, we have a flexible class of models on random functions with infinite-dimensional parameter vector, establishing a link with non-parametric inference.

Realising their potential, there have been several recent reviews of NNs from a statistical point of view, see e.g., Cheng and Titterington (1994), Ripley (1993), Stern (1996), Warner and Misra (1996), Neal (1996) or De Veaux and Ungar (1997). The general message they convey is that, although there has been a certain hype in the field, NNs may constitute a useful addition to the statistician's toolkit when dealing with complex nonlinear features in regression, classification or time series analysis. As an indication of their potential, note that some of the most popular statistical packages and libraries have recently included NN modules. Yet their impact is still to be fully perceived in statistical research practice.

Given that, it comes at no surprise that relatively little work on Bayesian analysis of NNs has taken place. Buntine and Weigend (1991) and MacKay in various papers, see e.g., (1992), develop the first approaches, based mainly on Gaussian approximations to the posteriors. Bishop (1996) reviews many of these earlier attempts. Of special relevance is the work of Neal, culminating in the above mentioned Neal (1996). From a theoretical point of view, he relates infinite hidden node FFNNs with Gaussian processes, when priors are appropriately chosen, hence sidestepping the issue of overfitting data. Other prior choices show that infinite FFNNs are actually richer than Gaussian processes, confirming their relevance as modeling tools. From a practical point of view, assuming a large, but fixed, number of hidden nodes, Neal introduces a hybrid Monte Carlo method, merging Metropolis algorithm with sampling techniques based on dynamical simulation, to perform inference and prediction with NNs. In previous work, Müller and Rios Insua (1998) introduced the idea of (parametric) variable architecture FFNNs, together with a very efficient Markov chain Monte Carlo method.

In this chapter, we build on this work providing analyses of nonparametric FFNN models, based on a novel reversible jump algorithm.

9.2 Feed Forward Neural Networks as Nonparametric Regression Models

A feedforward neural network model with p input nodes, one hidden layer with M hidden nodes, one output node and activation functions Ψ is a model relating p explanatory variables $x = (x_1, ..., x_p)$ and a response variable y of the form

$$\hat{y}(x) = \sum_{j=1}^{M} \beta_j \Psi(x'\gamma_j + \delta_j) \quad (1)$$

with $\beta_j \in \mathcal{R}, \gamma_j \in \mathcal{R}^{\,p}$. The terms δ_j are designated biases and may be assimilated to the rest of the γ_j vector if we consider an additional input with constant value one, say $x_0 = 1$.

The typical setup for FFNNs is: Given data $D = \{(x_1, y_1), ..., (x_N, y_N)\}$ and fixed M, choose $\beta = (\beta_1, ..., \beta_M), \gamma = (\gamma_1, ..., \gamma_M)$ according to a least squares criterion $\min_{\beta,\gamma} \sum_{i=1}^{N} (y_i - \hat{y}(x_i))^2$, either via *backpropagation* (Rumelhart and McLelland, 1986), an implementation of steepest descent, or other optimisation method such as quasi-Newton or simulated annealing. Hence, at least implicitly, we are assuming a normal error model and we are viewing a nonlinear parametric regression problem. It is also sometimes suggested to include a regularisation term in the objective function to avoid data overfitting.

Cybenko (1989) and others show that when Ψ is a sigmoidal function, finite sums of the form (1) are dense in $\mathcal{C}(I_p)$, the set of real continuous functions in the p-dimensional unit cube. For their proof, they assume that $M \to \infty$ as the approximation gets better. Hence, we may view FFNNs as nonparametric regression models.

9.3 Variable Architecture FFNNs

An important issue which has received comparatively little attention in the literature is the choice of architecture which, for the model we are considering, consists mainly of choosing the number M of hidden nodes. Typically, a choice is made by trial and error, though there are several heuristics helping in that task.

In Müller and Rios Insua (1998), we argue that architecture should be treated as an additional parameter M in the model: apart from simulated examples, there will always be uncertainty about the number of hidden nodes to include; a basic tenet of Bayesian inference suggests that uncertainty about architecture choice should be formalized in a prior probability model on M. We also argue that considering M as unknown parameter, rather than fixing it, will address some of the likelihood (and posterior) multimodality which makes inference in NN models notoriously difficult, as recognised by various authors.

Besides arguing for a variable architecture FFNN model, characterised as a mixture of M sigmoidal activation functions $\Psi(\cdot)$, with M unknown and ranging through the whole set of positive integers, we shall include a linear regression term $x'\lambda$. This corresponds to a standard model building strategy based on blocks, see West and Harrison (1996), where the linear term would model linear effects, and the FFNN term would take care of other effects. Hence, we shall deal with the following model

$$y_i = x_i'\lambda + \sum_{j=1}^{M} \beta_j \Psi(x_i'\gamma_j) + \epsilon_i,\ i = 1, ..., N,$$

where $\epsilon_i \sim N(0, \sigma^2)$. For the moment, we fix $\Psi(\eta) = \exp(\eta)/(1 + \exp(\eta))$, i.e., we use a logistic activation function. However, everything we shall say is valid for other sigmoidal functions. For contrast, our second example below will include a different type of activation functions.

The parameters in our model are the linear weights $\lambda = (\lambda_0, \lambda_1, ..., \lambda_p)$ and $(\beta_1, \beta_2, \ldots)$, the logistic parameters $(\gamma_1, \gamma_2, \ldots)$, the number M of terms and the error variance σ^2.

The prior over network parameters is:

$$\beta_j \sim N(\mu_\beta, \sigma_\beta^2),\quad \lambda_j \sim N(\mu_\beta, \sigma_\beta^2),\quad \gamma_j \sim N(\mu_\gamma, S_\gamma),$$
$$\sigma^{-2} \sim Gamma(s/2, sS/2). \tag{2}$$

The meaning and interpretation of parameters allow us to use informative priors. For example, the β's should reflect the order of magnitude of the data y_i. After standardisation, positive and negative values for β_j would be equally likely, calling for a symmetric prior around 0, with a standard deviation reflecting the range of plausible values for y_i. Similarly, a range of reasonable values for the logistic coefficients γ_j will be determined by the meaning of the data being modeled, to address smoothness issues.

We also need to provide a prior over the number M of hidden nodes. We shall keep it as a generic distribution, i.e., $p(M)$. Poisson distributions provide a convenient way of modelling various shapes for such distribution. Alternatively, we may choose a probability model which rewards parsimony by putting geometrically decreasing prior probability on larger networks, hence assuming a geometric prior with parameter α:

$$M \sim Geom(\alpha).$$

When there is non-negligible uncertainty about the prior hyperparameters, we may complete the prior model with a hyperprior over them. We shall use the following standard choices in hierarchical models:

$$\mu_\beta \sim N(a_\beta, A_\beta),\quad \mu_\gamma \sim N(a_\gamma, A_\gamma),$$
$$\sigma_\beta^{-2} \sim Gamma(c_b/2, c_b C_b/2),\quad S_\gamma^{-1} \sim Wish(c_\gamma, (c_\gamma C_\gamma)^{-1}), \tag{3}$$

Finally, as hyperprior for the geometric probability, we use $\alpha \sim Beta(a_\alpha, b_\alpha)$.

Since the likelihood is invariant to relabelings, and given our prior choice, we recommend including an order constraint to avoid trivial posterior multimodality due to permutations of indices. For example, we could use $\gamma_{1p} \leq \gamma_{2p} \cdots \leq \gamma_{Mp}$.

Note that the main difference with our previous model in Müller and Rios Insua (1998) resides on having removed the upper bound on the number of terms in the mixture of logistics, to allow for a nonparametric model. Since our hyperpriors allow for variances tending to infinity, we encompass the more general models described by Neal (1996).

9.4 Posterior inference with the FFNN model

Posterior inference in FFNNs is plagued by multimodality issues. Besides, trivial multimodality due to relabeling, which we mitigate with a constraint on logistic slopes, there is inherent multimodality due to nonlinearity and to nesting of models by inclusion of (approximately) duplicate or (approximately) irrelevant hidden nodes. As a consequence, there is little hope for normal approximations with these models, and we need to turn to MCMC methods. However, straightforward implementation of commonly used MCMC schemes is hindered by a variety of issues.

First, the variable architecture model defines a model with changing dimensionality parameter vector: posterior inference has to mix over models with hidden layers of different size. We introduce below an algorithm which addresses the changing dimensions by using an implementation of reversible jump (Green, 1995) with moves corresponding to "birth", "death", "thinning" and "seeding" of hidden nodes. Second, the high dimensional parameter vector is typically highly correlated *a posteriori*, necessitating MCMC strategies which use blocking to jointly update as many parameters as possible and marginalizing to partly avoid the random walk nature of Metropolis algorithms.

In Müller and Rios Insua (1998) we introduce a scheme which marginalizes over the weights β_j when updating the input weights γ_j. This is made possible by Lemma 1 which gives the marginal likelihood of γ, marginalizing over (β, λ). With this likelihood we use a Metropolis step to update values of the logistic regression parameters. The other key observation, which is actually used in the proof of the lemma, is that, once we have fixed M and γ, we have a standard linear model, hence facilitating, through Gibbs steps, the sampling of new linear weights and hyperparameters. For simplicity, we drop dependence on hyperparameters:

Lemma 1:

Let $z_{ij} = z_{ij}(\gamma) = \psi(x_i'\gamma_j)$, $Z = (z_{ij})_{i=1,\ldots,N}^{j=1,\ldots,M}$, $X = (x_{ij})_{i=1,\ldots,N}^{j=0,\ldots,p}$, $R = (X, Z)$. Define $A = R'R/\sigma^2$, $\rho = R'y/\sigma^2$, $C = 1/\sigma_\beta^2 I$, $\delta = \mu_\beta/\sigma_\beta^2(1, 1, \ldots, 1)'$, $m_{l,b}(\gamma) = (A+C)^{-1}(\rho+\delta)$ and $S_{l,b}(\gamma) = (A+C)^{-1}$.

Then,

$$\begin{aligned} p(D|\gamma, M) &= \frac{p[(\lambda,\beta) = m_{l,b}(\gamma)]}{p[(\lambda,\beta) = m_{l,b}(\gamma)|y,\gamma]} \prod_{i=1}^{N} p[y_i|(\lambda,\beta) = m_{l,b}(\gamma),\gamma] \\ &= p[(\lambda,\beta) = m_{l,b}(\gamma)]|S_{l,b}(\gamma)|^{1/2} \prod_{i=1}^{N} p[y_i|(\lambda,\beta) = m_{l,b}(\gamma),\gamma]. \end{aligned}$$

9.4.1 MCMC posterior simulation for FFNN models

We describe posterior simulation in the proposed model by stating the algorithm for one sweep, i.e., the transition from imputed parameter values at iteration t to those values at iteration $t+1$. The list below outlines the steps in one sweep. Details of each step are discussed afterwards. An item of the form $[a|b]$ indicates that parameter a is being updated using current values for parameters b. The absence of some parameter c in the conditioning set (to the right of the bar) indicates that either c is being marginalized over, or that a and c are conditionally independent given b. Let $\gamma_j = (\gamma_{j1}, \ldots, \gamma_{jp})'$, $\gamma = (\gamma_1, \ldots, \gamma_M)'$, $\beta = (\beta_1, \ldots, \beta_M)'$, $\lambda = (\lambda_0, ..., \lambda_p)$ and $\nu = (\mu_\beta, \sigma_\beta, \mu_\gamma, \sigma_\gamma, \sigma)$. Let γ_{-jk} indicate the list of all weights $\gamma_{j'k'}$ without γ_{jk}. The following steps define one sweep of the Markov chain:

(*i*) $[\gamma_{jk}|\gamma_{-jk}, M, \nu, D]$, $j = 1, \ldots, M$, $k = 0, \ldots, p$,
(*ii*) $[M|\gamma, \nu, D]$, "Birth/Death"
(*iii*) $[M|\gamma, \nu, D]$, "Seed/Thin"
(*iv*) $[\beta, \lambda|\gamma, M, \nu, D]$,
(*v*) $[\nu|\beta, \gamma, \lambda, D]$.

Step (i).

We update the logistic weights γ_{jk}, marginalizing over (λ, β). We use a random walk Metropolis proposal to propose a new value for γ_{jk}, and evaluate the Metropolis-Hastings acceptance probabilities using Lemma 1 to compute the required marginal likelihoods, $p(D|\tilde{\gamma}, M, \nu)$ and $p(D|\gamma, M, \nu)$. Denote with $\sigma_k = C_{\gamma,kk}$ the k-th element in the diagonal of C_γ. Let $\tilde{\gamma}$ denote the γ vector with γ_{jk} replaced by $\tilde{\gamma}_{jk}$ and, if necessary, the indices rearranged to satisfy the order constraint on the γ_{jp}. Let $x \wedge y$ denote the minimum of x and y. We then do:

$$\begin{aligned} \tilde{\gamma}_{jk} &\sim N(\gamma_{jk}, \sigma_k), \\ \pi &= 1 \wedge \frac{p(D|\tilde{\gamma}, M, \nu) p(\tilde{\gamma}|M, \nu)}{p(D|\gamma, M, \nu) p(\gamma|M, \nu)}, \end{aligned}$$

and

$$\gamma = \begin{cases} \tilde{\gamma} \text{ with probability } \pi \\ \gamma \text{ with probability } 1 - \pi. \end{cases}$$

The computation of $p(\gamma|M,\nu)$ is straightforward. See the note after (4) about the order constraint in $p(\gamma|M,\nu)$. Use Lemma 1 to evaluate $p(D|\gamma, M, \nu)$.

Step (ii).

A change in M by a reversible jump (Green 1995) type move is proposed. The algorithm follows the setup in Richardson and Green (1997). With probability $d_M = M/(M+1), M \neq 1$ and $d_1 = 0$, we propose to delete one hidden node ("death"); with probability $b_M = 1 - d_M$ we propose to add one hidden node ("birth"). If we decide to add a node, then we first generate a vector $\gamma_{M+1} \sim q(\gamma_{M+1}) = N(\mu_\gamma, S_\gamma)$. The random vector γ_{M+1} plays the role of the continuous random variable u in Richardson and Green's discussion of a birth move in the normal mixtures model. If we decide to delete a node, then we pick with equal probabilities $1/M$ one of the M hidden nodes to be proposed for deletion. Assume we choose index j. Then remove γ_j from the list of nodes; relabel the remaining indices to close the gap in the indeces; decrement M to $\tilde{M} := M - 1$; and record $\gamma_{\tilde{M}+1} = \gamma_j$.

To derive the appropriate acceptance probabilities, consider first the case of proposing a birth. Denote with $\tilde{\gamma}$ the vector γ augmented with the additional γ_{M+1}, relabeled if necessary to maintain the order constraint. Let $\tilde{M} = M + 1$. With probability $\pi = 1 \wedge A$, where

$$\begin{aligned} A &= \text{prior ratio} \times \text{likelihood ratio} \times \text{proposal ratio} \\ &= \frac{p(\tilde{\gamma}|\tilde{M},\nu)p(\tilde{M})}{p(\gamma|M,\nu)p(M)} \frac{p(D|\tilde{\gamma},\tilde{M},\nu)}{p(D|\gamma,M,\nu)} \frac{d_{\tilde{M}}\frac{1}{\tilde{M}}}{b_M q(\gamma_{M+1})} \\ &= \frac{\tilde{M}\, \alpha\, p(D|\tilde{\gamma},\tilde{M},\nu)\; 1/(\tilde{M}+1)}{p(D|\gamma,M,\nu)\; 1/(M+1)} \qquad (4) \end{aligned}$$

accept the proposal and set $\gamma := \tilde{\gamma}$. Otherwise keep γ unchanged. In verifying the expression for A, note that $p(\gamma|M,\nu) = (M!) \prod_{j=1}^{M} N(\gamma_j; \mu_\gamma, S_\gamma)$, with the factor $M!$ appearing because of the order constraint. The factor $\frac{1}{\tilde{M}}$ in the numerator of the proposal ratio corresponds to uniformly sampling one of the $\tilde{M}$ indices to decide which node should die. The α in the final expression derives from $p(\tilde{M})/p(M)$, if we use the geometric prior $M \sim Geom(\alpha)$. Note that the Jacobian appearing in the general expression for reversible jump moves equals one. The calculations for a death move can be derived by an analogous argument.

Step (iii).

As in the previous step, a change in the number of hidden nodes is proposed. Instead of generating an entirely new candidate node as in Step (ii), we propose a new node in relation to the current ones. We shall call this move "seeding a node". The matching move proposes to reduce M by "thinning" one current node. With probability $t_M = 0.5$ for $M > 1$ and $t_1 = 0$ we consider thinning and with probability $s_M = 1 - t_M$ we consider seeding.

Assume we decide to seed. With equal probabilities we choose an index $j \in \{1, \ldots, M\}$ and generate a new node $\tilde{\gamma}_{j+1}$ by first generating $u \sim N(0, S_\gamma)$ and then setting $\tilde{\gamma}_{j+1} = \gamma_j + \delta u$ for some factor δ. In the current implementation we use $\delta = 1.0$. Denote with $\tilde{\gamma}$ the augmented γ vector after relabeling the nodes $\gamma_{j+1}, \ldots, \gamma_M$ as $\tilde{\gamma}_{j+2}, \ldots, \tilde{\gamma}_{\tilde{M}}$, where $\tilde{M} = M + 1$. If $\tilde{\gamma}$ violates the order constraint imposed on the last coordinates $\gamma_{jp}, j = 1, \ldots, M$, then we immediately reject the proposal (if the order constraint is violated then the factor $p(\tilde{\gamma}|\tilde{M}, \nu)$ in the acceptance probabilities given below will evalute to zero).

For the matching thinning step, we choose with equal probabilities, $p = 1/(M-1)$, a pair of indices $(j, j+1) \in \{(j, j+1), i = 1, \ldots, M-1\}$; record $u = \delta^{-1}(\gamma_{j+1} - \gamma_j)$; reduce M by one, $\tilde{M} = M - 1$; remove γ_{j+1} from the list of nodes; and re-label the remaining indices to close the gap. Denote the proposal $\tilde{\gamma}$.

Again, we only derive the acceptance probability for a seed move. The expression for the thinning move is easily derived by a symmetric argument. Assume we decided to seed with respect to term j. With probability $\pi = 1 \wedge A$, where

$$\begin{aligned} A &= \text{prior ratio} \times \text{likelihood ratio} \times \text{proposal ratio} \\ &= \frac{p(\tilde{\gamma}|\tilde{M}, \nu)\, p(\tilde{M})}{p(\gamma|M, \nu)\, p(M)} \cdot \frac{p(D|\tilde{\gamma}, \tilde{M}, \nu)}{p(D|\gamma, M, \nu)} \cdot \frac{t_{\tilde{M}}\; 1/(\tilde{M}-1)}{q(u)\; |J|\; s_M/M} \\ &= \frac{N(\gamma_j; \mu_\gamma, S_\gamma)\tilde{M}\alpha\; p(D|\tilde{\gamma}, \tilde{M}, \nu)\; t_{\tilde{M}}}{N(u; 0, S_\gamma) p(D|\gamma, M, \nu)\; \delta^{-p}\; s_M}. \end{aligned}$$

The factor $1/M$ in the denominator, and the factor $1/(\tilde{M} - 1)$ in the numerator of the proposal ratio correspond to the step of sampling an index j from which we seed and sampling two indices $(j, j+1)$ from which we thin, respectively. The factor δ^p is the Jacobian appearing in the general formula for the reversible jump acceptance probability. The additional factor $\tilde{M}$ in the numerator stems from the order constraint on the $\gamma_{jp}, j = 1, \ldots, M$ (see the comment after the acceptance probability in step ii).

Step (iv).

Conditional on (γ, ν, M) the model becomes a normal linear regression problem with coefficients (β, λ). Resampling (β, λ) from $p(\beta, \lambda|\gamma, \nu, M, D)$ is a straightforward multivariate normal generation. See, for example, Bernardo and Smith (1994) for an explicit statement of the relevant moments.

Step (v).

We appeal to the same normal linear regression argument as before. It is then straightforward to resample the hyperparameters from their respective complete conditional posterior distributions.

Convergence of the algorithm to the desired stationary distribution comes from arguments in Tierney (1994) and Green (1995).

9.5 Examples

Example 1. Stochastic optimisation.

An application of interest is stochastic optimisation. Suppose we are interested in finding decisions of maximum expected utility, yet we do not have an explicit expression of such an objective function. Suppose however that we may approximate it quite accurately at several points, say via Monte Carlo integration. We could then fit a nonlinear surface and optimise its predictive expected value.

For an example, consider a reservoir management problem adapted from Rios Insua et al. (1997). We have to decide how much water should we release through turbines (u_1) and spillgates (u_2) from a reservoir. We evaluate them through their expected utility $\int g(u_1, u_2, i)h(i)di$, where g is the utility function, i is the inflow and h is the predictive density for inflows. At N design points, we approximate by standard Monte Carlo the expected utility. We then fit a neural network model to the data $\{(u_1^i, u_2^i, \Phi(u_1^1, u_2^1)), i - 1, ..., N\}$ where Φ designates the Monte Carlo approximation of the expected utility. Figure 1 shows the data and fitted FFNN model. Note that other smoothing methods would have problems in this example, due to the sharp edge on the front, which would typically be oversmoothed. Figure 2 shows the marginal posterior on M, suggesting bigger support for architectures with three or four hidden nodes. In ad-

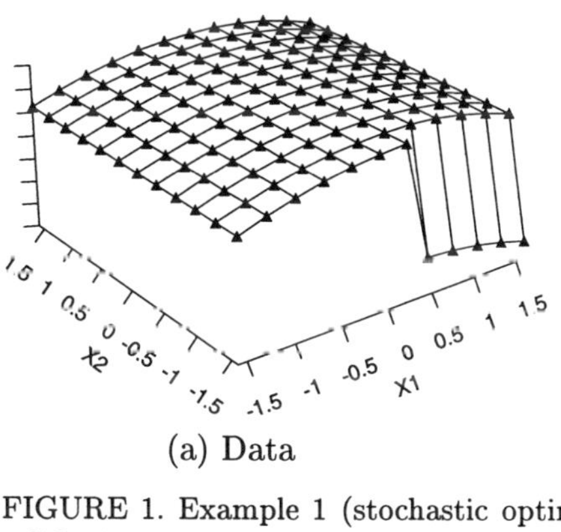

(a) Data

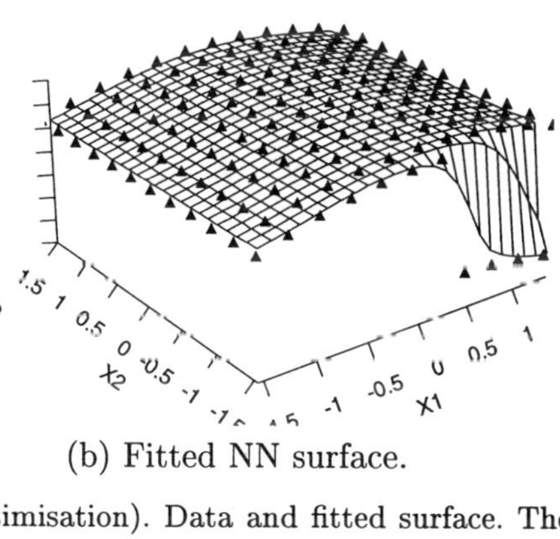

(b) Fitted NN surface.

FIGURE 1. Example 1 (stochastic optimisation). Data and fitted surface. The solid triangles indicate the data points.

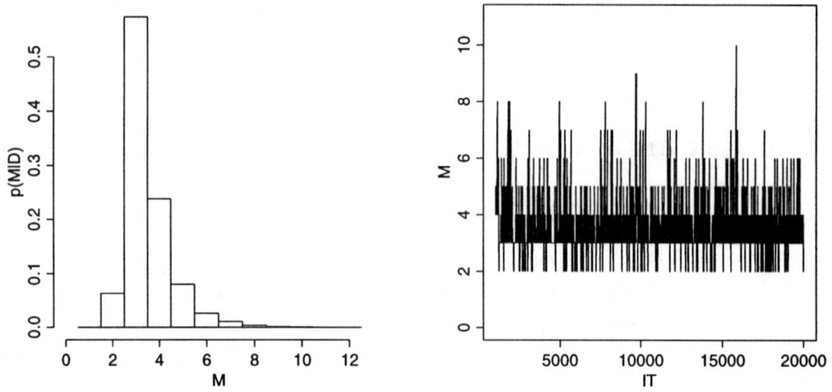

(a) (b)

FIGURE 2. Example 1 (stochastic optimisation). Posterior $p(M|D)$ on the size of the hidden layer. Panel (a) plots the estimated posterior distribution $p(M|D)$. Panel (b) shows the trajectory using the Markov chain.

dition to the normal prior (2), we constrained the γ_{jk} by $|\gamma_{jk}| < 10.0$ to avoid numerical problems. Otherwise proposals for the γ vector could lead to degenerate design matrices in the regression problems required for the evaluation of $p(\gamma|\nu)$ (Lemma 1).

Example 2. Robot arm.

In this example the response is bivariate and the activation functions are tanh instead of logistic. This is a standard example in the neural network literature called the robot arm problem, see e.g. MacKay (1992) and Neal (1996). We try to predict the robot arm position (y_1, y_2) from two joint angles (x_1, x_2), when they are actually related by the model

$$\begin{aligned} y_{i1} &= 2.0\cos(x_{i1}) + 1.3\cos(x_{i1} + x_{i2}) + \epsilon_{i1}, \quad i = 1, \ldots, N, \\ y_{i2} &= 2.0\sin(x_{i1}) + 1.3\sin(x_{i1} + x_{i2}) + \epsilon_{i2}, \\ & \epsilon_{ik} \overset{iid}{\sim} N(0, 0.05). \end{aligned}$$

The bivariate FFNN with tanh$(\cdot)$ activation functions we use is as follows:

$$\begin{aligned} y_{ik} &= x_i'\lambda_k + \sum_{j=1}^{M} d_j \beta_{jk} \Psi(x_i'\gamma_j) + \epsilon_{ik}, \quad i = 1, ..., N, \ k = 1, 2, \\ & \epsilon_{ik} \overset{iid}{\sim} N(0, \sigma^2), \quad \Psi(\eta) = \tanh(\eta). \end{aligned} \tag{5}$$

Note that we could change (β_j, γ_j) to $(-\beta_j, -\gamma_j)$ without changing the likelihood. Hence, with a prior symmetric around 0, the posterior distribution would remain invariant under such transformations. To avoid this identifiability problem we add a constraint $\gamma_{j1} > 0$.

In NN applications, it is frequent that the data set is broken in two subsets, one for estimation and the other for validation. We use MacKay's (1992) data set, splitting it into a training data set (first 200 observations), and a test data set (last 200 observations). Figure 3 shows the estimated surface, which is practically indistinguishable from the data. After around 300 iterations the mean squared error is 0.0064 and already close to the asymptotic value 0.0062 (note the theoretical minimum $2\sigma^2 = 0.0050$), indicating that short run-lengths of several hundred iterations are sufficient for predictive purposes. However, to monitor convergence diagnostics on some selected parameters, we needed 20,000 iterations to achieve practical convergence. The estimated marginal posterior probabilities $p(M|D)$ for the number of hidden nodes are 0.30, 0.53, 0.15 and 0.02 for $M = 6, 7, 8$ and 9, respectively.

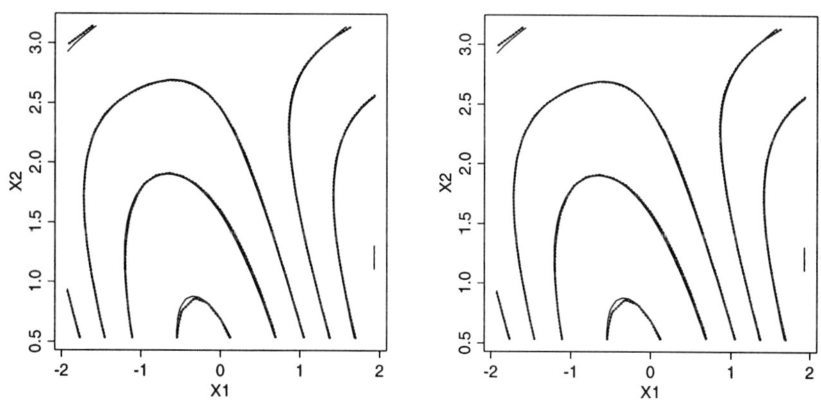

$E(y_{i1}|x_{i1}, x_{i2}, D), i = n + 1$ $\qquad$ $E(y_{i2}|x_{i1}, x_{i2}, D), i = n + 1$

FIGURE 3. Example 2 (robot arm). Estimated surface (dotted contours) and true contours (thin solid contours). The estimated and the true surface are almost undistinguishable. The average predictive mean squared error (MSE), i.e., the average of predictive mean squares, averaged over all 200 data points in the test data, is 0.0062. The theoretical optimum is $\sigma^2 = 0.005$.

9.6 Discussion

Several recent reviews suggest that neural networks constitute a useful enhancement to the statisticians' toolkit. Neal (1996), in particular, shows that FFNNs with infinite hidden nodes embrace a class of flexible models richer than Gaussian processes. We have shown here how a reversible jump algorithm permits routine analysis of FFNNs viewed as nonparametric regression models. By combining them, as we did, with other conventional models, such as linear regression, we have at hand a powerful modelling strategy for complex problems. Important applications include stochastic optimisation (Example 1), approximation (Example 2) or regression metamodels for simulation experiments, see Rios Insua et al (1997).

Many issues remain to be explored. For example, we should try other standard NN applications like classification, density estimation or time series analysis, e.g., via nonlinear autoregression models. Other reversible moves than birth, death, seed and thin, could be devised and should be compared. Also we have confined to FFNNs but many other NN models, analysed from a Bayesian point of view, might prove useful.

Acknowledgements

Research supported by grants from CICYT, the Government of Madrid, the Iberdrola Foundation and the National Science Foundation (NSF/DMS-9704934).

References

Bernardo, J.M. and Smith A.F.M. (1994). *Bayesian Theory*, Wiley, New York.

Bishop, C.M. (1996). *Neural Networks for Pattern Recognition*, Oxford University Press, Oxford.

Buntine, W.L. and Weigend, A.S. (1991). Bayesian back-propagation. *Complex Systems*, 5, 603-643.

Cheng, B. and Titterington, D.M. (1994). Neural networks: a review from a statistical perspective (with discussion). *Statistical Science*, 9, 2-54.

Cybenko, G. (1989). Approximation by superposition of sigmoidal functions, *Mathematics of Control Systems and Signals*, 2, 303-314.

De Veaux, R., Ungar, L. (1997). A brief introduction to neural networks, Technical Report, Williams College, Williamstown, MA.

Green, P. (1995). Reversible jump Markov chain Monte Carlo computation and Bayesian model determination. *Biometrika*, 82, 711-732.

Hornik, K., Stinchcombe, M. and White, H. (1989). Multilayer feedforward neural networks are universal approximators, *Neural Networks*, 2, 359-366.

Lippman, R. P. (1987). An introduction to computing with neural nets, *IEEE ASSP Magazine*, 4-22.

MacKay, D.J.C. (1992). A practical Bayesian framework for backprop networks. *Neural Computation*, 4, 448-472.

Müller, P. and Rios Insua, D. (1998). Issues in Bayesian Analysis of Neural Network Models, *Neural Computation*, 10, 571–592.

Neal, R. M. (1996). *Bayesian Learning for Neural Networks*, New York: Springer-Verlag.

Richardson, S., and Green, P. (1997). Modelling and Computation for Mixture Problems (with discussion) *Journal of the Royal Statistical Society, B*, 59, 731-792.

Rios Insua, D., Rios Insua, S. and Martin, J. (1997). *Simulacion*, RA-MA, Madrid.

Rios Insua, D., Salewicz, K.A., Müller, P. and Bielza, C. (1997). Bayesian methods in reservoir operations, in *The Practice of Bayesian Analysis* (eds: French, Smith), pp. 107–130, Wiley, New York.

Ripley, B.D. (1993). Statistical aspects of neural networks. In *Networks and Chaos* (eds: Barndorf-Nielsen, Jensen, Kendall), London: Chapman and Hall.

Rumelhart, D.E. and Mclelland, J.L. (eds) (1986). *Parallel Distributed Processing*, MIT Press, Cambridge.

Stern, H.S. (1996). Neural networks in applied statistics, *Technometrics*, 38, 205–220.

Tierney, L. (1994). Markov chains for exploring posterior distributions. *Annals of Statistics*, 22, 1701-1762.

West, M., Harrison, J. (1996). *Bayesian Forecasting and Dynamic Linear Models*, 2nd ed, New York: Springer-Verlag.

Warner, B., Misra, M. (1996). Understanding neural networks as statistical tools, *The American Statistician*, 50, 284-293.

10

Survival Analysis Using Semiparametric Bayesian Methods

Debajyoti Sinha
Dipak K. Dey

ABSTRACT This article presents the scope of nonparametric and semiparametric Bayesian methods for the analysis of survival data using models based on either the hazard or the intensity function. The nonparametric part of every model is assumed to have a suitable prior process. The parametric part, which may include a regression parameter or a parameter quantifying the heterogeneity of a population, is assumed to have a prior distribution with possibly unknown hyperparameters. Careful applications of some recently popular computational tools, including MCMC algorithms, are available to perform sophisticated Bayesian analyses even when we are dealing with complex models and unusual data structures.

10.1 Introduction

In medical studies interest often centers on the relation of the time to a particular event or 'failure' of interest (such as death or occurrence of a disease) to the explanatory variables (covariates such as treatments, design variables, etc.). There are several kinds of time to event data (popularly called survival data) encountered by statisticians in medical and reliability studies. The descriptions and examples of some of the different types of survival data are given in popular survival analysis texts such as Cox and Oakes (1984) and Andersen et al. (1993) (ABGK). Nonparametric and semiparametric models are very popular in survival analysis because they allow considerable generality and applicability but enough structure for useful physical interpretation and understanding particular applications at hand. In this article, we focus on *semiparametric Bayesian models* along with associated Bayesian statistical methods.

Following the path of Cox's (1972) article, we concentrate on survival models based on the so-called hazard functions and intensity functions. Though they may not include all the types of survival data encountered in practice, three major classifications among survival data problems are (1) univariate survival data, (2) multiple event-time data and (3) multivariate survival data.

If every subject of the study can experience the event at most once, the data are called *univariate survival data.* One example of univariate survival data is from a retrospective study which was carried out to compare 48 early breast cancer patients who had been treated with primary radiation therapy and adjuvant chemotherapy to those 46 patients treated with radiotherapy alone (Finkelstein and Wolfe, 1985). In this study, a patient is monitored at regular intervals (every 3-4 weeks) for cosmetic effect of the therapy. The survival time of a patient is the time to the cosmetic deterioration of her breasts from the time of the treatment and patients are assumed to be independent given their randomly assigned therapies.

Often in practice, one of the important and challenging aspects of survival data is censorings of different kinds depending on the particular design and method of data collection used for different studies. For the beast cancer study, a patient may miss some visits and return in a possibly changed state. For example, one patient has her survival time recorded as $(6, 18]$ which means that patient had one visit on the 6th week and had no cosmetic deterioration at that time and then the next visit was on the 18th week when she was found to have the cosmesis of breasts. We call this an interval censored data or an incompletely grouped data. Another patient had her last visit on 23rd week and she had no cosmesis by that time. The survival time for her is said to be right censored at the 23rd week.

For some studies, we may be interested in the intensity of the recurrent occurrences of an important but possibly non-fatal event to a subject. This intensity may serve as a measure of disease morbidity or physical status of the subject. An example of such *multiple event time data* is given by Gail et al. (1980). The observations are the times to occurrence of mammary tumors in 48 rats injected with a carcinogen and subsequently randomized to receive either the treatment or the control. This study also has regular follow-ups (only twice in each week) and hence for each rat only the number of new tumors between two consecutive follow up times is recorded. These highly grouped data contain many ties for event times within the same rat and across different rats. All observations are right censored after the 182nd day.

When some survival times are related, such as in litter-matched animal studies or in matched pair studies, then the data are called *multivariate survival data.* An example given by ABGK in Example I.3.6 presents remission lengths (survival times) of 21 matched pairs (clusters) of Leukemia patients. Survival times within the same cluster are likely to be related. Covariates such as treatment (either 6MP or Placebo for the Leukemia data) and remission status may influence the survival time.

These motivating examples indicate the challenges to the statisticians posed by typical modern day survival data problems with their need for complex models and computational tools sophisticated enough to deal with complex data structures. Semiparametric Bayesian methodologies, despite being comparatively new concepts, are very useful in survival analysis due to their potential to fit complex models and ability to deal with incomplete data.

The plan of this article is as follows. Section 2 reviews different semiparametric models to deal with different survival data structures and designs. Section 3 discusses the commonly used prior processes to model the nonparametric part. Section 4 presents the suitable semiparametric Bayesian models for the examples considered here using the prior processes from Section 3 and the analyses of different types of survival data using the described Bayesian models and implementing suitable computational tools. We are not going into the detailed descriptions and convergence criterions behind these computational tools such as MCMC, EM and MCEM algorithm. The general descriptions and convergence results of these tools can be found in Tanner (1996) and the references therein. In Section 4, we present key algorithmic steps for using MCMC algorithms on some of the representative models discussed here. Section 5 presents a brief survey of of some of the related works on these topics.

10.2 Models

We denote the survivor function of a single survival time T given a covariate vector x (the vector of explanatory and design variables) by $S(t|x)$ and the corresponding hazard function by $h(t|x)$ (when T is absolutely continuous). A popular semiparametric approach for analyzing univariate survival data makes use of Cox's (1972) proportional hazards model:

$$h(t|x) = h_0(t)\exp(\beta x) \tag{1}$$

where $h_0(\cdot)$ (the nonparametric part) is the unknown baseline hazard function and β is the unknown regression coefficient. Typically β is assumed to be constant over time t (a parametric part). If β is also a function of t, then we have a time-varying covariate effect; if x is a function of t, then we have a time-dependent covariate.

Suppose that the time axis $[0,\infty)$ is partitioned into a finite number of disjoint intervals, say $I_1, I_2, \cdots, I_{g+1}$ where $I_j = [a_{j-1}, a_j)$ for $j = 1, 2, \cdots, g+1$ with $a_0 = 0$ and $a_{g+1} = \infty$. For the breast cancer data, I_j can be the jth week. If the survival function is absolutely continuous, then we may assume an ordinary Cox's model $S(t|x) = \exp\{-H_0(t)\exp(\beta x)\}$, where β and x remain constant over time and $H_0(t) = \int_0^t h_0(u)\,du$ is the cumulative baseline hazard. Otherwise, we can still consider the discretized version of Cox's model, so that

$$S(a_j|\beta, x) = \prod_{k=1}^{j}(1-\alpha_k)^{\exp(\beta x)} \quad \text{for } j = 1, \cdots, g\ , \tag{2}$$

where α_k is the discretized baseline hazard rate for the interval I_k; that is, $\alpha_k = P(a_{k-1} < T \leq a_k | T > a_{k-1})$. In semiparametric Bayesian analysis, statisticians often use the partitioning of the time axis just described, thus reducing an infinite dimensional problem to a finite dimensional one. In practice, as data collection, patient monitoring, and measurement recording do almost always give rise to grouped data, so the grids of these partitioning are often decided either by the data or by the design of the study. For example, patients being monitored on a weekly schedule suggests the natural choice for the length of each I_j is one week.

A semiparametric model of multiple event time data, the so-called *proportional intensity model*, was presented in Sinha (1993a). In this model, tumors arise in the ith ($i = 1, \cdots, n$) rat as a conditional Poisson process with the conditional intensity function

$$b_i(t|\mathbf{x_i}, w_i) \;=\; b_0(t) \exp(\beta \mathbf{x_i}) w_i \tag{3}$$

given the unobserved random 'frailty' w_i and the observed covariate x_i for the ith rat (Oakes, 1992). The frailty random variables w_i's are assumed to be iid gamma (alternative distributions can be considered also) with unknown variance κ, quantifying the dependencies among inter-event times within each rat and the heterogeneity of the intensity processes in different rats. This model is useful when the focus is on the influence of the covariates (or treatments) and on the heterogeneity among subjects related to the rate of occurrence of events in each subject, and each subject may experience several events, but a moderate number of subjects are observed in a relatively short time span.

The most widely used semiparametric model for multivariate survival data is the *frailty model* (see Chapter IX of ABGK for a review). Denote the ith survival time in the lth cluster by T_{li} for $i = 1, \cdots, n_l$ where n_l (=2 for the Leukemia data) is the size of the lth cluster. The conditional hazard function of T_{li} given the unobserved frailty random variable w_l (random cluster effect) of the lth cluster and the fixed observed covariate x_{li} is given by

$$h_{li}(t|x_{li}, w_l) \;=\; h_0(t) \exp(\beta x_{li}) w_l \;, \tag{4}$$

where β is the unknown regression parameter and $h_0(\cdot)$ is the unknown baseline hazard function common to every survival time within every cluster. The frailty random variables, the w_l's, are assumed to be independently distributed for the clusters with some parametric distribution with unit mean (when the mean is assumed to be finite), usually gamma (Oakes, 1989; Clayton, 1991), where the unknown variance of the w_l's (say, κ) quantifies the amount of heterogeneity among patients. That is we may assume that $w_l \sim \tau_0(\cdot|\kappa)$ independently for $l = 1, \cdots, n$. It is important to keep in mind the interpretation of w_l in (4). The frailty rv w_l measures the random susceptibility of the lth cluster to the event after eliminating the effect of the covariate. That is why for finite mean frailty we need to assume unit mean (to assure identifiability) and we need to assume that the frailty distribution of subjects at different covariate levels have the same location (or mean) but may have different variability. So, if we like to incorporate the effect of covariate (x) on frailty then it is wise to incorporate that effect only on the variability of the frailty, possibly by taking κ as a function of x—that is $w_l \sim \tau_0(\cdot|\kappa(x_l))$ where x_l is the covariate vector shared by all the subjects within the lth cluster. For moderate sized data set, it is often difficult to separate the fixed effect of x_l on the hazard through β and the indirect random effect of x_l via $\kappa(x_l)$ to w_l, it is often assumed that w_l's are iid $\tau_0(\cdot|\kappa)$ where κ is an unknown variability parameter free of x_l.

Another important frailty distribution is the infinite mean positive stable distribution (see Oakes 1994, for a review) which has no closed form density function, but the laplace transformation is given by $E[\exp(sW)] = \exp(-s^\alpha)$ for the stable frailty parameter α. Positive stable frailty is discussed in more detail later in Section 4.3.

10.3 Prior Processes

Bayesian analysis of survival data using nonparametric and semiparametric models started immediately after Cox (1972) with work of Ferguson (1973) and Kalbfleisch (1978). A nonparametric Bayes approach requires a prior random process. In this section we explore different ways of Bayesian modeling for nonparametric parts of the semiparametric models. One example is the h_0 of the Cox's model. For the time being, we are considering only nonparametric models with no covariates.

The chapters by Wolpert and Ickstadt and by Walker and Damien in this volume use processes with positive independent increments in disjoint intervals (Lévy processes). One approach is to write the cdf of T as $P(T \leq t) = F(t) = 1-\exp\{-H(t)\}$, where $H(\cdot)$ is the realization of a Lévy process, i.e., $H(\cdot)$ has characteristic function of the form

$$E[\exp\{it(H(a) - H(b))\}] = \Psi(t)^{\Lambda^*(a) - \Lambda^*(b)}, \tag{5}$$

where $a \geq b$, $\Psi(t)$ is the characteristic function of an infinitely divisible r.v. with unit mean, and $\Lambda^*(\cdot)$ is an increasing function with only finitely many jumps (for most practical purposes in survival analysis). The cumulative hazard, $H(t)$, may be considered when $F(\cdot)$ is absolutely continuous and Λ^* is the prior mean of H. The most commonly used Lévy process for modeling $H(t)$ is the Gamma process where $\Psi(t) = \{c/(c-it)\}^c$ for $c > 0$ and $H(a) - H(b)$ has a gamma distribution with mean $[\Lambda^*(a) - \Lambda^*(b)]$ and variance $[\Lambda^*(a) - \Lambda^*(b)]/c$. The hyperparameter c can be interpreted as a measure of precision attached to the initial guess $\Lambda^*(\cdot)$ of $H(\cdot)$.

Burridge (1981) considered a gamma process model with with some possibly unknown hyperparameters for Λ^*. Burridge (1981) explained how this kind of model may give wrong inferences if applied to continuous survival data. Two unsatisfactory features of a gamma process are independent increments in adjacent intervals, and discrete realization of the sample path with probability 1, so the resulting marginal distribution implies a positive probability of tied survival data. Nonetheless, this process has found a vast number of applications in literature for modeling different types of survival data. Lévy prior process can also be used for multiple event-time data, as discussed in Section 4.2.

In a somewhat different approach, Dykstra and Laud (1981) defined an extended gamma process considering a function $H(t) = \int_0^t [\zeta(s)]^{-1} dG(s)$, where $\zeta(s)$ is a known parametric continuous function and $G(s)$ is a realization of a gamma process. The distribution of the survival time T is then given by $F(t) = 1 - exp[-\int_0^t h(s)ds]$. The extended gamma process has an independent increment and increasing hazard rate property, and the realization of the distribution is always absolutely continuous. This model has been generalized by several authors (see Ibrahim, Chen and MacEachern (1996) and the references therein).

A more general way of defining the cumulative hazard is

$$A(t) = \int_{[0,t]} \frac{dF(s)}{1 - F(s-)}, \quad \text{so that} \quad F(t) = 1 - \prod_{[0,t]} \{1 - dA(s)\}$$

(Hjort, 1990). This coincides with $H(t)$ in the absolutely continuous case and has jumps of size $P(T = t | T \geq t)$ when t is an atom of F. This cumulative hazard rate $A(t)$ can be modeled as a realization of a beta process with prior mean function $dA_0(s)$. For details, see the papeer by Laud et al. in this volume.

Leonard (1978) introduced a correlated prior process for modeling density functions with prior information on smoothness. A discrete survival data version of that model assumes a piecewise constant hazard function. Over a set of prespecified grouping intervals, $I_j = (a_{j-1}, a_j]$ for $j = 1, 2, \cdots, g$, the hazard function is given as $h(t) = \lambda_j$ for $t \in I_j$. Some smoothness for the values of the λ_j's in adjacent intervals can be introduced using a first order process for $\gamma_j = \log\lambda_j$ with

$$\gamma_{j+1} = \gamma_j + e_{j+1} \quad \text{for} \quad j = 1, \cdots, g-1 . \tag{6}$$

The e_j's are independent $N(0, \sigma^2)$ r.v. Higher values of σ imply less prior information on the smoothing of the λ_j's. The original model of Gamerman (1991) used the so-called *linear Bayesian method*, which requires only a partial specification (mean 0 and finite variance) of the distribution of e_{j+1} in (6). Also note that Gray (1994) used the prior process structure similar to (6) on λ_j's itself and used a 'flat' prior for λ_1. Despite some of their nice physical properties, correlated prior processes have so far enjoyed only limited applications for modeling survival data beyond the univariate survival analysis.

The correlated prior process is also related to the so-called *martingale process* introduced by Arjas and Gasbarra (1994). They modeled the hazard rate as a piecewise constant function with jumps. They have assumed that the jump times are realizations of a time-homogeneous poisson process with rate μ (say), and that the sequence of constant hazard rates in different jump intervals form a discrete martingale. Given $A_1 < A_2 < A_3 < \cdots$, the sequence of jump-times for the process, a martingale process may assume $\gamma_{j+1} | \{\gamma_k : 1 \leq k \leq j\} \sim N(\gamma_j, \sigma^2)$ where $\gamma_k = \log \lambda_k$ and λ_k is the constant hazard in the interval $[A_{k-1}, A_k]$. The realizations of a martingale process is a piecewise constant function, but the mean of the sample path is a smooth function and hence the mean function of the process is not a member of the class of possible sample paths. Martingale process is very similar to the model of (2) if we drop the condition of the jump times being random, and instead assume that the hazard rates change at the end of each grouping interval.

Gelfand and Mallick (1995) took a suitable known transformation (say, $\tilde{J}$) of an unknown cumulative hazard $H(t)$ to an unknown cdf $J(t)$, a continuous increasing function with range $(0, 1]$ and modeled this unknown $J(t)$ by a finite fixed size but random mixture of a set of known beta cdf's. This process has not been used for modeling other types of survival data except univariate data, possibly due to the intensive computation and difficulty in hyperparameter elicitation associated with this prior process.

There are so many choices of prior processes for semiparametric Bayesian analyses and the question is which one to use for a particular analysis. We think the selection of a prior process depend on practical considerations such as ease of elicitation, theoretical convenience and physical interpretation of the model. The prior elicitation for such processes are discussed in a separate paper in this volume.

10.4 Bayesian Analysis

10.4.1 Methods for univariate survival data

We begin with univariate survival data using the Cox model $S(t|x) = \exp(-e^{\beta x} H_0(t))$. The regression vector β has a specific (possibly noninformative) prior distribution $\tau(\cdot)$ and the cumulative baseline hazard $H_0(t)$ (or the baseline hazard $h_0(t)$) has a suitable prior process.

Suppose that, as described in Section 2, the time axis $[0, \infty)$ is partitioned into $g + 1$ disjoint intervals, $I_1, I_2, \cdots, I_{g+1}$ where $I_j = [a_{j-1}, a_j)$ for $j = 1, 2, \cdots, g + 1$, with $a_0 = 0$ and $a_{g+1} = \infty$. For the breast cancer data, I_j can be the jth week and $H_j = H_0(a_j) - H_0(a_{j-1})$. For data grouped by weeks, the process $H_0(\ )$ enters only through the g increments in consecutive g weeks, $H_1, H_2, \cdots, H_g$. Let $T_1, T_2, \cdots, T_N$ be survival times (i.e., times to cosmesis) for the N patients. Then $P(T_k \in I_j | \mathbf{H}, x_k)$ is

$\exp\left(-e^{\beta x_k}\sum_{m=1}^{j-1} H_m\right)\left[1-\exp(-e^{\beta x_k}H_j)\right]$. Given the data (denote it by $\mathbf{C}$) on time to cosmesis of each patient grouped by weeks, the complete-data likelihood of $(\beta, \mathbf{H})$ is

$$L(\beta, \mathbf{H}|\mathbf{C}) \propto \prod_{j=1}^{g} G_j \ , \tag{7}$$

where, $G_j = \exp(-H_j \sum_{k\in R_j} e^{\beta x_k}) \prod_{l \in D_j}[1-\exp(-e^{\beta x_l} H_j)]$ and R_j is the set of patients at risk and D_j is the set of patients having cosmesis in the jth week. In absence of any covariate x, G_j reduces to $\exp(-H_j r_j)[1 - \exp(H_j)]^{d_j}$, where r_j and d_j are respectively the numbers of patients in the sets R_j and D_j. When we use the discretized Cox model of (6), then G_j is equal to $(1-\alpha_j)^{\sum_{k\in R_j}\exp(\beta x_k)} \alpha_j^{\sum_{l\in D_j}\exp(\beta x_l)}$. Now we only need the joint distribution of H_j's (say, $\phi(\mathbf{H})$) and a prior for β (say, $\tau(\beta)$) to multiply with the likelihood in (7) to find the joint posterior

$$p(\beta, \mathbf{H}|\mathbf{C}) \ \propto \ L(\beta, \mathbf{H}|\mathbf{C}) \times \phi(\mathbf{H}) \times \tau(\beta) \ .$$

It is important to note that for grouped survival data, the dimension of $\mathbf{H}$ is often finite as for any survival experiment we observe only up to a certain time point and everything is right censored beyond that point. For example, for a gamma process with H_0, the priors for H_j's will be independent gamma with mean $\Lambda_j^* = \Lambda^*(a_j) - \Lambda^*(a_{j-1})$ and scale c, i.e.

$$\phi(\mathbf{H}) \ \propto \ \prod_j \phi_j(H_j) \ \propto \ \prod_j [c^{\Lambda_j^* c} H_j^{\Lambda_j^* c-1} \exp(-cH_j)/\Gamma(\Lambda_j^* c)] \ .$$

Now the form of the likelihood in (7), and mutual independence of the priors of the H_j's implies that the conditional posteriors of H_j's given β and grouped data $\mathbf{C}$ are also mutually independent. We denote the conditional density of V given U as $[V|U]$. Following this notation, we can show that $[H_j|\beta, \mathbf{C}] \propto \phi_j(H_j) \times G_j$. A discretized beta process prior for α_j's assumes the priors of α_j's to be independent beta and by a similar argument $[\alpha_j|\beta, \mathbf{C}]$'s are also mutually independent (see Sinha, 1997).

Burridge (1981) used a Lévy prior process for the $H_0(t)$ and applied it to the Cox model with grouped survival data to compute the marginal likelihood $L(\Lambda^*, c, \beta|\mathbf{C}) \ = \ \prod_{j=1}^{g} E(G_j|\Lambda^*, c)$. Expectation is taken w.r.t. the prior distribution of H_j and the integration can be achieved using either a Gauss-Laguerre quadrature formula (for small c) or by a Taylor series expansion (for large c). Burridge (1981) used this likelihood for empirical Bayesian methods to estimate hyperparameters of the Lévy process. In his seminal paper, Burridge (1981) demonstrated by example how an univariate survival model with a gamma process prior may give wrong inferences if applied to continuous survival data. Though it is not clear from his work what property of the gamma process causes this problem with continuous data, it may be due to the fact that the gamma process generates a discrete sample path with probability 1, so continuous survival data may contradict the properties of the gamma process. Keeping in mind the findings of Burridge (1981), we recommend using a gamma process prior only for grouped (or incompletely grouped) survival data.

For incompletely grouped (interval-censored data, denoted by $\mathbf{Y}$) survival data as in our example (1), suppose that the kth patient's censoring interval is $(6, 14]$. Then the augmented week of the occurrence of cosmesis, denoted by Z_k, for this patient will be one of the I_j's with $7 \leq j \leq 14$. The predictive distributions of $[Z_k|\beta, \mathbf{H}, \mathbf{Y}]$'s (independent multinomial for each patient), conditional posterior of $[\beta|\mathbf{H}, \mathbf{Z}, \mathbf{Y}]$ and the conditional posterior of $[H_j|\beta, \mathbf{Z}, \mathbf{Y}]$ can be obtained from (7) as $[H_j|\beta, \mathbf{C}] \propto \phi_j(H_j) \times G_j$. It is clear from (7), that for Lévy process prior, in every iteration of a MCMC chain, given the augmented data $(\mathbf{Z}, \mathbf{Y}) = \mathbf{C}$ each H_j can be sampled independent of the other H_k values $(k \neq j)$. Once all of the conditional posteriors and predictive distributions have been obtained, the steps of the MCMC algorithm are given as:

(1) sample $[Z_k|\beta, \mathbf{H}, \mathbf{Y}]$ using independent multinomial distribution for each subject (for $k = 1, \cdots, N$ patients)

(2) sample $[H_j|\beta, \mathbf{C}] \propto \phi_j(H_j) \times G_j$ independently for each I_j (for $j = 1, \cdots, g$)

(3) sample $[\beta|\mathbf{H}, \mathbf{C}] \propto \tau(\beta) \times \prod_j G_j$.

In the MCMC chain, some of the distributions may not be easy to sample from (e.g., conditional posterior of β), and so the Metropolis steps (or other suitable algorithm) can be used. Using MCMC algorithm, the marginal posterior of β and posterior estimate of any other relevant quantity can be computed. Explicit algorithms (with the detailed functional forms of the conditional posteriors) is available for interval-censored survival data using a gamma and a correlated prior process are available at Sinha et al. (1997). The fortran programs for these algorithms are available by request from the authors of that article.

A convenient way to incorporate a time-varying covariate effect—that is a time varying β in Cox model—is to assume that $\beta(t)$ is also piecewise constant—$\beta(t) = \beta_j$ for $t \in I_j$, $j = 1 \cdots, g$. A possible choice for the prior process for such a β function is the correlated (or martingale) prior process with

$$\tau(\beta) \propto \tau_1(\beta_1) \times \prod_{j=2}^{g} \tau_j(\beta_j; \beta_{j-1}, \sigma_j^2) ,$$

where $\tau_j(x; \mu, \sigma^2)$ is a known density (typicaly normal) with mean μ and variance σ^2. Now to use a MCMC algorithm for the joint posterior $[\beta, \mathbf{H}|\mathbf{C}]$

we need to use an additional step to sample β_j from

$$[\beta_j|\beta_{j+1}, \beta_{j-1}, H_j, \mathbf{Y}] \propto G_j \times \phi_j(\beta_j; \beta_{j-1}, \sigma_j^2) \times \phi_{j+1}(\beta_{j+1}; \beta_j, \sigma_{j+1}^2)$$

for $j = 1, \cdots, g$ where β_0 is the prior mean of β_1 and $\phi_{g+1}(x)$ is 1. When we use normal density for ϕ_j's this conditional density can be shown to be log-concave and hence the adaptive-rejection sampling algorithm of Gilks and wild (1992) can be used to sample from this density (see Sinha et al. (1997) for details).

Sinha (1997) discussed a time-discrete beta process model (Hjort, 1990) assuming that the discretized baseline hazard rates, the α_k's from the model in (2), have independent beta priors. This approximation to the time-continuous beta process model simplified the computation, because of the natural conjugacy of the beta priors with the grouped-data likelihood. The paper by Laud et al. in this volume describes the Bayesian analysis of survival data with time continuous right-censored data and using a Cox model. They use a time-continuous beta process for the cumulative hazard rate $A(t)$. Their method can be used to sample from any given time-continuous beta process thus we can use an MCMC algorithm to sample from the joint posterior distribution under an exactly beta process prior for $A(t)$ to perform a Bayesian analysis of the interval-censored survival data.

10.4.2 Analysis of multiple event-time data

Sinha (1993 a) described the Bayesian semiparametric model for multiple event time data. He used the conditional poisson process model of (3) to find the conditional likelihood given grouped data $\mathbf{Y} = \{d_{kj} : k = 1, \cdots n; j = 1, \cdots, g\}$ as

$$L(\beta, B_j|\mathbf{Y}, \mathbf{W}) \propto \prod_j \prod_k [B_j w_k \exp(\beta x_k)]^{d_{kj}} \exp\{-B_j w_k \exp(\beta x_k)\}$$

where g is the number of intervals of observations, n is the number of subjects (rats) and $B_j = \int_{a_{j-1}}^{a_j} b_0(u)du$ is the cumulative baseline intensity for the jth interval I_j. The likelihood is based on the fact that the conditional distribution of number of new tumors to ith rat in any interval is poisson, and thus the method is applicable to grouped data. Increments in common cumulative baseline intensity B_j's are assumed to have independent gamma priors with known hyperparameters (say, $Gamma(\Lambda_j c, c)$ where Λ_j is the prior mean of B_j and Λ_j/c is the variance), and the frailty random variable w has a gamma distribution (or some other known finite-mean positive distribution) with unit mean and unknown variance κ. For the gamma frailty model, the frailty r.v. has density $\tau_0(w|\kappa) = \eta^\eta w^{\eta-1} \exp(-w\eta)/\gamma(\eta)$, where $\eta = \kappa^{-1}$. An MCMC algorithm are used to compute the approximate marginal posterior distributions of (β, κ). The steps of the MCMC algorithm are as follows:

(1) sample $[B_j|\beta, \mathbf{W}, \mathbf{Y}]$ from $Gamma(\Lambda_j c + \sum_{k=1}^{n} d_{kj},\ c + \sum_{k=1}^{n} \theta_k w_k)$ independently for $j = 1, \cdots, g$

(2) sample $[w_k|\beta, \mathbf{B}, \mathbf{Y}]$ from $Gamma(\nu + d_{k*}, \nu + \theta_k \sum_{j=1}^{g} B_j)$ independently for $k = 1, \cdots, n$

(3) sample $[\kappa|\mathbf{W}]$ from density proportional to

$$\nu^{n\nu} \exp\{-\nu(S - \log P)\}\tau_2(\nu)/(\Gamma\nu)^n$$

(4) sample $[\beta|\mathbf{B}, \mathbf{W}, \mathbf{Y}]$ from a density proportional to $\exp[\sum_{k=1}^{n}\{w_k\theta_k + \beta x_k d_{k*}\}] \times \tau_1(\beta)$

where $\nu = \kappa^{-1}$, $\tau_1(\cdot)$ is the prior for β, $\tau_2(\cdot)$ is the prior for ν, $\theta_k = \exp(\beta x_k)$, $S = \sum_{k=1}^{n} w_k$, $P = \prod_{k=1}^{n} w_k$ and $d_{k*} = \sum_j d_{kj}$. Apparently, this method is not applicable to continuous data due to the theoretical difficulty in using a gamma process for $B_0(t)$ for the poisson process in (3.3).

The methodology of Sinha (1993 a) can be also adapted for autocorrelated prior process for $b_0(t)$ by using a suitable multivariate joint prior for the discretized $b_0(t)$.

10.4.3 Analysis of multivariate data

Clayton (1991) introduced the first semiparametric Bayesian model for analysis of multivariate survival data using the frailty model of (4), where β is the unknown regression parameter and $h_0(\cdot)$ is the unknown conditional baseline hazard common to every survival time (remission time) within every cluster (every mathced pair). The frailty random variables— the w_l's — are assumed to be iid for each cluster (each matched pair) with some known parametric distribution (gamma in particular) with unit mean but unknown variance κ. Clayton (1991) used a Lévy process, gamma process in particular, with known mean function and confidence parameter for modeling the prior information on $h_0(\cdot)$. But in many practical situations, an autocorrelated prior process may be more realistic for modeling $h_0(\cdot)$ (see later chapter by Ibrahim and Sinha). Clayton's work is applicable only for continuous data, and extensions of the methodology to the interval-censored data and a stable frailty are not immediate.

Aslanidou et al. (1998) also worked with the frailty model in (4). They assumed a discretized correlated prior process (though any discretized independent process can be easily accommodated) on the baseline hazard. The frailty distribution was assumed to be a gamma with unknown variance κ. The Metropolis-Hastings algorithm was used to sample from the joint posterior distribution of the paramcters, $(\beta, \Lambda, \mathbf{W}, \kappa)$, and then different posterior quantities and posterior predictions were computed using these samples. The algorithm described by Aslanidou et al. have been extended to stable frailty distribution by Qiou et al. (1997).

Suppose we use the piecewise constant baseline hazard for (4) and assume that the ith subject $(i = 1, \cdots, n_l)$ of the lth cluster $(l = 1, \cdots, m)$ has a constant hazard of $h_{li}(t|x_{li}, w_l) = \lambda_j \theta_{li} w_l$ in the jth interval $(j = 1, \cdots, g)$ given the unobserved frailty w_l.If the subject survived beyond the jth interval, i.e., $t_{li} > a_j$, where t_{li} is the observed survival time, the likelihood contribution for the jth interval is $\exp(-\lambda_j \Delta_j \theta_{li} w_l)$ where $\Delta_j = a_j - a_{j-1}$ and $\theta_{li} = \exp(\beta x_{li})$. If the subject failed or was censored in the jth interval, i.e., $a_{j-1} < t_{li} \le a_j$ then the likelihood contribution for that interval is $(\lambda_j \theta_{li} w_l)^{\delta_{li}} \exp\{-\lambda_j (t_{li} - a_{j-1}) \theta_{li} w_l\}$ where δ_{li} is the censoring indicator taking value 0 when right-censored or 1 otherwise. Given the complete data $(\mathbf{Y}, \mathbf{W})$ we derive the following complete data likelihood

$$
\begin{aligned}
& l(\beta, \Lambda|\mathbf{W}, \mathbf{Y}) \\
= & \prod_{l=1}^{m} \prod_{i=1}^{n_l} \Big[\prod_{j=1}^{g_{li}-1} \exp\{-\lambda_j \Delta_j \theta_{li} w_l\} \Big] \\
\times & \exp\{-\lambda_{g_{li}} (t_{li} - a_{g_{li}-1}) \theta_{li} w_l\} \, (\lambda_{g_{jk}} \theta_{jk} w_j)^{\delta_{jk}} ,
\end{aligned}
$$

where g_{li} is such that $t_{li} \in (a_{g_{li}-1}, a_{g_{li}}) = I_{g_{jk}}$. To get the joint posterior of $[\Lambda, \beta, \kappa|\mathbf{Y}]$ for the gamma frailty model we have to take

$$[\beta, \Lambda, \mathbf{W}, \kappa|\mathbf{Y}] \propto l(\beta, \Lambda|\mathbf{W}, \mathbf{Y}) \times \tau_0(\mathbf{W}|\kappa) \times \tau_1(\kappa) \times \tau_2(\beta) \times \phi(\Lambda|\Lambda^*) ,$$

and then integrate $\mathbf{W}$ by MCMC method. Here $\phi(\Lambda|\Lambda^*)$ denotes the joint prior of Λ given the hyperparameters Λ^* of the prior process.

For the stable frailty model we need to use

$$[\beta, \Lambda, \mathbf{W}, \mathbf{V}, \alpha|\mathbf{Y}] \propto l(\beta, \Lambda|\mathbf{W}, \mathbf{Y}) \times \tau_0(\mathbf{W}, \mathbf{V}|\alpha) \times \tau_1(\alpha) \times \tau_2(\beta) \times \phi(\Lambda|\Lambda^*) ,$$

where

$$\tau_0(\mathbf{W}, \mathbf{V}|\alpha) \propto \prod_{l=1}^{m} \frac{\alpha}{(1-\alpha) w_l} \Big(\frac{u_\alpha(v_l)}{w_l}\Big)^{\alpha/(1-\alpha)} \exp\Big\{-\Big(\frac{u_\alpha(v_l)}{w_l}\Big)^{\alpha/(1-\alpha)}\Big\} ,$$

and

$$u_\alpha(v) = \frac{\sin(\pi \alpha v + \psi_\alpha)}{\cos \pi v} \Big[\frac{\cos \pi v}{\cos\{\pi(\alpha - 1) v + \psi_\alpha\}}\Big]^{(\alpha-1)/\alpha}$$

for $v_j \in (-\frac{1}{2}, \frac{1}{2})$ and $\psi_\alpha = \min(\alpha, 2-\alpha)\pi/2$ (Buckle 1995, Qiou et al., 1997). So, for the stable frailty model, the steps of the MCMC algorithms are as follows:

(1) Sample from β

$$[\beta|\Lambda, \mathbf{W}, \mathbf{Y}] \propto l(\beta, \Lambda|\mathbf{W}, \mathbf{Y}) \times \tau_2(\beta)$$

(2) Sample Λ from

$$[\Lambda|\beta, \mathbf{W}, \mathbf{Y}] \propto l(\beta, \Lambda|\mathbf{W}, \mathbf{Y}) \times \phi(\Lambda|\Lambda^*)$$

(3) Sample w_l from

$$[w_l|v_l, \alpha, \Lambda, \beta, \mathbf{W}, \mathbf{Y}] \propto l(\beta, \Lambda|\mathbf{W}, \mathbf{Y}) \times \frac{\alpha}{(1-\alpha)w_l}(\frac{u_\alpha(v_l)}{w_l})^{\alpha/(1-\alpha)} \exp\{-(\frac{u_\alpha(v_l)}{w_l})^{\alpha/(1-\alpha)}\},$$

independently for $l = 1 \cdots, m$

(4) Sample v_l from

$$[v_l|w_l, \alpha] \propto \frac{\alpha}{(1-\alpha)w_l}(\frac{u_\alpha(v_l)}{w_l})^{\alpha/(1-\alpha)} \exp\{-(\frac{u_\alpha(v_l)}{w_l})^{\alpha/(1-\alpha)}\},$$

independently for $l = 1 \cdots, m$

(5) Sample α from

$$[\alpha|\mathbf{W}, \mathbf{V}] \propto \tau_1(\alpha) \times \prod_{l=1}^{m} \frac{\alpha}{(1-\alpha)w_l}\left(\frac{u_\alpha(v_l)}{w_l}\right)^{\alpha/(1-\alpha)} \exp\left\{-(\frac{u_\alpha(v_l)}{w_l})^{\alpha/(1-\alpha)}\right\}.$$

The exact expressions of these densities depend on the corresponding prior. For gamma frailty, there will be no step (4), and the step (5) will involve sampling κ from the density proportional to $\{(\prod_l \tau_0(w_l|\kappa)\} \times \tau_1(\kappa)$, where $\tau_0(w_l|\kappa)$ is the gamma density with variance κ and unit mean. For details on gamma, lognormal and stable frailty models with correlated prior process see Aslanidou et al. (1998), Sahu et al. (1997) and Qiou et al. (1997).

10.5 Further Readings

Gamerman (1991) presented a linear Bayesian method to analyze univariate survival data with possible right censoring using the autocorrelated prior process.

Sinha (1993 b) described a Monte Carlo version of the EM (MCEM) algorithm (Tanner, 1996) for empirical Bayes estimation of the hyperparameters of a Gamma process from interval-censored data. But, the convergence of MCEM algorithm for this case often will be very slow and each M-step requires intensive computation. A more practical and better Bayesian analysis would be to use a careful hyperprior for (Λ^*, c) and then use an appropriate MCMC algorithm to compute relevant posterior quantities.

Ghosh and Sinha (1995) described the adaptation of autocorrelated prior process methodology for interval censored data using the so-called posterior likelihood method to compute the discretized-maximum-posterior-likelihood-estimate (DMPLE) of the parameters. Ghosh and Sinha (1995) also estimated the dispersion matrix of the DMPLE and presented different competing methods of eliciting the hyperparameters of the autocorrelated prior process (so-called hyperparameters of smoothing).

Availability of the MCMC tools make it possible for us to consider Bayesian analyses of multiple event time data using other models, such as the modulated renewal process model and the m-site model (see Sinha and Dey, 1997, for a discussion).

Sinha (1998) used an autocorrelated incomplete prior process for the baseline hazard and a joint prior for (β, κ) (independent of the prior process) in the frailty model of (4). This model is analyzed using the posterior likelihood method, where a product of the data likelihood and prior likelihood (on smoothness of baseline hazard) is maximized to find Bayesian estimates of (β, κ). Sinha's (1998) posterior likelihood method can be used even when the w_l's in (4) follow the positive stable distribution. Walker and Mallick (1997) used a prior process on the nonparametric frailty distribution $\tau_0(\cdot)$ and even allowed the τ_0 to vary over covariate values. Say, for binary covariate ($x = 1$ or 0), Walker and Mallick (1997) allowed two separate frailty distributions, say τ_{00} for $x = 0$ and τ_{01} for $x = 1$, where both densities were modeled nonparametricaly. Identifiability may be a possible concern for such a model.

Another important recent research work for the Bayesian multivariate survival method is that of Gustafson (1995) who did not use any prior information on the baseline hazards h_{j0}'s. So the method used products of stratified partial likelihoods given the multivariate survival data. He assumed the frailty distribution to be gamma. Gustafson used the *hybrid Monte Carlo method* (Duane et al. 1987, Neal 1993) to sample from the joint posterior distributions of all of the parameters. Gustafson's algorithm is more intensive compared to some other MCMC algorithms, such as the Gibbs and the Metropolis. His method allows only the finite mean frailty distributions. The method may face problems if applied to tied or grouped multivariate survival data, as the common fixes of the partial likelihood for highly grouped survival data may have poor small sample properties. Also, the method cannot be used for most prediction purposes, such as predicting a future outcome of a particular patient at a given center.

Sinha and Dey (1997) reviewed the applications of semiparametric Bayes methods for univariate, multivariate, doubly censored and multiple event-time survival data problems. Sinha and Dey (1997) and the references therein complements this article for further references and details on some of the semiparametric models and Bayesian methods. This review article and the references therein are also sources for future research topics in this area.

Acknowledgments: Dr. Sinha's research was supported by the grant R29-CA69222 from the NCI.

References

Andersen, P. K., Borgan, O., Gill, R. D., and Keiding, N. (1993), *Statistical Models Based on Counting Processes*, Springer-Verlag: New York.

Arjas, E., and Gasbarra, D. (1994), "Nonparametric Bayesian inference from right censored survival data, using the Gibbs sampler", *Statistica Sinica,* **4**, 505-524.

Aslanidou, H., Dey, D. K., and Sinha, D. (1998), "Bayesian analysis of multivariate survival data using Monte Carlo methods", *Canadian Journal of Statistics*, **26**, to appear.

Buckle, D. J. (1995), "Bayesian inference for stable distribution". *Journal of the American Statistical Association*, **90**, 605-613.

Burridge, J. (1981), "Empirical Bayes analysis of survival time data". *Journal of the Royal Statistical Society, Series B*, **43**, 65-75.

Clayton, D. (1991), "A Monte Carlo Method for Bayesian inference in frailty models". *Biometrics*, **47**, 467-485.

Cox, D. R. (1972), "Regression models and life tables", *Journal of the Royal Statistical Society, Series B*, **34**, 187-220.

Cox, D. R., and Oakes, D. (1984), *Analysis of Survival Data.* London: Chapman & Hall.

Duane, S., Kennedy, A. D., Pendleton, B. J., and Roweth, D. (1987) "Hybrid Monte Carlo", *Physics Letters, Series B*, **195**, 216-222.

Dykstra, R. L., and Laud, P. (1981), "A Bayesian nonparametric approach to reliability". *Annals of Statististics*, **9**, 356-367.

Ferguson, T. S. (1973), "A Bayesian analysis of some nonparametric problems". *Annals of Statistics*, **1**, 209-230.

Finkelstein, D. M., and Wolfe, R. A. (1985), "A semiparametric model for regression analysis of interval-censored failure time data". *Biometrics*, **41**, 933-945.

Gail, M. H., Santner, T. J., and Brown, C. C. (1980), "An analysis of comparative carcinogenesis experiments with multiple times to tumor". *Biometrics*, **36**, 255-266.

Gamerman, D. (1991), "Dynamic Bayesian models for survival data", *Applied Statistics*, **40**, 63-79.

Gelfand, A. E., and Mallick, B. (1995), "Bayesian analysis of proportional hazards model built from monotone functions". *Biometrics*, **51**, 843-852.

Ghosh, S., and Sinha, D. (1995), "Analysis of interval-censored survival data using posterior likelihood". *Technical Report*, Department of Statistics, University of Connecticut, USA.

Gilks, W.R., and Wild, P. (1992), "Adaptive Rejection Sampling for Gibbs Sampling," *Applied Statistics*, **41**, 337-348.

Gray, R. J. (1994), "A Bayesian analysis of institutional effects in multicenter cancer clinical trial", *Biometrics*, **50**, 244-253.

Gustafson, P. (1995), "Hierarchical Bayesian analysis of clustered survival data", *Technical Report #144*, Department of Statistics, University of British Columbia, Canada.

Hjort, N. L. (1990), "Nonparametric Bayes estimators based on beta processes in models for life history data". *Annals of Statistics*, **18**, 1259-1294.

Hougaard, P. (1986), "A class of multivariate failure time distribution", *Biometrika*, **73**, 671-678.

Ibrahim, J., Chen, M-H. and MacEachern, S. N. (1996), "Bayesian variable selection for proportional hazards model", *Submitted for publication.*

Kalbfleisch, J. D. (1978), "Nonparametric Bayesian analysis of survival time data", *Journal of the Royal Statistical Society, Series B*, **40**, 214-221.

Leonard, T. (1978), "Density estimation, stochastic processes and prior information", *Journal of the Royal Statistical Society, Series B*, **40**, 113-146.

Neal, R. M. (1992), "Bayesian training of backpoint propagation using hybrid Monte Carlo methods", *Technical Report, CRG-TR-92-1*, Department of Computer Science, University of Toronto, Canada.

Oakes, D. (1989), "Bivariate survival models induced by frailties". *Journal of the American Statistical Association*, **84**, 487-493.

Oakes, D. (1992), "Frailty models for multiple event times", *Survival Analysis: State of the Art* (eds. J.P. Klein and P.K. Goel), 371-379, Kluwer Academic Publisher: Netherlands.

Oakes, D. (1994), "Use of frailty models for multivariate survival data". *Proceeding of XVIIth International Biometric Conference*, Hamilton, Ontario, Canada.

Qiou, Z., Ravishanker, N., and Dey, D. K. (1997), "Multivariate survival analysis with positive stable frailties", *Technical Report*, Department of Statistics, University of Connecticut, USA.

Sahu, S. K., Dey, D. K., Aslanidou, H., and Sinha, D. (1997), "A Weibull Regression model with Gamma frailties for multivariate survival data". *Lifetime Data Analysis*, **3**, 123–137.

Sinha, D. (1993 a), "Semiparametric Bayesian analysis of multiple time data". *Journal of the American Statistical Association*, **88**, 979-983.

Sinha, D. (1993 b), "Semiparametric Bayesian analysis of event time data". *PhD. Thesis*, University of Rochester, Rochester, NY, USA.

Sinha, D., Chen, M-H., and Ghosh, S. K. (1996), "Bayesian analysis and model selection for interval-censored survival data". *Submitted for publication.*

Sinha, D. (1997), "Time-discrete beta process model for interval censored survival data". *Canadian Journal of Statistics*, **25**, 445–456.

Sinha, D. (1998), "Posterior likelihood method for multivariate survival data". *Biometrics*, to appear.

Tanner, M. A. (1993), "Tools for Statistical Inference: Observed Data and Data Augmentation Methods". *Lecture Notes in Statistics*, Springer-Verlag, New York.

Walker, S. G., and Mallick, B. K. (1996), "Hierarchical Generalized Linear models and Frailty Models with Bayesian Nonparametric Mixing". *Journal of the Royal Statistical Society, Series B*, **59**, 845–860.

11

Bayesian Nonparametric and Covariate Analysis of Failure Time Data

Purushottam W. Laud
Paul Damien
Adrian F.M. Smith

ABSTRACT A Bayesian analysis of the semi-parametric regression model of Cox (1972) is given. The cumulative hazard function is modelled as a beta process. The posterior distribution of the regression parameters and the survival function are obtained using a combination of recent Monte Carlo methods. An illustrative analysis within the context of survival time data is given.

11.1 Introduction

Cox (1972) proposed a model for survival time data in the presence of covariates. To state it, let T represent the time to the event of interest for an individual and $\mathbf{z} = (z_1, \ldots, z_p)'$ the vector of covariates. Then the survival distribution of T is taken to be

$$P(T > t|\mathbf{z}) = exp\{-H(t)exp(\mathbf{z}'\beta)\} \tag{1}$$

where $H(t) = \int_0^t h(s)ds$ is the baseline cumulative hazard function (CHF), $h(.)$ the corresponding hazard rate and $\beta = (\beta_1, \ldots, \beta_p)'$ is the vector of regression coefficients. This model is often called the proportional hazards model since the ratio of hazard rates of two individuals with differing $\mathbf{z}$'s is constant in time. Its use and success are now so widespread [see, for example, Klein and Moeschberger (1996) and the references therein] as to make it the primary tool in the analysis of lifetime data.

In his original article Cox used a factor of the full likelihood, which he later justified and termed partial likelihood in Cox (1975), to estimate β. Kalbfleisch (1978) considered a Bayesian analysis using a gamma process prior on the logarithm of the baseline survival function and Clayton (1991) discussed a Monte Carlo method for related frailty models. Designed for use in various survival analysis problems, the beta process was advanced by Hjort (1990) as a prior on the space of CHF's. He also briefly treated the Cox model and gave some descriptions of the resulting posterior distributions.

In this paper, we follow Hjort (1990) and model the cumulative hazard using a beta process. We give a full Bayesian solution in that the posterior distributions of the baseline CHF $H(t)$ and the regression coefficient vector β are obtained. This is accomplished by using and extending a combination of Monte Carlo methods described in Smith and Roberts (1993), Damien et.al (1995) and Walker (1995). By straightforward transformations, samples obtained from the full posterior distribution also allow inference for other relevant quantities such as the baseline survival function, the relative risk and survival probabilities at given values of covariates and times.

Section 2 below describes the use of the beta process with the Cox model and the attendant posterior results available from Hjort (1990). In Section 3, we develop the computational form of the model and give details for its implementation. Section 4 provides an illustrative analysis of the leukemia data previously analysed, using Bayesian methods, by Kalbfleisch (1978). Some extensions are discussed in Section 5.

11.2 Cox Model with Beta Process Prior

Much of nonparametric Bayesian inference has proceeded by modelling the unknown cumulative distribution function (CDF) as a stochastic process. One of the exceptions appears in Hjort (1990) where an alternative is introduced by placing a (beta) stochastic process on the space of cumulative hazard functions. This has several attractive features: it offers a broader class of models in the context of life history data; there is an accessible interpretation of the prior process in terms of hazard functions; and it provides a natural generalisation of neutral to the right processes, such as the Dirichlet (Ferguson, 1973).

In the discussion that follows, we will consider the general time-continuous version of the beta process. Let T be a random variable with CDF $F(t) = Pr(T \leq t)$ on $[0, \infty)$ and $F(0) = 0$. The CHF for F or T is a nonnegative, nondecreasing, right continuous function A on $[0, \infty)$. As in Hjort (1990) we use the symbol A in three different ways : $A[a, b)$, $A(t)$ and $A\{t\}$ denote the increment of the function A in the interval $[a, b)$, the value of the function A at a point t, and the increment of the function A at t, respectively. Now let A be such that it satisfies

$$dA(s) = A[s, s + ds) = Pr\{T \in [s, s + ds)|T \geq s\} = dF(s)/F[s, \infty)$$

so that

$$A[a, b) = \int_{[a,b)} \frac{dF(s)}{F[s, \infty)}.$$

Recovering F from A requires product integrals (Gill and Johansen, 1990) as

$$F(t) = 1 - \prod_{[0,t]} \{1 - dA(s)\},\ t \geq 0.$$

If F is continuous, this can be shown to reduce to the usual formula

$$A(t) = -log(1 - F(t)).$$

In general, however, $F(t)$ does not equal $1 - exp\{-A(t)\}$. As explicated in Andersen et al.(1993), this definition of the cumulative hazard function unifies the treatment of discrete, continuous and mixed random variables while retaining the appropriate meaning of hazard as used in the actuarial and survival analysis literature.

While we omit the mathematical details of its construction, the beta process has independent increments that, infinitesimally speaking, have beta distributions. More precisely, it is a Lévy process with Lévy measure concentrated on $[0, 1]$. Using this process as a prior on the space of cumulative hazard functions, the posterior distribution is, in the absence of covariates, again a beta process. Relevant definitions and results are stated in the Appendix. Full Bayesian inference using the Gibbs sampler for this situation is given in Damien et al (1996).

To include covariates as prescribed in the Cox model (1), let A_i and $\mathbf{z}_i$ denote the cumulative hazard function and the covariate vector for individual i, $i = 1, \cdots, n$. Then,

$$1 - dA_i(s) = \{1 - dA(s)\}^{exp(\mathbf{z}_i'\beta)}.$$

Suppose $X_1, \cdots, X_n$ are the times to the event of interest for the n individuals and the data are in the form $t_i = min(X_i, c_i)$, $\delta_i = I(X_i \leq c_i)$, where I denotes the indicator function. The c_i's here are censoring times assumed to be independent of the event times. Now the discussion in Section 6 of Hjort (1990) leads to the following likelihood for A and β :

$$\mathcal{L}(A, \beta; data) = \left[\prod_{[0,\infty)} \{1 - dA(s)\}^{R(s,\beta)} \right] \Pi_{i=1}^{n} \left[\{1 - A\{t_i\}\}^{-exp(\mathbf{z}_i'\beta)} - 1 \right]^{\delta_i} \tag{2}$$

where $R(s, \beta) = \sum_{j=1}^{n} exp(\mathbf{z}_i'\beta) I(t_j \geq s)$. Using a beta process prior $A \sim beta(c(.), A_0(.))$ for A (see Appendix for notation) and an independent prior for β, the conditional posterior of A given β and the data is a beta process between points where actual event times are observed. At such event times, there are positive jumps in A with distributions

$$f_{A\{s\}}(u) \propto u^{-1}(1-u)^{c(s)+R(s,\beta)-d(s,\beta)-1} \Pi_{i=1}^{n} \left\{ 1 - d_i(s)(1-u)^{exp(\mathbf{z}_i'\beta)} \right\} \tag{3}$$

where $d_i(s) = I(t_i = s, \delta_i = 1)$ and $d(s, \beta) = \sum_{j=1}^{n} d_i(s) exp(\mathbf{z}_i'\beta)$. A prior specification for β and the likelihood in (2) above suffice to yield a conditional posterior for β given A. In the next section we employ these two conditionals and several auxiliary variables to develop a Gibbs sampler.

11.3 The Computational Model

Let $\Delta = (\Delta_1 \cdots \Delta_N)$ denote increments in the $A(\cdot)$ process between the t_i's; i.e., the grid is taken to be all distinct censoring or event times in the data. Call these time points $s_1 < s_2 < \cdots < s_N$, among which are event times $x_1 < x_2 < \cdots < x_m$ occuring with multiplicities $k_1, k_2, \cdots, k_m$ repectively. Let $U = (U_1 \cdots U_m)$ stand for the jumps in $A(\cdot)$ at these x's. Finally, let $i(j)$ be the index of the jth event observation at x_i, $j = 1, \cdots, k_i$ in some fixed order for each $i = 1, \cdots, m$. This means that $i(j)$ is a unique number between 1 and n such that $t_{i(j)} = x_i$ and $\delta_{i(j)} = 1$. We use the notation $[X|Y]$ to denote the conditional probability distribution of X given Y. Likewise, $[X]$, denotes the marginal distribution of X.

With this notation, the Gibbs sampler proceeds by simulating from $[\Delta, U|\beta, data]$ and $[\beta|\Delta, U, data]$. As outlined in the previous section, $[\Delta|\beta, data]$ is the distribution of the increments of a beta process with parameters $\{c(\cdot) + R(\cdot, \beta)\}$ and $\int_0^{(\cdot)} \frac{c(s)dA_0(s)}{c(s)+R(s,\beta)}$. The components of Δ are independent, each with an infinitely divisible distribution. Simulation of these is detailed in Damien et al.(1996). Moreover, U and Δ are independent, U having independent components with distributions

$$f(u_i) \propto u_i^{-1}(1-u_i)^{c(x_i)+R(x_i,\beta)-\{\sum_{j=1}^{k_i} exp(\mathbf{z}_{i(j)}'\beta)\}-1} \Pi_{j=1}^{k_i}\{1-(1-u_i)^{exp(\mathbf{z}_{i(j)}'\beta)}\} .$$

Simulation from such a distribution can be effected using auxilliary variables. Transforming u to $v = -\log(1-u)$, the task is to simulate from

$$[v] \propto exp(-av)(1-exp(-v))^{-1}\Pi_{i=1}^{k}(1-exp(-b_i v)) .$$

Using $\frac{[v|\mathbf{w},y]}{[\mathbf{w},y|v]} \propto [v]$ and taking y and w_i to be geometric and truncated exponential variates respectively as

$$[y|v] = (1-exp(-v))exp(-vy), \quad y = 0, 1, \cdots$$

and

$$[w_i|v] = b_i v\ exp(-b_i v w_i)(1-exp(-b_i v))^{-1} I_{(0,1)}(w_i) ,$$

we get

$$[v|\mathbf{w}, y] \propto v^k\ exp\{-v(a + y + \sum_{i=1}^{k} b_i w_i)\}$$

which is a gamma variate.

It only remains to simulate $\beta|\Delta, U, data$. Adapting the likelihood (2) to the notation of the current section yields

$$[\beta|\Delta, U, data] \propto [\beta] \left\{\Pi_{i=1}^{N}(1-\Delta_i)^{R(s_i,\beta)}\right\}$$
$$\Pi_{i=1}^{m}(1-u_i)^{R(x_i,\beta)-d(x_i,\beta)}\Pi_{j=1}^{k_i}\{1-(1-u_i)^{exp(\mathbf{z}'_{i(j)}\beta)}\}$$

where $[\beta]$ is the prior distribution of the regression coefficients. Using $v_i = -\log(1-u_i)$ and incorporating the conditionals of the auxiliary variables $y_i, w_{ij}, i = 1, \cdots, m, j = 1, \cdots, k_i$, we arrive at

$$[\beta|\Delta, V, Y, W, data] \propto [\beta] exp \left[\sum_{i=1}^{N} R(s_i, \beta)\log(1-\Delta_i)\right.$$
$$\left. -\sum_{i=1}^{m} v_i \left\{R(x_i,\beta) - \sum_{j=1}^{k_i}(1-w_{ij})exp(\mathbf{z}'_{i(j)}\beta)\right\} + \sum_{i=1}^{m}\sum_{j=1}^{k_i}\mathbf{z}'_{i(j)}\beta\right].$$

It is straightforward to show that this is logconcave in each component of β if the prior is so, leading to efficient simulation of β by the algorithm of Gilks and Wild (1992). To summarize, we use

$$\begin{aligned}
[Y|\beta, \Delta, V, W, data] &\propto \textit{Independent Geometrics}\\
[W|\beta, \Delta, V, Y, data] &\propto \textit{Independent Truncated Exponentials}\\
[V|\beta, \Delta, W, Y, data] &\propto \textit{Independent Gammas}\\
[\Delta|\beta, V, W, Y, data] &\propto \textit{Independent Beta Process Increments}\\
[\beta|\Delta, V, W, Y, data] &\propto \textit{Componentwise Logconcave}
\end{aligned}$$

as the full conditionals in a Gibbs sampler.

11.4 Illustrative Analysis

To demonstrate the types of inference one can make using the techniques of this article, we consider the leukemia data analyzed by Cox (1972) and Kalbfleisch (1978), among others. These data, listed in Table 1 as reported by these authors, consist of remission times (in weeks) of leukemia patients assigned to treatment with a drug or a placebo during remission maintenance therapy. Although Andersen et al.(1993) point out that the data actually consisted of 21 pairs matched by status of initial remission attained, we have chosen to disregard this aspect here in view of the present purposes of illustration and comparison.

In the analyses below, the prior distribution for β is taken to be normal with mean 1.5 and standard deviation 5.0. Runs with varying means in the

TABLE 1. Leukemia Remission Time Data

Group 0 (drug)	$6^*, 6, 6, 6, 7, 9^*, 10^*, 10, 11^*, 13, 16, 17^*, 19^*, 20^*, 22, 23, 25^*, 32^*, 32^*, 34^*, 35^*$ *Censored
Group 1 (placebo)	$1, 1, 2, 2, 3, 4, 4, 5, 5, 8, 8, 8, 11, 11, 12, 12, 15, 17, 22, 23$

TABLE 2. Posterior Summaries for β

	k				
Summary	.01	.10	1.0	10.0	100.0
Mean	1.69	1.71	1.71	1.67	1.62
Std. Deviation	.438	.421	.422	.419	.381
Cntr. 90% limits	(.99,2.43)	(1.06,2.44)	(1.02,2.42)	(.98,2.36)	(.97,2.25)

range -5.0 to 10.0 gave similar results not reported here. The prior for the baseline cumulative hazard function is a beta process with $A_0(t) = 0.05t$ and $c(t) = ke^{-0.05t}$. This simple choice corresponds to a prior mean survival function that is exponential with mean 20 while the prior variability can be controlled via the number k. Our illustrative calculations use a wide range for k from 0.01 to 100.0. The beta process, however, admits a much richer expression of prior opinion as designed in Hjort (1990) and demonstrated in Damien et al.(1996). One can specify different time-dependent degrees of prior variance in $A(t)$ since $c(t)$ can be interpreted as the number at risk at t in an imagined prior sample. Such flexibility is especially welcome in constructing prior specifications from past data sets and reported studies.

With these priors and the data in Table 1 the Gibbs sampler of the previous section was implemented and run with each setting of k, saving 2000 iterations after discarding the first 3000. The results for β are summarized in Table 2. Values of k near zero represent little prior input whereas large values correspond to a strong prior belief that the baseline hazard is that of an exponential distribution with mean 20. These posterior summaries are remarkably stable over a large range of k, a phenomenon observed also by Kalbfleisch (1978) regarding the posterior mode and an approximation to the posterior standard deviation when using a gamma process prior. In Figure 1(a) we show the posterior distributions of β.

For comparison with results based on maximizing the partial likelihood, calculations were carried out using four different methods including those advanced by Efron (1977) and Kalbfleisch and Prentice (1980) to accomodate tied observations. Details of these are given, for example, in Section 8.3 of Klein and Moeschberger (1996). Estimates of β varied from 1.51 with the method of Breslow (1974) to 1.63 by that of Cox (1972). The discrepancy is about 28% of the standard error. In contrast, the Bayesian analysis in this paper uses the full likelihood and calculates the appropriate posterior, with or without multiplicities. The Monte Carlo error in the posterior mean of β is less than 0.01. Figure 1(b) shows posterior distributions of the more easily interpreted hazard ratio of placebo to drug treatment. These

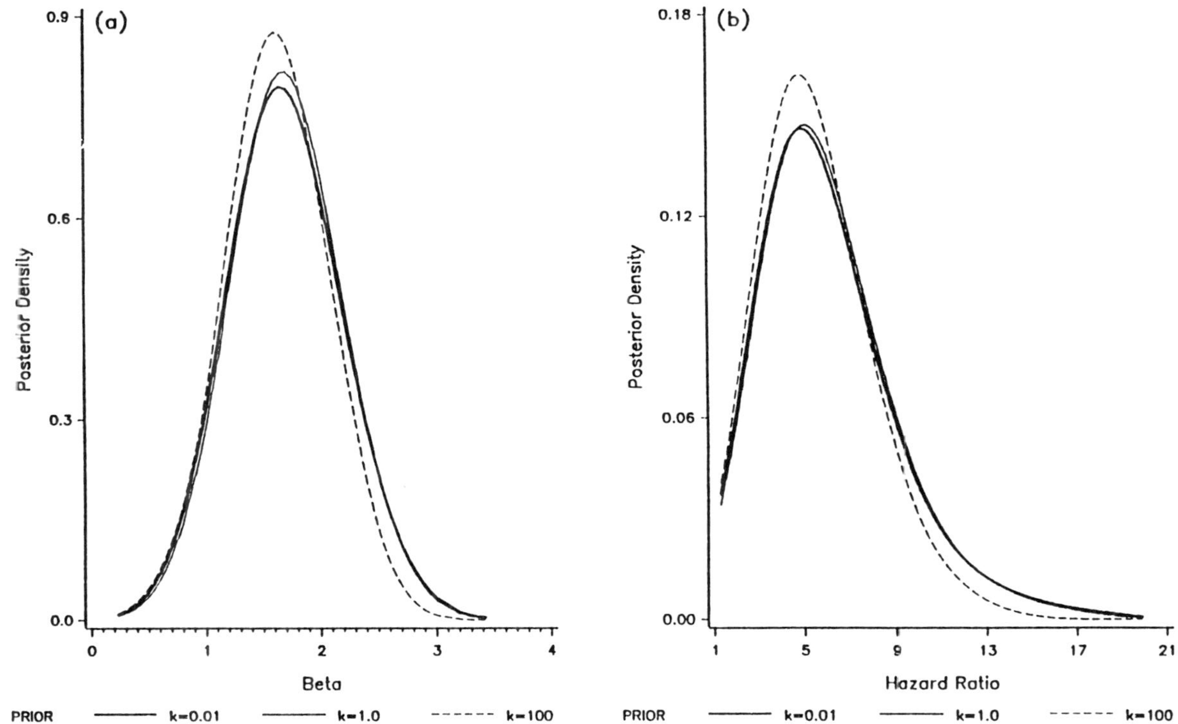

Figure 1. Posterior distributions of (a) the regression coefficient and (b) the hazard ratio.

are readily obtained by transformation of the posterior samples of β.

The posterior mean of the baseline survival function and pointwise 90% limits for it are plotted in Figure 2. As k increases, the shape approaches that of the exponential as chosen in the prior. Values of k upto 10 give remarkably stable posteriors except in the right tail where a different $c(\cdot)$ function could be chosen to express higher prior uncertainty. In general, such stable ranges of k will depend on the size of the data.

Figure 3 demonstrates the types of inference that can be easily obtained from the full posterior samples. In addition to the mean survival functions for the two groups, part (b) shows how all parameter uncertainty can be propagated to useful quantities such as the difference in the survival functions. A more detailed description at $t = 10$ weeks is shown in part (c). It clearly conveys the effectiveness of the drug in maintaining remission past 10 weeks.

The calculations for this analysis were performed on a Hewlett-Packard 9000 Series 700 workstation using code written in SAS-IML. Generation of 5000 posterior samples was accomplished in 28 minutes. Code in FORTRAN or C could be expected to speed this up considerably. Convergence diagnostics not reported here showed a burn-in of 1000 would suffice.

11.5 Conclusion

A semiparametric model using Cox regression and the beta process was described and exemplified. The complex form of the posterior distribution, following latent variable transformations, was shown to reduce to a computational model that could be sampled easily. The beta process is a rich family of prior distributions on the space of cumulative hazards with easy to interpret parameters. However, in some contexts, it may be necessary or easier to model the hazard rate nonparametrically. One class of prior distributions for hazard rates is called the extended gamma process (Dykstra and Laud, 1981). A semiparametric analysis similar to the one in this paper for increasing, decreasing and U-shaped hazard rates can be developed. In another direction, the methods of this paper can be adapted for use with various frailty models.

There are alternative approaches to modelling grouped survival data. The article by Sinha and Dey (1997) contains and excellent survey. It would be interesting to attempt a deatiled comparison of the alternative approaches in the presence of ties in the data. Yet another extension would be to incorporate time-dependent covariates into the model, and investigate the subsequent implications for assesing the proportional hazards assumtion.

Acknowledgments: We thank Nils Hjort for personal communication clarifying the distribution of jumps in the beta process and Craig Cumbus for

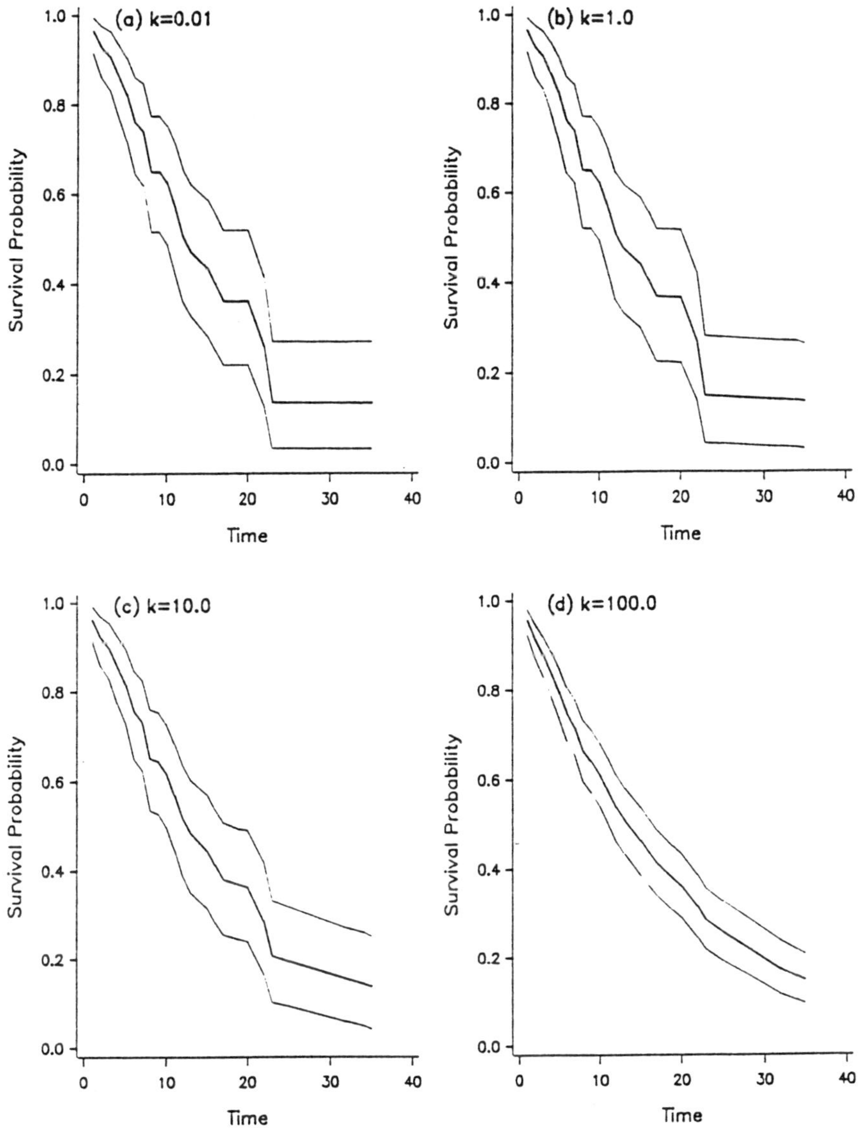

Figure 2. Posterior mean and central 90% limits for baseline survival function.

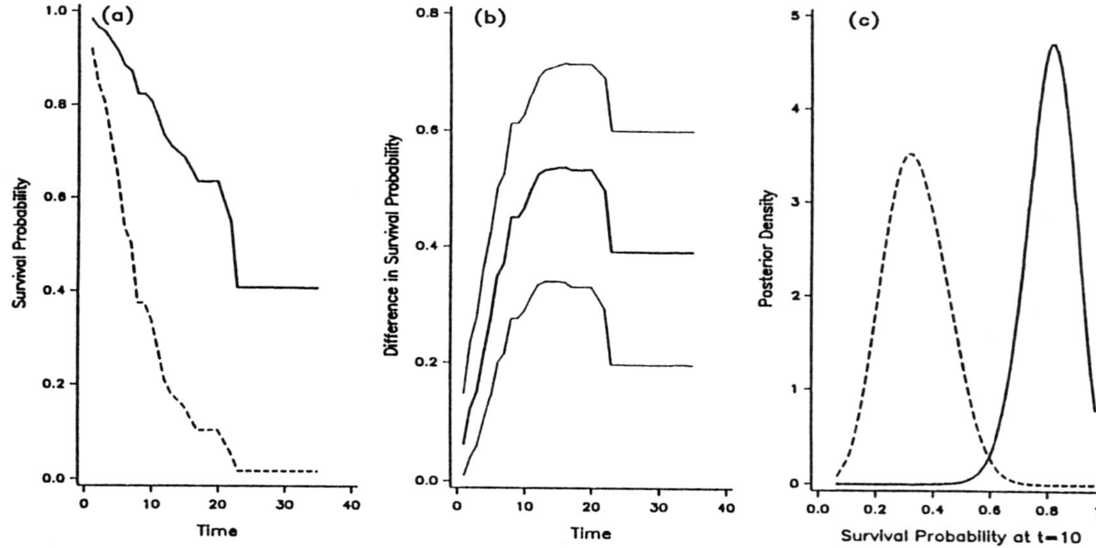

Figure 3. Posterior comparison of groups, drug ———, placebo - - - - - -, k=0.01
(a) mean survival functions
(b) mean difference in survival functions with 90% limits
(c) distributions of $P(X>10)$

SAS-IML code for the Gilks-Wild algorithm. Research reported here was partially funded by a grant from the National Science Foundation.

Appendix – Beta Process Summary

With Beta(α, β) denoting the beta distribution proportional to $x^{\alpha-1}(1-x)^{\beta-1}$ on $[0,1]$, we have,

Definition: Let A_0 be a cumulative hazard function with a finite number of jumps taking place at $t_1, t_2, \cdots, t_m$ and let $c(\cdot)$ be a piecewise continuous, nonnegative function on $[0, \infty)$. A process A is called a beta process with parameters $c(\cdot)$ and $A_0(\cdot)$, denoted

$$A \sim beta(c(\cdot), A_0(\cdot)),$$

if it has Lévy representation

$$E(e^{-\theta A(t)}) = \left[\prod_{j:t_j \leq t} E(e^{-\theta S_j})\right] exp\left\{-\int_0^1 (1-e^{-\theta s}) dL_t(s)\right\},$$

with

$$S_j = A\{t_j\} \sim Beta(c(t_j)A_0\{t_j\}, c(t_j)(1 - A_0\{t_j\}))$$

and

$$dL_t(s) = \int_0^t \left[c(z)s^{-1}(1-s)^{c(z)-1} dA_{0,c}(z)\right] ds$$

for $t \geq 0$ and $0 < s < 1$, where $A_{0,c}(t) = A_0(t) - \sum_{t_j \leq t} A_0\{t_j\}$.

A_0 is the prior guess at the cumulative hazard, and $c(t)$ can be interpreted as the number at risk at t in an imagined prior sample. The above definition implies that the beta process has independent increments and at fixed points of discontinuity the increments have beta distributions. More general choices of $dL_t(s)$ are possible with attendant results given in Hjort (1990) and simulation techniques in Damien et al.(1996).

Let $X_1, \cdots, X_n$ be independent and identically distributed with cumulative hazard function A, and $A \sim beta(c(\cdot), A_0(\cdot))$. Suppose $(T_1, \delta_1), \cdots, (T_n, \delta_n)$ is observed, where $T_i = min(X_i, c_i)$, $\delta_i = I\{X_i \leq c_i\}$ and $c_1, \cdots c_n$ are censoring times. Define the counting process N and the left continuous at-risk process Y as

$$N(t) = \sum_{i=1}^n I\{T_i \leq t \;\&\; \delta_i = 1\}, \quad Y(t) = \sum_{i=1}^n I\{T_i \geq t\},$$

where I is the indicator function. Then

$$A|(T_1,\delta_1),\cdots,(T_n,\delta_n) \sim beta\left\{c(\cdot)+Y(\cdot),\int_0^{(\cdot)} \frac{c(s)dA_0(s)+dN(s)}{c(s)+Y(s)}\right\}.$$

The posterior process contains fixed points of discontinuity even if the prior does not. These extra points occur at observations with $\delta_i = 1$. The distribution of any jump is given by

$$A\{t\}|data \sim Beta(c(t)A_0\{t\}+dN(t), c(t)(1-A_0\{t\})+Y(t)-dN(t)).$$

References

Andersen, P.K., Borgan, O., Gill, R.D. and Keiding, N. (1993). *Statistical Models Based on Counting Processes.* New York : Springer-Verlag.

Besag, J. and Green, P.J. (1993). Spatial Statistics and Bayesian Computation. *J. R. Statist. Soc.* **B 55**, 25-37.

Breslow, N.E. (1974). Covariate Analysis of Censored Survival Data. *Biometrics* **30**, 89-99.

Cox, D.R. (1972). Regression Models and Life Tables (with Discussion). *J. R. Statist. Soc.* **B 34**, 187-220.

Damien, P., Laud, P.W., and Smith, A.F.M., (1995). Approximate Random Variate Generation from Infinitely Divisible Distributions with Applications to Bayesian Inference. *J. R. Statist. Soc.* **B 57**, 547-563.

Damien, P., Laud, P.W., and Smith, A.F.M., (1996) Implementation of Bayesian Nonparametric Inference Based on Beta Processes. *Scand. J. Statist.* **23**, 27-36.

Dykstra, R. L. and Laud, P. W. (1981). Bayesian Nonparametric Approach Toward Reliability, *Ann. Statist.* **9**, 356-367.

Efron, B. (1977). The Efficiency of Cox's Likelihood Function for Censored Data. *J. Am. Statist. Assoc.* **72**, 557-565.

Ferguson, T. S. (1973). A Bayesian Analysis of Some Nonparametric Problems, *Ann. Statist.* **1**, 209-230.

Gilks, W.R. and Wild, P. (1992). Adaptive Rejection Sampling for Gibbs Sampling. *Appl. Statist.* **41**, 337-348.

Gill, R.D. and Johansen, S. (1990). A Survey of Product-Integration with a View toward Application in Survival Analysis. *Ann. Statist.* **18**, 1501-1555.

Hjort, N.L. (1990). Nonparametric Bayes Estimators Based on Beta Processes in Models for Life History Data. *Ann. Statist.* **18**, pp 1259-1294.

Kalbfleisch, J.D. (1978). Non-parametric Bayesian Analysis of Survival Time Data. *J. R. Statist. Soc.* **B 40**, 2, 214-221.

Kalbfleisch, J.D. and Prentice, R.L. (1980). *The Statistical Analysis of Failure Time Data.* New York : John Wiley.

Klein, J.P. and Moeschberger, M.L. (1996). *Survival Analysis : Methods for Censored and Truncated Data.* New York : Springer-Verlag.

Sinha, D. and Dey, D.K. (1997). *J. Am. Statist. Assoc.* **92**, 1195-1212.

Smith, A.F.M. and Roberts, G.O. (1993). Bayesian Computations via the Gibbs Sampler and Related Markov Chain Monte Carlo Methods. *J. R. Statist. Soc.* **B 55**, 3-23.

Walker, S.G. (1996). Random Variate Generation from an Infinitely Divisible Distribution via Gibbs Sampling. *Technical Report*, Imperial College, London.

12

Simulation of Lévy Random Fields

Robert L. Wolpert
Katja Ickstadt

ABSTRACT An efficient recently developed method, the Inverse Lévy Measure (ILM) algorithm, is presented for drawing random samples from gamma, skewed stable and other nonnegative independent-increment random fields, which we call Lévy random fields. The method is useful for computing posterior distributions in nonparametric hierarchical Bayesian statistical analysis. Algorithms are illustrated through prototype implementations in S-PLUS.

12.1 Introduction and Overview

Following the emergence of Gibbs sampling and more generally Markov chain Monte Carlo methods (Gelfand and Smith, 1990; Tierney, 1994; Gilks et al., 1996), Bayesian inference is now commonly based on simulations from the posterior distributions of model parameters. In nonparametric Bayesian analysis the "parameter" is typically a function or distribution so the simulation-based approach entails drawing random samples of functions (stochastic processes) or measures (random fields). Stationary independent-increment (i.e. homogeneous) random fields can serve as building blocks for Bayesian nonparametric priors: the Dirichlet process is simply the gamma random field normalized to have unit mass, for example, while the distributions called "neutral to the right" (Ferguson, 1973, 1974; Ferguson and Phadia, 1979) are simply those whose hazard functions have independent increments.

The extreme ease with which such samples may be drawn from the Dirichlet process distribution (Antoniak, 1974; Ferguson, 1973) is largely responsible for its frequent use in early Bayesian nonparametric analysis, despite its limitations (negative correlation for probabilities assigned to disjoint sets, absence of ways to express spatial correlation or varying precision, etc.).

Here we present a new and very efficient scheme, the Inverse Lévy Measure (ILM) algorithm first introduced in Wolpert and Ickstadt (1998), based on an idea of (Lévy, 1937), for sampling from a wide class of random measures assigning positive, independent, infinitely divisible random variables to disjoint sets. The method is useful in Bayesian hierarchical models, is sufficiently flexible to allow expression of a wide range of possible prior beliefs, and is nearly as simple to implement as are the Dirichlet process methods—our prototype implementations require only five lines of S-PLUS code (MathSoft, 1997). The new method's flexibility allows a modeler to construct and make inference from hierarchical Bayesian nonparametric models without making unnecessary assumptions or approximations.

The underlying idea is presented in section 2. In section 3 it is illustrated for gamma random variables, processes, and fields; in section 4 it is presented in full generality for possibly inhomogeneous Lévy processes and fields. The method is compared with others in section 5 and discussed in section 6. Two appendices includes some S-PLUS display functions and the proof of Theorem 2.

12.2 Increasing Independent-Increment Processes: A New Look at an Old Idea

12.2.1 Infinitely-Divisible Random Variables

The celebrated Lévy-Khinchine formula (Jacod and Shiryaev, 1987, p. 75) asserts that every infinitely-divisible distribution $\mu(dx)$ has a characteristic function $\hat{\mu}(t) \equiv \int_{\mathbb{R}} e^{itx} \mu(dx)$ of the form

$$\hat{\mu}(t) = \exp\left\{ ita - t^2\sigma^2/2 + \int_{\mathbb{R}} \left[e^{itu} - 1 - it\,h(u)\right] \nu(du) \right\}$$

for some numbers $a \in \mathbb{R}$, $\sigma^2 \in \mathbb{R}_+ \equiv [0, \infty)$, and some positive measure $\nu(du)$ (called the "Lévy measure") on the real line satisfying the integrability condition $\int_{\mathbb{R}} (1 \wedge u^2)\nu(du) < \infty$, where the notation $1 \wedge u^2$ denotes the minimum of 1 and u^2 and $h(u)$ denotes an arbitrary bounded continuous function satisfying $h(u) = u + \mathcal{O}(u^2)$ near $u = 0$. For distributions of positive random variables, necessarily $a \geq 0$, $\sigma^2 = 0$, $\nu(\mathbb{R}_-) = 0$ and $\int_{\mathbb{R}_+} (1 \wedge u)\nu(du) < \infty$, so we may take $h(u) \equiv 0$ (absorbing its effect into a) and simplify this to the form

$$\hat{\mu}(t) = \exp\left[ita + \int_0^\infty (e^{itu} - 1)\nu(du) \right].$$

For example, the log characteristic functions of the Poisson (with mean λ), gamma (with shape α and scale β) and fully skewed stable (with exponent ξ) distributions are given by

$$\begin{aligned}
\log \hat{\mu}_P(t) &= (e^{it} - 1)\lambda &&= \int (e^{itu} - 1)\, \lambda\delta_1(du) \\
\log \hat{\mu}_\Gamma(t) &= -\log(1 - i\beta t)\alpha &&= \int (e^{itu} - 1)\, u^{-1} e^{-u/\beta} \alpha du \\
\log \hat{\mu}_S(t) &= -\alpha|t|^\xi \Gamma(1 - \xi) e^{-it\xi\pi/2|t|} &&= \int (e^{itu} - 1)\, u^{-1-\xi} \xi\alpha du
\end{aligned}$$

with Lévy measures on $\mathbb{R}_+$ given by $\nu_P(du) = \lambda\delta_1(du)$ (where $\delta_x(du)$ denotes the unit point mass at locus $u = x$), $\nu_\Gamma(du) = u^{-1}e^{-u/\beta}\alpha du$ and $\nu_S(du) = u^{-1-\xi}\xi\alpha du$, respectively.

12.2.2 Increasing Independent-Increment Processes

Each nonnegative infinitely-divisible distribution may be thought of as the distribution at time $t = 1$ of an increasing stochastic process X_t with stationary independent increments, or a *subordinator* (Rogers and Williams, 1994, p. 78); in this context the Lévy measure of a Borel set $E \subset \mathbb{R}_+$ may be interpreted as the Poisson arrival rate $\nu(E)$ of jumps for the process of sizes $[X_t - X_{t-}] \in E$. For example, the Poisson process has jumps only of size $u = 1$, arriving at constant rate λ, while for each $\epsilon > 0$ the gamma process has jumps of size $u \geq \epsilon$ arriving as a Poisson process with rate $\nu_\Gamma\big([\epsilon,\infty)\big) = \int_\epsilon^\infty u^{-1}e^{-u/\beta}\alpha\, du$ and the one-sided stable process has jumps of size $u \geq \epsilon$ arriving as a Poisson process with rate $\nu_S\big([\epsilon,\infty)\big) = \alpha\epsilon^{-\xi}$. Other independent-increment processes have similar representations: the beta process (Hjort, 1990), with $\nu_\beta(du) = u^{-1}(1-u)^{\beta-1}\beta\alpha\, du$ on $[0,1]$, for example, or the simple homogeneous process (Ferguson and Phadia, 1979), with $\nu(du) = e^{-\beta u}(1-e^{-u})^{-1}\alpha\, du$ on $\mathbb{R}_+$.

12.2.3 The Key Idea

Implicit in the Lévy-Khinchine formula (and in Lévy (1937, ch. VII)) is the representation of a subordinator X_t as a stochastic integral $X_t = \iint_{\mathbb{R}_+\times[0,t]} uH(du\, ds)$ of a Poisson measure on $\mathbb{R}_+^2$ with mean measure $\mathrm{E}[H(du\, ds)] = \nu(du)\, ds$. This perspective suggests an exact method for simulating a subordinator X_t on the time interval $0 \leq s \leq T$: generate independent jump times σ_m from the uniform distribution on $[0,T]$; generate successive jumps τ_m of a standard unit-rate Poisson process (perhaps as partial sums of independent standard exponential variates) and set $v_m \equiv \inf\{u \geq 0 : T\,\nu\big([u,\infty)\big) \leq \tau_m\}$; then set $X_t \equiv \sum[v_m : \sigma_m \leq t]$. The sum will have finitely many terms if $\nu(\mathbb{R}_+) < \infty$, but will have infinitely many if $\nu(\mathbb{R}_+) = \infty$, as in the stable and gamma examples above. If we set $L(u) \equiv \nu\big([u,\infty)\big)$ we can write $v_m = L^{-1}(\tau_m/T)$, justifying the name Inverse Lévy Measure (ILM) for this algorithm Wolpert and Ickstadt (1998).

12.3 Example: Gamma Variates, Processes and Fields

For the gamma process the jump sizes $v_m \equiv \inf\{u \geq 0 : T\nu\big([u,\infty)\big) \leq \tau_m\}$ are found by solving for $u = x$ the equation $\tau/T = \nu\big([x,\infty)\big) =$

$\int_x^\infty u^{-1} e^{-u/\beta} \alpha\, du = \alpha E_1(x/\beta)$, where $E_1(x) \equiv \int_x^\infty e^{-u}\, u^{-1}\, du$ is the exponential integral function (Abramowitz and Stegun, 1964, *p.* 228). This function, included in many special function subroutine libraries (IMSL, NSWC, etc.) and mathematical environments (Mathematica, MatLab, etc.), is the limit as $a \to 0$ of the complementary incomplete gamma function

$$E_1(x) = \lim_{a \to 0} \left[\Gamma(a, x) \equiv \int_x^\infty e^{-t}\, t^{a-1}\, dt \right]$$

(Abramowitz and Stegun, 1964, *p.* 260) and so can also be computed as the limit of rescaled χ^2-tail probabilities as the degrees of freedom $d = 2a$ tend to zero,

$$E_1(x) = \lim_{d \to 0} \Gamma(d/2) \Pr[\chi_d^2 > 2x] = \lim_{d \to 0} (2/d) \Pr[\chi_d^2 > 2x].$$

Thus $E_1(x)$ and its inverse $E_1{}^{-1}(y)$ can be approximated in S-PLUS by the functions

```
"e1"    <- function(x, d = 1e-9) {(2/d) * (1-pchisq(2*x, d))}
"e1inv"<- function(y, d = 1e-9) {qchisq(1-(d/2)*y, d)/2}
```

12.3.1 Gamma Random Variates

The first M jumps of a standard Poisson process can also be simulated easily in S-PLUS, as the partial sum `cumsum(rexp(M))`, so a gamma random variate with shape and scale parameters a and b can be constructed approximately using the ILM algorithm with the function

```
"gv"  <- function(a, b, M = 1000) {
      return(sum(b * e1inv(cumsum(rexp(M)/a)))); }
```

Of course gamma random variates may be generated directly in S-PLUS with the command `b*rgamma(1,a)`; the strength of the present method lies in its extensibility to arbitrary infinitely-divisible distributions and to independent increment random processes and fields.

12.3.2 Gamma Random Processes

The ILM method may also be used for generating a gamma process (Ferguson and Klass, 1972); now we must additionally generate the jump times uniformly on the interval $[0, T]$, and sort the jumps in increasing temporal order. Figure 1, generated by the command `disp.1d(gp(10,2,M=250), "Gamma", c("a","b"), c(10,2));`, illustrates this with plots of both the gamma process and its jumps, using the display function `disp.1d()` given in Appendix 1 and the following function implementing the ILM algorithm for a gamma process:

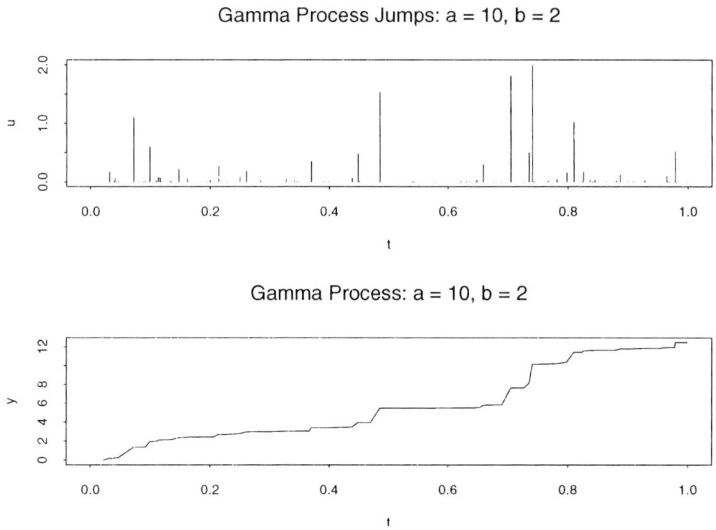

FIGURE 1. Gamma Process realization with $\alpha = 10$, $\beta = 2$.

```
"gp"<- function(a, b, tlim=1, M = 1000) {
    x <- runif(M)*tlim;                         # times x
    u <- b * e1inv(cumsum(rexp(M)/(tlim*a))); # jumps u
    ox<- order(x);  x <- x[ox];  u <- u[ox];  # sort x,u on x
    invisible(list(M=M,x=x,u=u,y=cumsum(u),t=tlim)); }
```

12.3.3 Gamma Random Fields

A gamma random field is a continuous (in probability) linear assignment of random variables $X[\phi]$ to bounded compactly-supported Borel functions $\phi(\cdot)$ on some set $\mathcal{S}$ in Euclidean space such that indicator functions $\phi = 1_A$ will be assigned gamma-distributed random variables $X[1_A] \sim \text{Ga}(\alpha|A|, \beta)$ with shape parameter proportional to volume. With the ILM approach we can generate gamma random fields in any number of dimensions. Figure 2, produced with the command `disp.2d(grf(a=10,b=2), "Gamma", c("a","b"), c(10,2));`, shows representations of the gamma random field and its jumps on the unit square $\mathcal{S}$ in two dimensions:

```
"grf"<- function(a, b, M=10000, g=51) {
  x  <- runif(M);  y <- runif(M);        # jump locations
  u  <- b * e1inv(cumsum(rexp(M)/a));    # jump sizes
  xx <- seq(0,1,length=g);  yy <- seq(0,1,length=g);
  zz <- matrix(0,nrow=g,ncol=g);         # g*g grid
  for(i in 2:g) {
```

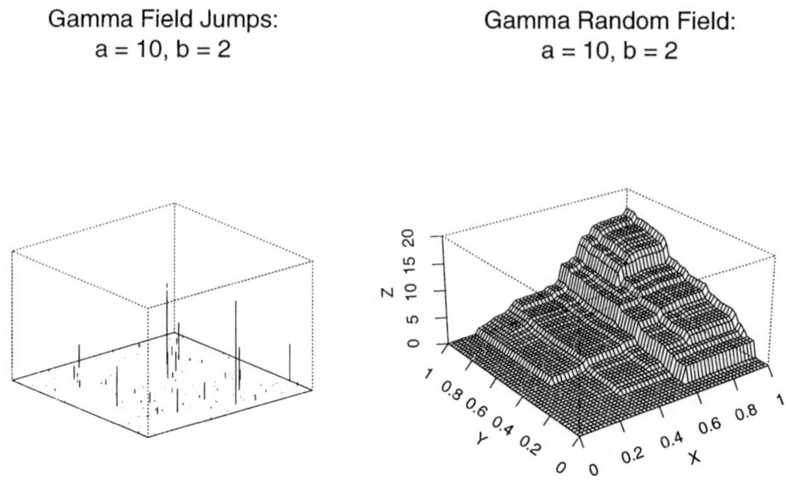

FIGURE 2. Gamma Random Field realization with $\alpha = 10$, $\beta = 2$.

```
    xok <- x < xx[i];
    for(j in 2:g) {
      yok <- y < yy[j];
      zz[i,j] <- sum(u[xok&yok]);           # sum of jumps SW
    }                                       # of (x[i],y[j])
  }
  invisible(list(x=x,y=y,u=u, xx=xx,yy=yy,zz=zz)); }
```

The ILM simulation method is applicable to more general gamma random fields: the shape measure may be any measure $\alpha(ds)$, not necessarily a constant times Lebesgue measure; the scale parameter $\beta(s)$ need not be constant; and the jump locations need not be drawn from a uniform distribution, but may be drawn from any convenient sampling distribution $\Pi(ds)$; see Wolpert and Ickstadt (1998) for the proof of:

Theorem 1 *Let $\alpha(s) \geq 0$ and $\beta(s) > 0$ be measurable functions on a space $\mathcal{S}$. Let $\{\sigma_m\}$ be independent identically distributed draws from any probability distribution $\Pi(ds)$ on $\mathcal{S}$, and let $\tau_m \geq 0$ be the successive jump times of a standard Poisson process. Set $\tau(u,s) \equiv E_1\big(u/\beta(s)\big)\alpha(s)$ and $\upsilon_m \equiv \inf[u \geq 0 : \tau(u,\sigma_m) \leq \tau_m]$, that is,*

$$\upsilon_m \equiv {E_1}^{-1}\big(\tau_m/\alpha(\sigma_m)\big)\,\beta(\sigma_m),$$

or $\upsilon_m \equiv 0$ if $\alpha(\sigma_m) = 0$. Then the random field $\Gamma[\phi] \equiv \sum_{m<\infty} \upsilon_m \phi(\sigma_m)$ has the gamma process distribution $\Gamma(ds) \sim \mathrm{Ga}\big(\alpha(ds), \beta(s)\big)$ for the measure $\alpha(ds) \equiv \alpha(s)\Pi(ds)$.

12.4 Inhomogeneous Lévy Random Fields

Independent increment processes and fields that are *not* homogeneous nevertheless have a representation similar to that of Lévy and Khinchine. From (Jacod and Shiryaev, 1987, Chapter II, §4c) it follows that a nonnegative independent-increment random field must have a characteristic functional of the form

$$\hat{\mu}[\phi] = \mathrm{E}\left\{e^{iX[\phi]}\right\} = \exp\left\{i\eta[\phi] + \iint_{\mathbb{R}_+\times\mathcal{S}} [e^{iu\phi(s)} - 1]\,\nu(du\,ds)\right\}, \quad (1)$$

where now $\eta(ds)$ is a sigma-finite positive measure on $\mathcal{S}$, $\eta[\phi] \equiv \int_{\mathcal{S}} \phi(s)\,\eta(ds)$, and the inhomogeneous Lévy measure $\nu(du\,ds)$ on $\mathbb{R}_+ \times \mathcal{S}$ satisfies the integrability condition $\iint_{\mathbb{R}_+\times K}(1 \wedge u)\nu(du\,ds) < \infty$ for each compact $K \subset \mathcal{S}$. The inhomogeneous gamma random field of Theorem 1 is of this form with $\eta(ds) = 0$ and $\nu(du\,ds) = \exp\big(-u/\beta(s)\big)\,u^{-1}\,\alpha(s)\,du\,\Pi(ds)$, and inhomogeneous versions of the Poisson, stable, and other processes may be generated in the same way. The heart of the idea is the same as before: represent the Lévy field in the form $X[\phi] = \eta[\phi] + \iint_{\mathbb{R}_+\times\mathcal{S}} \phi(s)uH(du\,ds)$, where now the Poisson measure $H(du\,ds)$ on $\mathbb{R}_+ \times \mathcal{S}$ has mean $\mathrm{E}[H(du\,ds)] = \nu(du\,ds)$, leading to the following algorithm (in which we use the same letter to denote both the Lévy measure $\nu(du\,ds)$ and its density $\nu(u,s)$ with respect to the measure $du\,\Pi(ds)$):
Algorithm ILM:

- Choose any probability measure $\Pi(ds)$ on a Borel set $\mathcal{S}$ from which samples may be drawn, any sigma-finite drift measure $\eta(ds)$ on $\mathcal{S}$ and any nonnegative Borel function $\nu(u,s)$ on $\mathbb{R}_+ \times \mathcal{S}$ that satisfies the condition $\iint_{\mathbb{R}_+\times K}(1 \wedge u)\,\nu(u,s)\,du\,\Pi(ds) < \infty$ for compact $K \subset \mathcal{S}$.
- Set $\nu(du\,ds) \equiv \nu(u,s)\,du\,\Pi(ds)$ and $\tau(u,s) \equiv \int_u^\infty \nu(x,s)\,dx$.
- Generate independent draws σ_m from $\Pi(ds)$ and construct the event times of a standard Poisson process by generating independent unit-mean exponential random variables T_i and setting $\tau_m \equiv \sum_{i\le m} T_i$.
- Set $v_m \equiv \inf[u \ge 0 : \tau(u,\sigma_m) \le \tau_m]$ (typically by setting v_m to the inverse Lévy measure $\tau(\cdot,\sigma_m)^{-1}(\tau_m)$).
- Set $X[\phi] = \eta[\phi] + \sum_m v_m\phi(\sigma_m)$ for bounded compactly-supported Borel $\phi(s)$ on $\mathcal{S}$.

Of course in practice only a finite number M of jumps (v_m, σ_m) can be sampled, but the convergence in distribution of the truncated version $X_M[\phi]$ as $M \to \infty$ is an immediate consequence of the following theorem, whose proof appears in Appendix 2; truncation error bounds and suggestions for choosing M appear in Wolpert and Ickstadt (1998).

Theorem 2 *Let $\Pi(ds)$ be any Borel probability measure on $\mathcal{S}$ and $\nu(u,s)$ a nonnegative Borel function on $\mathbb{R}_+ \times \mathcal{S}$ for which the measure $\nu(du\,ds) = \nu(u,s)\,du\,\Pi(ds)$ satisfies the condition $\iint_{\mathbb{R}_+ \times K}(1 \wedge u)\nu(du\,ds) < \infty$ for each compact $K \subset \mathcal{S}$, and let $\eta(ds)$ be a sigma-finite measure on $\mathcal{S}$. Let $\{\sigma_m\}$ be independent identically distributed draws from $\Pi(ds)$ and let $\{\tau_m\}$ be the successive jump times of a standard Poisson process. Set $\tau(u,s) \equiv \int_u^\infty \nu(x,s)\,dx$ and, for $m \in \mathbb{N}_+$, set $\upsilon_m \equiv \inf[u \geq 0 : \tau(u,\sigma_m) \leq \tau_m]$. Then the random field defined by $X[\phi] \equiv \eta[\phi] + \sum_{m<\infty} \upsilon_m \phi(\sigma_m)$ for compactly-supported bounded measurable $\phi(s)$ has characteristic functional (1).*

Note that there is no condition prohibiting fixed points of discontinuity—if $\Pi(ds)$ assigns positive mass to some point $s_0 \in \mathcal{S}$, as typically happens for posterior distributions in Bayesian analysis, then $X[\cdot]$ will be discontinuous at s_0. It is usually more efficient numerically to sample the discontinuity points separately, however, and use the ILM algorithm only for the continuous portion of the random process or field.

12.4.1 Examples: Stable, Beta, Simple Homogeneous Fields

Lévy processes are of emerging importance in Bayesian hierarchical nonparametric and semiparametric modeling (Sinha and Dey, 1997). The one-sided stable processes have been used as time series innovations (Qiou and Ravishanker, 1997) and as survival analysis frailties (Qiou et al., 1997), the beta for survival analysis (Hjort, 1990), and the gamma process for point process intensities Wolpert and Ickstadt (1998), for example.

The ILM algorithm can be used to construct inhomogeneous analogues of all these Lévy processes: the one-sided stable distribution, with inhomogeneous Lévy measure $\nu(du\,ds) = u^{-1-\xi(s)}\xi(s)\,du\,\alpha(s)\,\Pi(ds)$ whose exponent $0 < \xi(s) < 1$ and intensity $\alpha(s)$ may vary with $s \in \mathcal{S}$, for example, or of the beta process with Lévy measure $\nu_\beta(du\,ds) = u^{-1}\beta(s)(1-u)^{\beta(s)-1}\,du\,\alpha(s)\,\Pi(ds)$ whose at-risk parameter $\beta(s)$ and mean $\alpha(s)\,\Pi(ds)$ need not be uniform. Evidently $\tau(u,s) = u^{-\xi(s)}\alpha(s)$ for the stable process and so the ILM jumps are available in closed form as $\upsilon_m \equiv \inf[u \geq 0 : \tau(u,\sigma_m) \leq \tau_m] = \left(\alpha(\sigma_m)/\tau_m\right)^{1/\xi(\sigma_m)}$, while the jumps for the beta process can be computed from the incomplete beta function (Abramowitz and Stegun, 1964, p. 263) using $\tau(u,s) = \alpha(s)\beta(s)B_{1-x}(\beta,0)$.

The ILM approach is illustrated below for a stable process with constant exponent ξ and intensity α and for a beta process with constant at-risk parameter β and mean α with the prototype S-PLUS functions `sp()` and `bp()`. Figures 3 and 4 give the outputs, generated using the commands `disp.1d(sp(a=3,xi=0.9), "Stable", c("a","xi"), c(3,0.9));` and `disp.1d(bp(a=3, b=0.9),"Beta", c("a","b"), c(3,0.9));`.

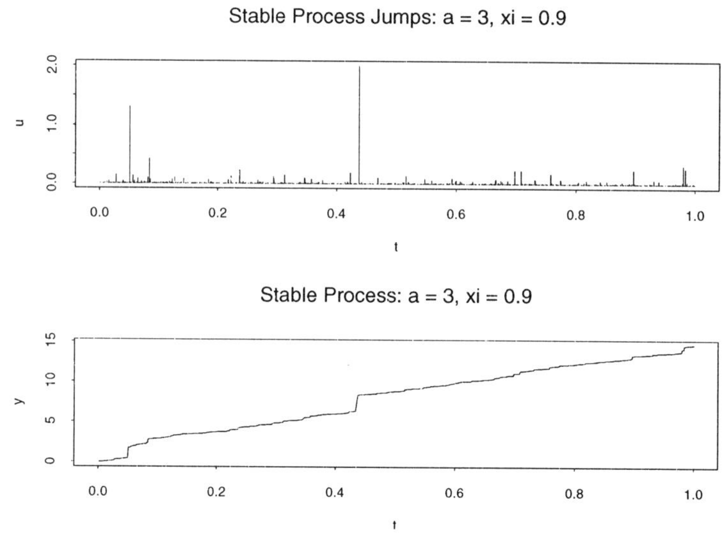

FIGURE 3. Stable Process realization with $\alpha = 3$, $\xi = 0.9$.

```
"sp"<- function(a, xi, tlim=1, M = 1000) {
    x <- runif(M)*tlim;                            # times x
    u <- (tlim*a/cumsum(rexp(M)))^(1/xi);          # jumps u
    ox<- order(x);  x <- x[ox];  u <- u[ox];  # sort x,u on x
    invisible(list(M=M,x=x,u=u,y=cumsum(u),t=tlim)); }


"bp"<- function(a, b, tlim=1, M = 1000, eps=1e-9) {
    x <- runif(M)*tlim;                            # times x
    u <- 1-qbeta(eps*cumsum(rexp(M))/(a*b),b,eps);
    ox<- order(x);  x <- x[ox];  u <- u[ox];  # sort x,u on x
    invisible(list(M=M,x=x,u=u,y=cumsum(u),t=tlim)); }
```

The simple homogeneous processes (Ferguson and Phadia, 1979) has a similar construction:

```
"sh"<- function(a, b, tlim=1, M = 1000, eps=1e-9) {
    x <- runif(M)*tlim;                            # times x
    u <- -log(qbeta(eps*cumsum(rexp(M))/a,b,eps));
    ox<- order(x);  x <- x[ox];  u <- u[ox];  # sort x,u on x
    invisible(list(M=M,x=x,u=u,y=cumsum(u),t=tlim)); }
```

In all cases the random field versions offer no additional difficulties, in any number of dimensions.

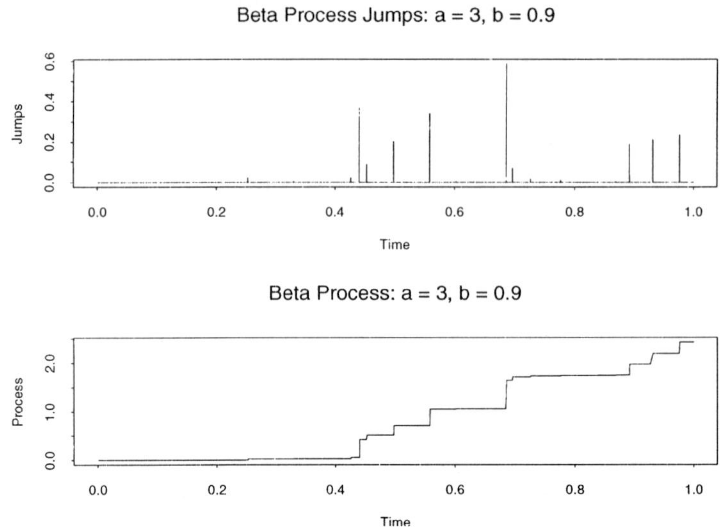

FIGURE 4. Beta Process realization with $\alpha = 3$, $\beta = 0.9$.

12.5 Comparisons with Other Methods

12.5.1 Method of Bondesson

For the gamma process in one dimension with $\alpha(\mathcal{S}) < \infty$ and constant $\beta(s) \equiv \beta$ the choice $\Pi(ds) = \alpha(ds)/\alpha(\mathcal{S})$ makes $\alpha(s) = \alpha(\mathcal{S})$ constant too, and the ILM algorithm will generate the jumps v_m in strictly decreasing order starting at the largest mass point of $\Gamma(ds)$. In this special case, in one dimension, $\mathcal{S} = [0, 1] \subset \mathbb{R}^1$, with Lebesgue measure for $\alpha(ds)$ and $\Pi(ds)$, the ILM method reduces to that of Bondesson (1982).

12.5.2 Method of Damien, Laud, Smith

A related method for generating Lévy processes was described in (Damien et al., 1995). These authors generate a finite number n of jump levels u_i from some distribution $\gamma(du)$ (they use $\gamma(du) \propto \frac{u}{1+u}\nu(du)$), take n Poisson processes P_i with rates $\lambda_i = \frac{1}{n}\frac{d\nu}{d\gamma}(u_i)$, and approximate the Lévy process X_t with Lévy measure $\nu(du)$ by $X_t \approx \Sigma_i u_i P_i(t)$. In our context this amounts to replacing the exact Poisson measure $H(du\,ds)$ on $\mathbb{R}^2_+$ with mean $\mathrm{E}[H(du\,ds)] = \nu(du\,ds)$ by an approximate one concentrated on the n lines $[(u, s) : u \in \{u_i\}]$, with mean $\mathrm{E}[H(du\,ds)] = \frac{1}{n}\sum \delta_{u_i}(du)\lambda_i\,ds$. For sufficiently large n (or small T) the restriction to a finite set of jump levels is scarcely noticeable; figure 5 illustrates this aspect of the method by requesting $n = 100$ levels drawn from an exponential distribution $\gamma(du)$ and drawing dotted horizontal lines at the 13 distinct levels that actually appear

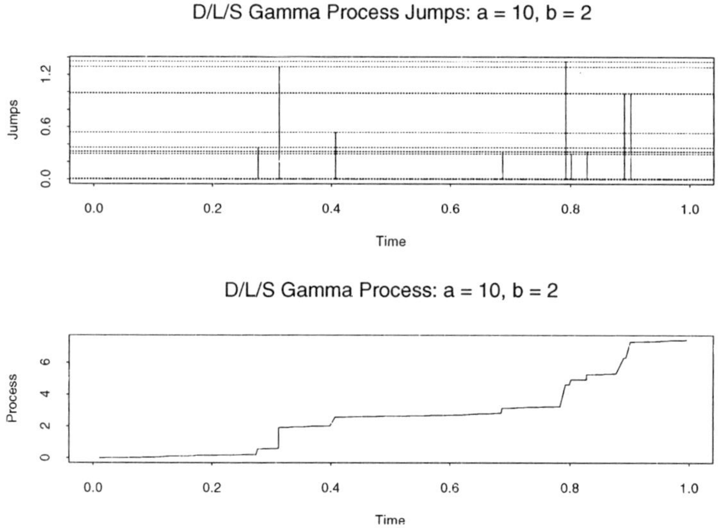

FIGURE 5. Method of Damien, Laud, and Smith.

in the realization (87 of the $n = 100$ Poisson processes had no jump during the time interval $[0, 1]$). This realization generated approximately the same number of mass points (245) as used by our ILM method illustrated in Figure 1 (250). Figure 5 was generated in S-PLUS with the commands

```
gam <- dls(a=10,b=2,n=100);
disp.1d(gam,"D/L/S Gamma",c("a","b"), c(10,2),h=gam$u);
```

using the function `dls()` given by

```
"dls"<- function(a, b, tlim=1, n = 100, aa=1, bb=b) {
    uvec <- rgamma(n,aa)*bb;
    lam  <- a*bb/(n*uvec*dgamma(uvec,aa)*exp(b*uvec));
    M <- sum(mvec <- rpois(n,lam*tlim));
    x <- runif(M)*tlim;                         # times x
    u <- rep(uvec,mvec);                        # jumps u
    ox<- order(x); x <- x[ox]; u <- u[ox]; # sort on x
    invisible(list(n=n,x=x,u=u,y=cumsum(u),M=M,t=tlim)); }
```

The ILM method is more flexible in that it allows inhomogeneous processes (in the gamma example, both $\alpha(s)$ and $\beta(s)$ may be nonconstant), applies in any number of dimensions, does not require sampling from an inconvenient distribution (but when sampling from a distribution $\Pi(ds)$ that makes $\nu(u, s)$ constant in s will offer high efficiency by generating the biggest jumps v_m first) and will generate samples from the exact target distribution without any approximation error other than the unavoidable truncation error that arises from taking only finitely many jumps v_m.

Elsewhere in this volume Walker and Damien (1998) present an improvement on the method of (Damien et al., 1995) which appears to be mathematically equivalent to a special case (one dimensional $\mathcal{S}$ and a specific choice of $\Pi(ds)$) of the earlier ILM method. There the method is described differently and with a different choice of truncation technique (they draw a Poisson-distributed number of points (σ_m, v_m) from a distribution proportional to $\nu(du\,ds)$ on $[\epsilon, \infty) \times \mathcal{S}$, while we give a specific method of drawing a fixed number M of points from the same distribution), but they are otherwise identical.

12.5.3 Dirichlet Processes

Dirichlet processes and their mixtures frequently are chosen to represent nonparametric prior information in Bayesian analysis due to the extreme simplicity with which they can be simulated and manipulated (MacEachern, 1998). The present method offers an alternative that is almost as simple to implement: generate a gamma random process $\Gamma(s)$ with shape measure $\alpha(ds)$ and constant scale $\beta(s) \equiv 1$ and set $\mathcal{D}(s) = \Gamma(s)/\Gamma(1)$ on the unit interval $s \in \mathcal{S} = [0, 1]$. The appeal of the ILM approach is the ease with which the Dirichlet process limitations may be overcome: prior opinion can be more informed at some locations $s \in \mathcal{S}$ than at others, by allowing $\beta(s)$ to be nonconstant; spatial and spatio-temporal problems can be handled without any additional difficulty, using random fields $\mathcal{D}[\phi] \equiv \Gamma[\phi]/\Gamma[1]$ instead of processes; more robustness against prior misspecification can be obtained, by using a flatter-tailed distribution such as a stable process rather than the gamma process; mixtures are still available, just as for the Dirichlet process (a gamma random field mixture example appears in Ickstadt et al. (1998)).

12.6 Conclusions

The ILM algorithm enables the efficient generation of both homogeneous and inhomogeneous random processes and fields, in any number of dimensions, without any approximation other than truncation. It only requires sampling from convenient distributions (the exponential and the investigator's choice of an arbitrary distribution $\Pi(ds)$ on $\mathcal{S}$). It is nearly as simple to implement as Dirichlet process methods, but offers far more flexibility to enable an investigator to express a wide range of possible prior beliefs in simulation-based approaches to Bayesian nonparametric statistical analysis, especially in spatial and spatio-temporal applications where random fields are required.

The ILM method has been applied to spatial hierarchical mixture models in which observed point processes are treated as Poisson random fields whose latent intensity is given by a kernel mixture of Lévy random fields in a variety of applications areas: bioabundance (Ickstadt and Wolpert, 1997), transportation engineering (Ickstadt et al., 1998), and, in ongoing work, to problems in epidemiology and ecology where the use of inhomogeneous fields is critical to allow prior specifications that reflect spatial demographic and environmental collateral information.

Appendix 1: Display Functions

The algorithms described in this paper are straightforward to implement in any numerical programming language, and for production work should be implemented in a fast compiled language like C, FORTRAN, or JAVA. We include prototype implementations in the widely available interpreted language S-PLUS (MathSoft, 1997) in order to illustrate them and to encourage readers to experiment with them. The short functions needed to generate the processes and fields are included with the text discussion; here are the one- and two-dimensional display routines `disp.1d()` and `disp.2d()` used to generate the figures:

```
"disp.1d" <- function(p=list(), name="", pn=c("",""),
             pv=c(NA,NA), h=NULL) {
    oldpar <- par(mfrow=c(2,1));
    plot(p$x, p$u, type = "h",xlim=c(0,p$t),
         xlab="Time", ylab="Jumps");
    if(length(h)) {abline(h=h,lty=2)}
    title(paste(name, " Process Jumps: ", pn[1]," = ", pv[1],
          ", ", pn[2]," = ", pv[2], sep=""));
    plot(p$x, p$y, type = "l",xlim=c(0,p$t),
         xlab="Time", ylab="Process");
    title(paste(name, " Process: ", pn[1]," = ", pv[1],
          ", ", pn[2]," = ", pv[2], sep=""));
    par(mfrow=oldpar$mfrow);
    invisible(); }

"disp.2d" <- function(p=list(),name="",pn=c("",""),
             pv=c(NA,NA)) {
    par(mfrow=c(1,2));
    pincushion(p);                # see note below
    title(paste(name," Field Jumps:\n", pn[1], " = ", pv[1],
          ", ", pn[2]," = ", pv[2], sep=""));
    persp(p$xx,p$yy,p$zz);
    title(paste(name," Random Field:\n", pn[1], " = ", pv[1],
          ", ", pn[2]," = ", pv[2], sep=""));
    invisible(); }
```

The following function `pincushion()` is called from within `disp.2d()` to generate a plot of the Lévy random field jumps as impulses (see Figure 2):

```
"pincushion" <- function(p=stop("Need list of x,y,u"),
```

```
n=length(x), screen=list(z=40, x=-60, y=0), hori=F) {
co <- cos((screen$z * pi)/180);
si <- sin((screen$z * pi)/180);
xmat <- matrix(c(co,si,0,  -si,co,0, 0, 0,1),
                  nr = 3);          # Transformation matrix
co <- cos((screen$x * pi)/180);
si <- sin((screen$x * pi)/180);
xmat <- matrix(c(1,0,0, 0,co,si, 0,-si,co),
                  nr = 3) %*% xmat;
co <- cos((screen$y * pi)/180);
si <- sin((screen$y * pi)/180);
xmat <- matrix(c(co,0,-si, 0,1,0, si,0,co),
                  nr = 3) %*% xmat;
n  <- min(n,length(p$x));
or <- rev(order(p$u))[1:n];         # Sort by jump sizes:
x  <- p$x[or]; y <- p$y[or]; z <- p$u[or];
zmax <- max(z);  z <- 0.8*z/zmax; # Rescale jumps
box3d <- xmat %*% matrix(c(0,0,0, 0,1,0, 0,1,1, 1,1,1,
      1,0,1, 0,0,1, 0,0,0, 1,0,0, 1,0,1, 0,0,1, 0,1,1,
      0,1,0, 1,1,0, 1,1,1, 1,1,0, 1,0,0), nrow = 3)
box2d <- xmat %*% matrix(c(0,0,0, 0,1,0, 1,1,0,
                        1,0,0, 0,0,0), nrow = 3)
xl <- c(min(box3d[1,  ]), max(box3d[1,  ]));
yl <- c(min(box3d[2,  ])-.75, 0.75 + max(box3d[2,  ]));
plot(x = c(0,0), type = "n", xlim = xl, ylim = yl,
    axes = F, xlab = "", ylab = "");
lines(t(box2d[-3,]),lty=1); lines(t(box3d[-3,]),lty=2);
xy0 <- xmat %*% rbind(x, y, 0);    # Bases of pins
xyz <- xmat %*% rbind(x, y, z);    # Heads of pins
segments(xy0[1,], xy0[2,], xyz[1,], xyz[2,]); }
```

Appendix 2: Proof of Theorem 2

Let $\epsilon > 0$, and, for $i \in \mathbb{N}$, set $a_i \equiv i\epsilon$, $A_i = [a_i, a_{i+1})$ and, for $u \geq 0$, set $A^\epsilon(u) \equiv \sum a_i 1_{A_i}(u) = \epsilon\lfloor u/\epsilon \rfloor$, u rounded down to the nearest integral multiple of ϵ. Let $\{B_j\}$ be any Borel partition of $\mathcal{S}$ and take $\phi(s) = \sum \phi_j 1_{B_j}(s)$ and $\nu(u,s) = \sum \nu_j(u) 1_{B_j}(s)$ to be nonnegative and constant on partition elements, with $\phi(s) \leq M < \infty$ bounded. Set $\pi_j \equiv \Pi(B_j)$ and denote by N_{ij} the cardinality of the set $\{m : v_m \in A_i, \sigma_m \in B_j\}$. Then for $i \geq 1$ the N_{ij} have independent Poisson distributions with means $\mathsf{E}[N_{ij}] = \nu(A_i \times B_j) = \pi_j \int_{A_i} \nu_j(u)\,du$, and the approximation $X_\epsilon[\phi] \equiv \eta[\phi] + \sum_m A^\epsilon(v_m)\phi(\sigma_m)$ has characteristic function

$$
\begin{aligned}
\mathsf{E}\left(e^{iX_\epsilon[\phi]}\right) &= e^{i\eta[\phi]} \prod_{ij} \mathsf{E}[e^{ia_i\phi_j N_{ij}}] \\
&= \exp\left(i\eta[\phi] + \sum_{ij} [e^{ia_i\phi_j} - 1]\nu(A_i \times B_j)\right) \\
&= \exp\left(i\eta[\phi] + \iint_{\mathbb{R}_+ \times \mathcal{S}} [e^{iA^\epsilon(u)\phi(s)} - 1]\nu(du\,ds)\right).
\end{aligned}
$$

By the dominated convergence theorem we can take $\epsilon \to 0$ to find $\mathrm{E}[e^{iX[\phi]}] = \exp\big(i\eta[\phi] + \iint_{\mathbb{R}_+\times\mathcal{S}}[e^{iu\phi(s)} - 1]\nu(du\,ds)\big)$, as claimed. Now refine the partition $\{B_j\}$ to complete the proof for any bounded compactly-supported Borel function $\phi(s)$ and $(1 \wedge u)du\Pi(ds)$–integrable $\nu(u,s)$. □

Acknowledgments: This analysis was supported by U.S. E.P.A. grant CR822047-01-0, U.S. N.S.F. grant DMS-9626829, and the Deutsche Forschungsgemeinschaft.

References

Abramowitz, M. and Stegun, I. A., eds. (1964), *Handbook of Mathematical Functions With Formulas, Graphs, and Mathematical Tables*, volume 55 of *Applied Mathematics Series*, Washington, D.C.: National Bureau of Standards.

Antoniak, C. E. (1974), "Mixtures of Dirichlet Processes with Applications to Bayesian Nonparametric Problems," *Annals of Statistics*, 2, 1152–1174.

Bondesson, L. (1982), "On simulation from infinitely divisible distributions," *Advances in Applied Probability*, 14, 855–869.

Damien, P., Laud, P. W., and Smith, A. F. M. (1995), "Approximate random variate generation from infinitely divisible distributions with applications to Bayesian inference," *Journal of the Royal Statistical Society, Series B (Methodological)*, 57, 547–563.

Ferguson, T. S. (1973), "A Bayesian analysis of some nonparametric problems," *Annals of Statistics*, 1, 209–230.

Ferguson (1974), "Prior distributions on spaces of probability measures," *Annals of Statistics*, 2, 615–629.

Ferguson, T. S. and Klass, M. J. (1972), "A representation of independent increment processes without Gaussian components," *Annals of Mathematical Statistics*, 43, 1634–1643.

Ferguson, T. S. and Phadia, E. G. (1979), "Bayesian nonparametric estimation based on censored data," *Annals of Statistics*, 7, 163–186.

Gelfand, A. E. and Smith, A. F. M. (1990), "Sampling-based approaches to calculating marginal densities," *Journal of the American Statistical Association*, 85, 398–409.

Gilks, W. R., Richardson, S., and Spiegelhalter, D. J., eds. (1996), *Markov Chain Monte Carlo in Practice*, New York, NY, USA: Chapman & Hall.

Hjort, N. L. (1990), "Nonparametric Bayes estimators based on beta processes in models for life history data," *Annals of Statistics*, 18, 1259–1294.

Ickstadt, K. and Wolpert, R. L. (1997), "Multiresolution Assessment of Forest Inhomogeneity," in *Case Studies in Bayesian Statistics, Volume III* (eds: C. Gatsonis, J. S. Hodges, R. E. Kass, R. E. McCulloch, P. Rossi, and N. D. Singpurwalla), volume 121 of *Lecture Notes in Statistics*, 371–386, New York: Springer-Verlag.

Ickstadt, K., Wolpert, R. L., and Lu, X. (1998), "Modeling travel demand in Portland, Oregon," in *Practical Nonparametric and Semiparametric Bayesian Statistics* (eds: Dey, D. and Müller, P. and Sinha, D.), New York: Springer-Verlag.

Jacod, J. and Shiryaev, A. N. (1987), *Limit Theorems for Stochastic Processes*, volume 288 of *Grundlehren der mathematischen Wissenschaften*, Berlin: Springer-Verlag.

Lévy, P. (1937), *Théorie de L'Addition des Variables Aléatoires*, Paris: Gauthier-Villars.

MacEachern, S. N. (1998), "Computational Methods for Mixture of Dirichlet Process Models," in *Practical Nonparametric and Semiparametric Bayesian Statistics* (eds: Dey, D. and Müller, P. and Sinha, D.), New York: Springer-Verlag.

MathSoft, D. A. P. D. (1997), *S-Plus User's Guide*, Seattle: MathSoft.

Qiou, Z. and Ravishanker, N. (1997), "Bayesian inference for time series with stable innovations," To appear in *Journal of Time Series Analysis*.

Qiou, Z., Ravishanker, N., and Dey, D. K. (1997), "Multivariate survival analysis with positive stable frailties," Under revision for Biometrics.

Rogers, L. and Williams, D. (1994), *Diffusions, Markov Processes, and Martingales*, volume 1, New York, NY, USA: John Wiley & Sons, second edition.

Sinha, D. and Dey, D. (1997), "Semiparametric Bayesian Analysis of Survival Data," *Journal of the American Statistical Association*, 92, 1195–1212.

Tierney, L. (1994), "Markov chains for exploring posterior distributions," *Annals of Statistics*, 22, 1701–1762, (With discussion).

Walker, S. and Damien, P. (1998), "Sampling methods for Bayesian nonparametric inference involving stochastic processes," in *Practical Nonparametric and Semiparametric Bayesian Statistics* (eds: Dey, D. and Müller, P. and Sinha, D.), New York: Springer-Verlag.

Wolpert, R. L. and Ickstadt, K. (1998), "Poisson/gamma random field models for spatial statistics," *Biometrika*, to appear.

13

Sampling Methods For Bayesian Nonparametric Inference Involving Stochastic Processes

Stephen Walker
Paul Damien

ABSTRACT An algorithm for simulating the class of neutral to the right processes is described. The mixture of Dirichlet processes (MDP) model has proved to be successful in a variety of contexts. A new sampling method, using auxiliary variables, for MDP models is developed and exemplified.

13.1 Introduction

A large section of the Bayesian nonparametric literature is concerned with the construction of prior distributions on the space of probability distributions on $(0, \infty)$. One particular application which has received a great deal of attention includes survival analysis. An important and rich class of prior distributions is the *neutral to the right* (NTR) processes (Doksum, 1974; Ferguson, 1974), including the Dirichlet process (Ferguson, 1973). Applications using NTR processes other than the Dirichlet appear in Damien et al. (1995).

There have only been limited applications involving NTR process priors in the statistics literature (besides the well known Dirichlet process). The reason for this is the difficulty in obtaining anything more than an expression for the posterior expectation of the distribution function (Ferguson and Phadia, 1979, Section 6). In Section 2, we discuss sampling such a process, providing the basis for a full Bayesian analysis.

Another class of stochastic process models is the mixture of Dirichlet processes (MDP); see, for example, Escobar, 1994; Escobar & West, 1995; West et al., 1994; Mueller et al., 1996. These authors provide illustrations of the usefulness of MDPs in a variety of contexts. In Section 3, we develop a new algorithm to sample an MDP model, based on a result of Damien et al. (1998). Illustrative examples are given. A brief discussion in Section 4 concludes the paper.

13.2 Neutral to the right processes

DEFINITION (Ferguson, 1974). A random distribution function $F(t)$ on $(0,\infty)$ is neutral to the right (NTR) if for every m and $0 < t_1 < \ldots < t_m$ there exist independent random variables $V_1, \ldots, V_m$ such that

$$\mathcal{L}(1 - F(t_1), \ldots, 1 - F(t_m)) = \mathcal{L}\left(V_1, V_1V_2, \ldots, \prod_{j=1}^{m} V_j\right).$$

If F is NTR then $Z(t) = -\log[1 - F(t)]$ has independent increments. The converse is also true: if $Z(t)$ is an independent increments process, nondecreasing almost surely (a.s.), right continuous a.s., $Z(0) = 0$ a.s. and $\lim_{t\to\infty} Z(t) = +\infty$ a.s. then $F(t) = 1-\exp(-Z(t))$ is NTR. For the process $Z(.)$ there exist at most countably many fixed points of discontinuity at $t_1, t_2, \ldots$ with jumps $J_1, J_2, \ldots$ (independent) having densities $f_{t_1}, f_{t_2}, \ldots$. Then $Z_c(t) = Z(t) - \sum_{t_j \le t} J_j$ has no fixed points of discontinuity and therefore has a Lévy representation with log-Laplace transform given by

$$-\log E \exp\left(-\phi Z_c(t)\right) = \int_0^\infty (1 - \exp(-\phi z))dN_t(z),$$

where $N_t(.)$, a Lévy measure, satisfies: $N_t(B)$ is nondecreasing and continuous for every Borel set B; $N_t(.)$ is a measure on $(0,\infty)$ for each t and $\int_{(0,\infty)} z(1+z)^{-1} dN_t(z) < \infty$. In short, we have

$$F(t) = 1 - \exp[-Z(t)],$$

with Z an independent increments (Lévy) process.

THEOREM 1 (Ferguson, 1974). *If F is NTR and $X_1, \ldots, X_n$ is a sample from F, including the possibility of right censored samples (where X_i represents the censoring time if applicable), then the posterior distribution of F given $X_1, \ldots, X_n$ is NTR.*

The prior distribution for $Z(.)$ is characterised by

$$M = \{t_1, t_2, \ldots\}, \quad \{f_{t_1}, f_{t_2}, \ldots\},$$

the set of fixed points of discontinuity with corresponding densities for the jumps, and $N_t(.)$, the Levy measure for the part of the process without fixed points of discontinuity. The posterior distribution is now given for a single observation X. The case for n observations can then be obtained by repeated application. In the following, we assume the Lévy measure to be of the type $dN_t(z) = dz \int_{(0,t]} K(z,s)ds$; the beta-Stacy process with parameters $\alpha(.)$ and $\beta(.)$ arises when $K(z,s)ds = (1-e^{-z})^{-1} \exp[-z\beta(s)]d\alpha(s)$.

THEOREM 2 (Ferguson, 1974; Ferguson and Phadia, 1979). Let F be NTR and let X be a random sample from F.

(i) Given $X > x$ the posterior parameters (denoted by a $*$) are $M^* = M$,

$$f^*_{t_j}(z) = \begin{cases} c \cdot e^{-z} f_{t_j}(z) & \text{if } t_j \leq x \\ f_{t_j}(z) & \text{if } t_j > x \end{cases}$$

and $K^*(z,s) = \exp[-zI(x \geq s)]K(z,s)$ (here c is the normalising constant).

(ii) Given $X = x \in M$ the posterior parameters are $M^* = M$,

$$f^*_{t_j}(z) = \begin{cases} c \cdot e^{-z} f_{t_j}(z) & \text{if } t_j < x \\ c \cdot (1-e^{-z}) f_{t_j}(z) & \text{if } t_j = x \\ f_{t_j}(z) & \text{if } t_j > x \end{cases}$$

and, again, $K^*(z,s) = \exp[-zI(x \geq s)]K(z,s)$.

(iii) Given $X = x \notin M$ the posterior parameters are $M^* = M \cup \{x\}$, with $f_x(z) = c \cdot (1-e^{-z})K(z,x)$,

$$f^*_{t_j}(z) = \begin{cases} c \cdot e^{-z} f_{t_j}(z) & \text{if } t_j < x \\ f_{t_j}(z) & \text{if } t_j > x \end{cases}$$

and, again, $K^*(z,s) = \exp[-zI(x \geq s)]K(z,s)$.

With this general characterisation of a posterior NTR process, a procedure for sampling the posterior process is developed in the first sub-section.

13.2.1 Simulating the posterior process

We are now concerned with simulating from the posterior distribution of $F[a,b)$. Since $F(t) = 1 - \exp[-Z(t)]$ we have

$$F[a,b) = \exp\big[-Z(0,a)\big]\big[1 - \exp\big(-Z[a,b)\big)\big]$$

and note that $Z(0,a)$ and $Z[a,b)$ are independent. Therefore we only need to consider sampling random variables of the type $Z[a,b)$. This will involve sampling from densities $\{f^*_{t_j} : t_j \in [a,b)\}$, corresponding to the jumps in the process, and secondly, the continuous component, the random variable $Z^*_c[a,b)$.

The type of densities corresponding to points in M^*, the jumps, are defined, up to a constant of proportionality, by

$$f^*(z) \propto \big(1 - \exp(-z)\big)^\lambda \exp\big(-\mu z\big) f(z), \quad \lambda, \mu \geq 0 \ \text{(integers)}$$

and

$$f^*(z) \propto \big(1 - \exp(-z)\big)^\lambda K(z, x), \quad \lambda > 0 \ \text{(integer)}$$

where the subscripts t_j have been omitted.

Simulating the jump component. The second of these two types of densities for the random variable corresponding to the jump in the process will most likely be a member of the Smith-Dykstra (SD) family of densities (Damien et al., 1995) and algorithms for sampling from such densities are given in Walker (1995).

For the first of these two types we can proceed via the introduction of latent variables: we define the joint density of z, $u = (u_1, \dots, u_\lambda)$ and v by

$$f(z, u, v) \propto e^{-\mu v} I(v > z) \left\{ \prod_{l=1}^{\lambda} e^{-u_l} I(u_l < z) \right\} f(z).$$

Clearly the marginal density of z is as required. The full conditional densities for implementing Gibbs sampling (Smith and Roberts, 1993) are given by

$$f(u_l | u_{-l}, v, z) \propto e^{-u_l} I(u_l < z), \quad l = 1, \dots, \lambda,$$

$$f(v | u, z) \propto e^{-\mu v} I(v > z)$$

and

$$f(z | u, v) \propto f(z) I\big(z \in (\max_l \{u_l\}, v)\big).$$

So, provided it is possible to sample from $f(z)$, this algorithm is easy to implement; see, for example, Damien et al. (1998).

Simulating the continuous component. The random variable $Z = Z_c^*[a, b)$, corresponding to the continuous component, is *infinitely divisible* (id) with log-Laplace transform given by

$$-\log E\big(e^{-\phi Z}\big) = \int_0^\infty \big(1 - e^{-\phi z}\big) \int_a^b \exp\Big(-z \sum_{i=1}^n I(X_i \geq s)\Big) K(z, s) ds dz.$$

Damien et al. (1995) introduced an algorithm for sampling from id laws. An alternative idea is based on the fact that

$$\mathcal{L}(Z) = \mathcal{L}\left(\int_0^\infty z dP(z) \right),$$

where $P(.)$ is a Poisson process with intensity measure $dz \int_{[a,b)} K^*(z,s)ds$; see, for example, Bondesson (1982).

Here we consider sampling an infinitely divisible variable Z with log-Laplace transform given by

$$\log E \exp(-\phi Z) = \int_0^\infty (e^{-z\phi} - 1)dN(z),$$

for some appropriate Lévy measure $N(.)$. For a fixed $\epsilon > 0$, $Z = X_\epsilon + Y_\epsilon$, with X_ϵ and Y_ϵ being independent, where X_ϵ is the sum of the jumps of $P(.)$ in (ϵ, ∞) and Y_ϵ is the sum of the jumps in $(0, \epsilon]$. The aim is to approximate a random draw Z by sampling an X_ϵ. The 'error' variable Y_ϵ has mean $E(Y_\epsilon) = \int_{(0,\epsilon]} z dN(z)$ and variance $\text{var}(Y_\epsilon) = \int_{(0,\epsilon]} z^2 dN(z)$. Now ϵ can be taken as small as one wishes and for appropriate $N(.)$ we have, for $j = 1, 2$, $\int_{(0,\epsilon]} z^j dN(z) \to 0$ as $\epsilon \to 0$. This is the case for all Lévy measures appearing in practical applications. The approach just described was originally suggested by Bondesson (1982), but rejected by him because of the difficulty in sampling some of the required distributions to implement this strategy. However, we can now do this using, in particular, the Gibbs sampler.

Let $\lambda_\epsilon = N(\epsilon, \infty) < \infty$ and take $\nu \sim \text{Poisson}(\lambda_\epsilon)$. Here ν is the (random) number of jumps in (ϵ, ∞) and therefore the jumps $\tau_1, \ldots, \tau_\nu$ are taken iid from $G_\epsilon(s) = \lambda_\epsilon^{-1} N(\epsilon, s]$, for $s \in (\epsilon, \infty)$. Then X_ϵ is given by $X_\epsilon = \sum_{k=1}^{\nu} \tau_k$. To implement the algorithm it is required to (a) simulate from $G_\epsilon(.)$, and (b) calculate λ_ϵ. These are available via the construction of a Gibbs sampler and the use of auxiliary variables (see Damien et al., 1998). The specific algorithm for the beta-Stacy process (Walker and Muliere, 1997) is presented in Walker and Damien (1998).

An alternative approach to (approximately) sampling Z, also based on the Poisson process representation, called the Inverse Lévy Measure (ILM) method, has been developed by Wolpert and Ickstadt (1998). A comparative discussion with Damien et al. (1995) is provided by these authors in the current volume.

13.3 Mixtures of Dirichlet processes

A key aspect to the sampling method developed below is that even though we will be working with a Dirichlet process, the process itself is *not* sampled; from a practical perspective, this will lead to an user-friendly algorithm. Consider, then, the following general hierarchical model involving the NTR process — the Dirichlet process — commonly known as the mixture of Dirichlet processes (MDP) model:

$$z_i|\theta_i \sim f(.;\theta_i), \quad i = 1, \ldots, n,$$

$$\theta_1, \ldots, \theta_n | G \sim_{\text{iid}} G$$

and

$$G \sim \mathcal{D}(\alpha G_0).$$

Briefly, the probability model assumes z_i given θ_i $(i = 1, \ldots, n)$ is an observation from the known distribution $f(.;\theta_i)$. Here $\mathcal{D}$ represents a Dirichlet process (Ferguson, 1973) and G is a discrete random distribution (taken from the Dirichlet process) conditional on which the mixing parameters, $\{\theta_i\}$, are independent and identically distributed (iid); whence the title, "mixture of Dirichlet processes." The parameters of the Dirichlet process are the distribution G_0 which centers the process, in that $EG = G_0$, and $\alpha > 0$, which loosely represents the *strength of belief* in G_0 via a prior sample size; note also that the marginal prior distribution of θ_i is G_0. Throughout the discussion we will use G and G_0 to denote both distributions and densities.

Inference is undertaken via Markov chain Monte Carlo (MCMC) methods (Tierney, 1994) and, in particular, the Gibbs sampler (Smith and Roberts, 1993). This requires sampling sequentially from the full conditional distributions. For us there is no loss of generality in assuming that both α and G_0 are fixed.

The full conditional distribution for θ_i can pose the problem for the Gibbs sampler (usually the others are of standard types). This distribution has density given, up to a constant of proportionality, by

$$f^*(\theta_i) \propto f(z_i|\theta_i)\{\alpha G_0(\theta_i) + \sum_{j \neq i} \delta_{\theta_j}(\theta_i)\},$$

where $\delta_\theta(.)$ is the measure putting mass 1 at θ. Here we will always use a $*$ to represent a full conditional distribution. If $q_{i0} = \alpha \int f(z_i|\theta_i) dG_0(\theta_i)$ is analytically tractable then $f^*(.)$ can be sampled directly. That is, θ_i is taken from the distribution proportional to $f(z_i|\theta_i)G_0(\theta_i)$ with probability proportional to q_{i0} or is taken to be θ_j $(j \neq i)$ with probability proportional to $q_{ij} = f(z_i|\theta_j)$. An alternative way to understand the above is to note that the Dirichlet process,G, has been integrated out in the conditonal distribution of θ_i.

The only difficulties with implementing a Gibbs sampler for the MDP model are when the integral $\int f(z_i|\theta_i)dG_0(\theta_i)$ is intractable and hence q_{i0} is not available; see, for example, MacEachern & Mueller (1998); The method of MacEachern & Mueller to solving the nonconjugate model relies on the introduction of a latent model, described in Section 3 of their paper. We omit the details of their algorithm but note that their approach requires sampling from the non-standard density

$$f^*(\theta) \propto f(.|\theta)G_0(\theta).$$

Our approach avoids this and instead the only 'non-standard' sampling is from a truncated version of G_0.

In the next sub-section we introduce a latent model which uses a Gibbs sampler in which all the full conditional distributions are of known types and can be sampled directly. Essentially, we are introducing latent variables which means that the resultant q_{i0}s can be evaluated.

13.3.1 A new latent MDP model

Recall the problem is sampling from the distribution given by

$$f^*(\theta_i) \propto l(\theta_i) \left\{ \alpha G_0(\theta_i) + \sum_{j \neq i} \delta_{\theta_j}(\theta_i) \right\},$$

where we have written $f(z_i|\theta_i)$ as $l(\theta_i)$; in the following, $l(\theta_j) > u_i$ means $j \neq i$ with $l(\theta_j > u_i$. Here we introduce the latent variable u_i and construct the joint density with θ_i given, up to a constant of proportionality, by

$$f^*(\theta_i, u_i) \propto I\left(u_i < l(\theta_i)\right) \left\{ \alpha G_0(\theta_i) + \sum_{j \neq i} \delta_{\theta_j}(\theta_i) \right\},$$

where $I(.)$ represents an indicator function. Clearly the marginal distribution for θ_i is $f^*(.)$. Note that the full posterior distribution for (θ, u) is given by

$$f^*(\theta, u) \propto \prod_i \left\{ I\left(u_i < l(\theta_i)\right) \left(\alpha G_0(\theta_i) + \sum_{j<i} \delta_{\theta_j}(\theta_i) \right) \right\}.$$

Therefore, the Gibbs sampler now runs over the additional full conditionals given by

$$u_i|\theta_i \sim U\left(0, l(\theta_i)\right),$$

where $U(a, b)$ is the uniform distribution on the interval (a, b), and the full conditional for θ_i is now in the more friendly form,

$$f^*(\theta_i|u_i) \propto \alpha G_0(\theta_i) I\left(l(\theta_i) > u_i\right) + \sum_{l(\theta_j) > u_i} \delta_{\theta_j}(\theta_i).$$

The full conditional for u_i is obviously trivial to sample. The full conditional for θ_i should not pose any problem either.

Example 1. Here we consider a one-dimensional nonconjugate normal/uniform MDP model. Then $l(\theta_i) = \exp[-0.5(z_i - \theta_i)^2/\sigma^2]$, $G_0 = U(\mu_L, \mu_H)$ and

$$f^*(\theta_i|u_i) \propto \frac{\alpha}{\mu_H - \mu_L} I\left(\max\{\mu_L, a_{ui}\} < \theta_i < \min\{\mu_H, b_{ui}\}\right) + \sum_{l(\theta_j) > u_i} \delta_{\theta_j}(\theta_i),$$

where $a_{ui} = z_i - \sigma\sqrt{-2\log u_i}$ and $b_{ui} = z_i + \sigma\sqrt{-2\log u_i}$ (note that $u_i < 1$). Thus θ_i is taken from the uniform distribution on the interval

$$(\max\{\mu_L, a_{ui}\}, \min\{\mu_H, b_{ui}\})$$

with probability proportional to

$$q_{i0} = \alpha(\mu_H - \mu_L)^{-1}\left(\min\{\mu_H, b_{ui}\} - \max\{\mu_L, a_{ui}\}\right),$$

or is equal to θ_j, for which $l(\theta_j) > u_i$, with probability proportional to 1. Note, therefore, that if $\{\theta_j : l(\theta_j) > u_i\} = \emptyset$ then θ_i is automatically taken from the uniform distribution.

For the general nonconjugate model, let $A_i = \{\theta_i : l(\theta_i) > u_i\}$ so we can write

$$f^*(\theta_i|u_i) = \frac{\alpha G_0(A_i)G_0(\theta_i, A_i) + \sum_{l(\theta_j)>u_i}\delta_{\theta_j}(\theta_i)}{\alpha G_0(A_i) + \sum_{l(\theta_j)>u_i} 1},$$

where $G_0(\theta_i, A_i)$ is the normalised $G_0(\theta_i)I(\theta_i \in A_i)$. Therefore, we take θ_i from $G_0(\theta_i, A_i)$ with probability

$$\frac{\alpha G_0(A_i)}{\alpha G_0(A_i) + \sum_{l(\theta_j)>u_i} 1},$$

or take $\theta_i = \theta_j$, for which $l(\theta_j > u_i)$, with probability $[\alpha G_0(A_i) + \sum_{l(\theta_j)>u_i} 1]^{-1}$. Therefore, to know which component to sample from we only need to know $G_0(A)$.

Example 2. Here we consider a binomial model with logit link;

$$z_i|p_i \sim \text{binomial}(p_i, n_i) \;\text{ and }\; \text{logit}p_i = \theta_i.$$

Therefore

$$f(z_i|\theta_i) \propto \frac{\exp(\theta_i z_i)}{(1+\exp(\theta_i))^{n_i}}.$$

For this model we introduce the latent variables (u_i, v_i) so that the joint posterior with θ_i is given by

$$f^*(\theta_i, u_i, v_i|z) \propto e^{-u_i z_i - v_i(n_i - z_i)} I\left(u_i > \log\{1+e^{-\theta_i}\}, v_i > \log\{1+e^{\theta_i}\}\right) G_0(\theta_i),$$

where $G_0(\theta_i) = N(\theta_i|\mu, \sigma^2)$ is the prior. Clearly the marginal posterior for θ_i is as required. With the inclusion of the Dirichlet prior the full conditional for θ_i becomes

$$f^*(\theta_i|u_i, v_i) \propto I(a_{ui} < \theta_i < b_{vi}) \left\{ \alpha G_0(\theta_i) + \sum_{j \neq i} \delta_{\theta_j}(\theta_i) \right\},$$

where $a_{ui} = -\log(e^{u_i} - 1)$ and $b_{vi} = \log(e^{v_i} - 1)$. Therefore, we take $\theta_i = \theta_j$, for which $a_{ui} < \theta_j < b_{vi}$, or from a truncated normal distribution, according to easily available probabilities.

Provided we can implement the algorithm for the parametric model $z_i|\theta_i \sim f(.|\theta_i)$ with $\theta_i \sim_{\text{iid}} G_0$ using the latent variable idea, then we can also implement it for the corresponding MDP model. Damien et al. (1998) show that such an algorithm is applicable for a large class of nonconjugate models. This is detailed in the following Theorem, whose proof is omitted since it is similar to the theorem appearing in Damien et al. (1998), from which further details and examples can be found.

THEOREM 3. *If*

$$l(\theta) = \prod_{k=1}^{K} l_k(\theta),$$

where the $l_k(\theta)$ are nonnegative invertible functions then it is possible to obtain the set $A_{ku} = \{\theta : l_k(\theta) > u\}$, and it is possible to implement a Gibbs sampler for generating random variates from the MDP model in which all the full conditionals are of known types.

13.3.2 Numerical example

We use the same example as that used by MacEachern et al. (1998) who demonstrate their "next generation" sequential importance sampling technique in that paper. The data comes from Beckett and Diaconis (1994) and consists of 320 binomial experiments where 320 thumbtacks were flicked nine times each. The data for each experiment was the number of times the tack landed point up. That is, for $i = 1, \ldots, 320$, $y_i|p_i \sim \text{binomial}(p_i, 9)$. Therefore taking $G_0 = U(0,1)$ and introducing latent variables (u_i, v_i) we have

$$f^*(p_i, u_i, v_i) \propto$$

$$I\left(u_i < p_i^{y_i}, v_i < (1-p_i)^{9-y_i}\right) \left\{ \alpha I(0 < p_i < 1) + \sum_{j \neq i} \delta_{p_j}(p_i) \right\}.$$

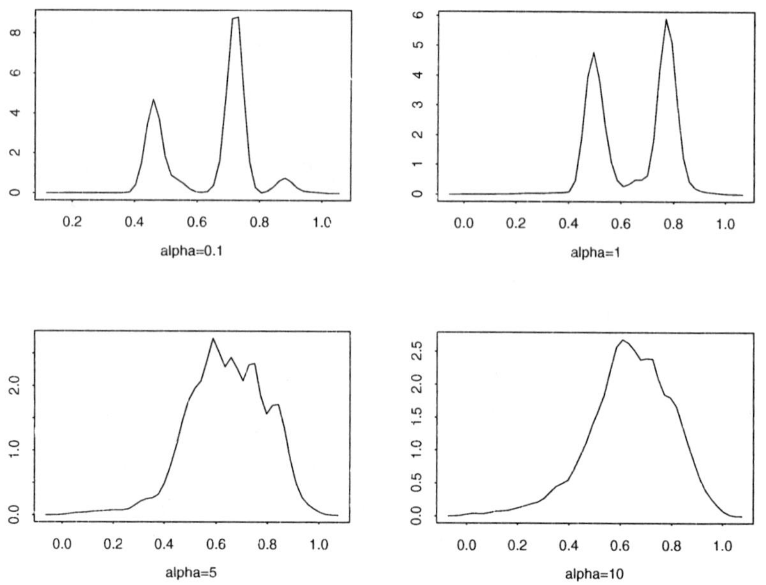

FIGURE 1. Estimates of the densities for p_{n+1}, given the data, for four values of α.

Therefore

$$f^*(p_i|u_i, v_i) \propto \alpha I\,(a_i < p_i < b_i) + \sum_{a_i < p_j < b_i} \delta_{p_j}(p_i),$$

where

$$a_i = \begin{cases} u_i^{1/y_i} & \text{if } y_i > 0 \\ 0 & \text{otherwise,} \end{cases}$$

and

$$b_i = \begin{cases} 1 - v_i^{1/(9-y_i)} & \text{if } y_i < 9 \\ 1 & \text{otherwise.} \end{cases}$$

Also $f^*(u_i) = U(0, p_i^{y_i})$ and $f^*(v_i) = U(0, (1-p_i)^{9-y_i})$. These full conditionals are sampled straightforwardly. We implement the Gibbs sampler with α fixed at four different values $(0.1, 1, 5, 10)$. Within each iteration we generate a sample from the distribution of $p_{n+1}|$data to compare with the analysis of MacEachern et al. (1998). We collect 5000 samples for each of the four analyses which takes several seconds. The estimates for the predicitve density of $p_{n+1}|$data for the four values of α are shown in Figure 1; these alpha values are the same as those used by MacEachern et al. The graphs in Figure 1 are similar to the graphs obtained by MacEachern et al. (1998, Figure 2).

13.4 Conclusions

We have described an algorithm to simulate from any NTR process; the well-known Dirichlet process is NTR. In particular, simulating such processes involves random variate generation from infinitely divisible laws. A straightforward algorithm was described in this paper. The MDP class of models is a natural extension of the basic NTR model. This is because the MDP models are hierarchical models that involve the Dirichlet process. A new method, using auxiliary variables, to sample an MDP model was developed and exemplified.

References

Beckett, L. and Diaconis, P. (1994). "Spectral analysis for discrete longitudinal data," *Advances in Mathematics*, 103, 107-128.

Bondesson, L. (1982). On simulation from infinitely divisible distributions. *Advances in Applied Probability* **14**, 855-869.

Damien, P., Laud, P.W. and Smith, A.F.M. (1995). "Random variate generation approximating infinitely divisible distributions with application to Bayesian inference," *Journal of the Royal Statistical Society, Series B*, 57, 547-564.

Damien, P., Wakefield, J.C. and Walker, S.G. (1998). "Gibbs sampling for Bayesian non-conjugate and hierarchical models using auxiliary variables." *Journal of the Royal Statistical Society, Series B*, to appear.

Doksum, K.A. (1974). "Tailfree and neutral random probabilities and their posterior distributions," *The Annals of Probability*, 2, 183-201.

Escobar, M.D. (1994). "Estimating normal means with a Dirichlet process prior," *Journal of the American Statistical Association*, 89, 268-277.

Escobar, M.D. and West, M. (1995). "Bayesian density estimation and inference using mixtures," *Journal of the American Statistical Association*, 90, 577-588.

Ferguson, T.S. (1973). "A Bayesian analysis of some nonparametric problems," *The Annals of Statistics*, 1, 209-230.

Ferguson, T.S. (1974). "Prior distributions on spaces of probability measures," *The Annals of Statistics*, 2, 615-629.

Ferguson, T.S. and Phadia, E.G. (1979). "Bayesian nonparametric estimation based on censored data," *The Annals of Statistics*, 7, 163-186.

MacEachern, S.N. and Mueller, P. (1998). "Estimating mixtures of Dirichlet process models," *Journal of Computational and Graphical Statistics*, to appear.

MacEachern, S.N., Clyde, M. and Liu, J.S. (1998). "Sequential importance sampling for nonparametric Bayes models: The next generation," *Canadian Journal of Statistics*, to appear.

Mueller, P., Erkanli, A. and West, M. (1996) "Bayesian curve fitting using multivariate normal mixtures," *Biometrika*, 83, 67-79.

Smith, A.F.M. and Roberts, G.O. (1993). "Bayesian computations via the Gibbs sampler and related Markov chain Monte Carlo methods," *Journal of the Royal Statistical Society, Series B*, 55, 3-23.

Tierney, L. (1994). "Markov chains for exploring posterior distributions," *The Annals of Statistics*, 22, 1701-1762.

Walker, S.G. (1995). "Generating random variates from D-distributions via substitution sampling," *Statistical Computing*, 5, 311-315.

Walker, S.G. and Muliere, P. (1997). "Beta-Stacy processes and a generalisation of the Polya-urn scheme," *The Annals of Statistics*, 25, 1762-1780.

Walker, S.G. and Damien, P.(1998). "A full Bayesian nonparametric analysis involving a neutral to the right process." *Scandinavian Journal of Statistics*, to appear.

Wolpert, R. L. and Ickstadt, K. (1998). "Poisson/gamma random field models for spatial statistics," *Biometrika*, to appear.

West, M., Müller, P., and Escobar, M.D. (1994), "Hierarchial Priors and Mixture Models, with Applications in Regression and Density Estimation," *Aspects of Uncertainty: A Tribute to D. V. Lindley* (eds: A.F.M. Smith and P.R. Freeman), London: John Wiley and Sons, 363-386.

14

Curve and Surface Estimation Using Dynamic Step Functions

Juha Heikkinen

ABSTRACT This chapter describes a nonparametric Bayesian approach to the estimation of curves and surfaces that act as parameters in statistical models. The approach is based on mixing variable dimensional piecewise constant approximations, whose 'smoothness' is regulated by a Markov random field prior. Random partitions of the domain are defined by Voronoi tessellations of random generating point patterns. Variable dimension Markov chain Monte Carlo methods are proposed for the numerical estimation, and a detailed algorithm is specified for one special case. General applicability of the approach is discussed in the context of density estimation, regression and interpolation problems, and an application to the intensity estimation for a spatial Poisson point process is presented.

14.1 Introduction

Statistical modelling is often concerned with finding functions that describe some features of a given data set. Perhaps the most common examples are density estimation, where the object is a probability density function, and regression analysis, where the function of interest usually returns the expected value of the response variable corresponding to given value(s) of the predictor(s). This chapter addresses the general problem of estimating such functions nonparametrically.

The method described here was motivated by Arjas and Gasbarra (1994), where a new approach was introduced to the nonparametric Bayesian curve estimation. Curves defined on a finite interval were parametrised by the function values at a finite number of points (to be called generating points here). In the prior (and hence also in the posterior) the number and locations of generating points were random, i.e. they varied from one realisation to another. This lead to an infinite-dimensional parameter space, whereby the approach can be honestly called nonparametric. Deterministic interpolation was applied between the generating points by the simplest possible means, piecewise constant functions with jumps at the generating points. More sophisticated forms of interpolation can easily be applied in the curve estimation context; a concrete demonstration was provided by Denison,

Mallick and Smith (1998), where sequences of piecewise polynomials were used.

Since Bayesian inference is not concerned with selecting a point estimate (here a single curve) from the postulated model class, the precise functional form of its individual members is not as crucial as in the frequentist approach. More important is that the integrals of test functions of interest (e.g. predictive densities or probabilities) w.r.t. the posterior distribution obtained from the approximate model are close to those obtained from the 'true' model (see Arjas and Andreev, 1996). Furthermore, a 'Bayesian point estimate', the posterior mean, does not necessarily belong to the model class. Hence, in the model of Arjas and Gasbarra (1994) pointwise posterior means don't need to form a piecewise constant curve since the jump points are variable, and indeed the posterior mean is typically a smooth continuous function. Further discussion on the topic can be found in Section 14.7, in the papers cited above, and in Arjas (1996).

In Arjas and Gasbarra (1994) a prior distribution for the marked generating points was specified in terms of the corresponding local characteristics. A martingale structure was assumed, which penalises large differences between nearby function values. The aim was, besides smoothing the oscillations, to have the change points concentrated on the places of most rapid changes.

Markov chain Monte Carlo (MCMC) methods were proposed to carry out the computations required for the actual inferences. It is not possible, however, to apply standard MCMC algorithms, such as the Gibbs sampler, when the length of the parameter vector is variable. Arjas and Gasbarra (1994) modified the Gibbs sampler in a rather *ad hoc* way that is practical only in curve estimation. A general framework for handling variable dimension was provided by the reversible jump MCMC approach of Green (1995).

Two very similar extensions of this 'dynamic step function approach' to surface estimation, using Voronoi tessellations to define the regions of constant function value, were presented in Green (1995) and in Heikkinen and Arjas (1998). The main concern in the example of Green (1995) was in finding change-curves in a surface that is truly discontinuous, and accordingly independence was assumed between step function values in different regions. Heikkinen and Arjas (1998) followed the approach of Arjas and Gasbarra (1994) also in building a smoothing prior across the tiles. Locally dependent Markov random fields, widely used in Bayesian image analysis, were applied for this purpose. A related approach, using triangulations, was taken by Nicholls (1998) in the context of image segmentation.

Arjas and Heikkinen (1997) presented the one-dimensional (curve estimation) version of the method of Heikkinen and Arjas (1998), and Heikkinen and Arjas (1997) demonstrated the flexibility of the approach through an application, where the model contains three unknown functions to be estimated, non-parametrically and simultaneously. Based on the experiences from these works the purpose of this chapter is to suggest a rather simple, but in many cases sufficient, 'prototype algorithm' for the nonparametric Bayesian function estimation, and to discuss its possible modifications and extensions. Although the proposed method is theoretically applicable in any dimension, it is likely to be practical mainly in the estimation of uni- and bivariate functions, i.e. curves and surfaces. Both the computational effort and the amount of data required for useful nonparametric inference is usually prohibitive in higher dimensions.

Section 14.2 introduces the general setup and some common statistical problems, for which the proposed method might be useful. Concrete demonstrations can be found in the papers cited above; a brief description of one of these examples is given in Section 14.6.

Section 14.3 reviews some concepts from spatial statistics to be used in Section 14.4, where a basic version of the proposed prior model is defined. MCMC inference based on the corresponding posterior distribution is considered in Section 14.5 with a detailed algorithm specified in the Appendix. Possible modifications and extensions are discussed in Section 14.7.

14.2 Some statistical problems

Suppose that a statistical model is specified for data $\mathbf{x}$ by a likelihood expression $p(\mathbf{x} \mid \theta)$ containing an unknown parameter function

$$\theta : E \to \mathbf{R}$$

defined on a bounded domain $E \subset \mathbf{R}^d$, $d \in \{1, 2, \dots\}$. The general aim here is to describe a method for the nonparametric estimation of θ. Although the method is in principle applicable in any dimension d, it has been implemented in practice only for the estimation of curves ($d = 1$, Arjas and Heikkinen, 1996) and surfaces ($d = 2$, Arjas and Heikkinen, 1997a).

The form of data $\mathbf{x}$ and likelihood function p may be arbitrary as long as likelihood ratios can be easily evaluated. In *density estimation*, for example, θ would be a probability density function on E, $\mathbf{x} = (x_1, \dots, x_n)$ a sample of n realisations $x_i \in E$, and

$$p(\mathbf{x} \mid \theta) = \prod_{i=1}^{n} \theta(x_i). \tag{1}$$

Arjas and Heikkinen (1997) and Heikkinen and Arjas (1998) considered the estimation of the *intensity function* of a non-homogeneous Poisson process (see Section 14.3.1), a problem essentially analogous to density estimation. An example from Heikkinen and Arjas (1998) is briefly reviewed in Section 14.6 to demonstrate the method.

Another common example is *regression analysis* with a d-dimensional predictor, whose values lie within E. Each data item x_i consists of a predictor value $s_i \in E$ and a related observation z_i of a real valued response variable. The unknown regression function $\theta : E \to \mathbf{R}$ describes the relationship between the variables, usually $\theta(s_i)$ is the expected value of z_i. Given θ, the records z_i, $i = 1, \ldots, n$, are assumed conditionally independent with conditional density functions $p(z_i \,|\, \theta(s_i))$. The likelihood function of this model is defined by

$$p(\mathbf{x} \,|\, \theta) = \prod_{i=1}^{n} p(x_i \,|\, \theta) = \prod_{i=1}^{n} p(z_i \,|\, \theta(s_i)). \tag{2}$$

Using the same notation, the problem of *interpolating* data $\mathbf{x}$ can be formulated as a degenerate case of likelihood (2) with $p(z_i \,|\, \theta(s_i))$ having all its mass at $\theta(s_i)$. Both cases are demonstrated in Heikkinen and Arjas (1997), where the model contains a regression curve and an interpolated surface.

14.3 Some spatial statistics

This section reviews some concepts, originating from spatial statistics, that are needed to introduce the suggested prior distribution for parameter functions θ (Section 14.4).

14.3.1 Poisson point processes

A point process on a bounded region $E \subset \mathbf{R}^d$ is a statistical model for point patterns $\boldsymbol{\xi} = \{\xi_1, \ldots, \xi_K\} \subset E$ with (usually) a random number $K = K(\boldsymbol{\xi})$ of points. For $K = 1, 2, \ldots$, let Ω_K be the space of all such patterns on E which have K points, and let $\Omega_0 = \{\emptyset\}$, where $\emptyset$ denotes an empty configuration. Formally, a point process may be defined as a random variable taking values in $\Omega = \bigcup_{K=0}^{\infty} \Omega_K$.

The *Poisson process* on E is specified by an intensity function $\lambda : E \to [0, \infty)$, or by the corresponding intensity measure $\Lambda(A) = \int_A \lambda(s)\, \nu(ds)$, where ν is the Lebesque measure on $\mathbf{R}^d$. The number of points, K, follows the Poisson distribution with expectation $\Lambda(E)$, whereas conditional on K, the locations $\xi_1, \ldots, \xi_K$ are distributed on E as i.i.d. random variables with density $\lambda/\Lambda(E)$. The distribution P of the Poisson process is the probability measure on Ω defined by

$$P(F) = e^{-\Lambda(E)} \left[\mathbf{1}(\emptyset \in F) + \sum_{K=1}^{\infty} \frac{1}{K!} \int_{E^K} \mathbf{1}(\{\xi_1, \ldots, \xi_K\} \in F) \prod_{k=1}^{K} \Lambda(d\xi_k) \right].$$

A Poisson process with a constant intensity function λ is called homogeneous. The density function p of a Poisson process with distribution P can be defined as a Randon-Nikodym derivative of P w.r.t. the distribution of a homogeneous reference process, yielding

$$p(\boldsymbol{\xi}) \propto e^{-\Lambda(E)} \prod_{k=1}^{K(\boldsymbol{\xi})} \lambda(\xi_k).$$

For more detailed discussion on the Poisson process and point processes in general, see e.g. Karr (1991), Stoyan, Kendall and Mecke (1995), or chapter 8 of Cressie (1993).

14.3.2 Voronoi tessellations

The *Voronoi tessellation* of E generated by point pattern $\boldsymbol{\xi}$ is the partition $E = \bigcup_{k=1}^{K} E_k(\boldsymbol{\xi})$ into non-overlapping *tiles*

$$E_k(\boldsymbol{\xi}) = \{s \in E : \|s - \xi_k\| \le \|s - \xi_j\| \text{ for all } j\}.$$

Thus the interior of $E_k(\boldsymbol{\xi})$ consists of all points in E which are closer to ξ_k than to any other point of $\boldsymbol{\xi}$.

On the real line, $E = [T_1, T_2] \subset \mathbf{R}$, Voronoi tiles are intervals, whose end-points lie half way between successive points of $\boldsymbol{\xi}$:

$$E_k(\boldsymbol{\xi}) = \begin{cases} [T_1, \frac{1}{2}(\xi_1 + \xi_2)] & k = 1, \\ [\frac{1}{2}(\xi_{k-1} + \xi_k), \frac{1}{2}(\xi_k + \xi_{k+1})] & k = 2, \ldots, K-1, \\ [\frac{1}{2}(\xi_{K-1} + \xi_K), T_2] & k = K. \end{cases}$$

On the plane, $d = 2$, Voronoi tiles are convex polygons, whose edges are half way between two points of $\boldsymbol{\xi}$ and perpendicular to the line segment joining them (see Figure 1, left).

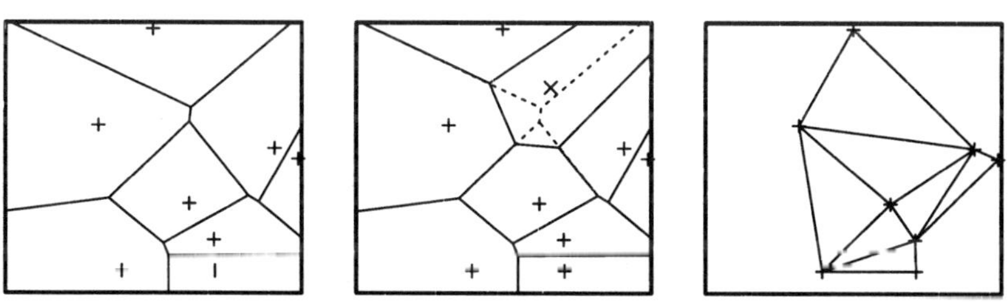

FIGURE 1. Voronoi tessellation generated by the +'s (left); point × is added to the pattern, its Voronoi tile consists of parts 'conquered' from the neighbours, other tiles are not affected (middle), Delaunay graph of the +'s (right).

A natural neighbourhood relation $\underset{\boldsymbol{\xi}}{\sim}$ between the points of $\boldsymbol{\xi}$ can be derived from the Voronoi tessellation by letting those pairs be neighbours, whose tile boundaries intersect. Hence on the real line successive points are neighbours, and on the plane those whose Voronoi polygons share a common boundary segment. This neighbourhood relation defines the *Delaunay graph* (Figure 1, right). On the plane, if $\boldsymbol{\xi}$ is a realisation from such point process which is absolutely continuous w.r.t. the Poisson process, then (with probability 1) the Delaunay graph is a triangulation of $\boldsymbol{\xi}$. For regular patterns, such as lattices, this is not necessarily the case.

In order to avoid unnecessarily complex notation for Voronoi tiles and neighbours, the references to the underlying point pattern $\boldsymbol{\xi}$ will be omitted whenever possible without ambiguity. Shorthand notation $j \sim k$ will also be applied for $\xi_j \sim \xi_k$. The index set $\{j : j \sim k\}$ will be denoted by $\partial_k = \partial_k(\boldsymbol{\xi})$, and its cardinality, i.e. the number of neighbours of point ξ_k, by $N_k = N_k(\boldsymbol{\xi})$.

A particularly useful property of the Voronoi tessellations in the current context is that they can be locally updated to small changes in the generating pattern $\boldsymbol{\xi}$: When a new pont is added to a tessellated pattern (or when one is removed), only the neighbours of the new (or removed) tile need to be modified (see Figure 1, middle).

For a theoretical treatment of random Voronoi tessellations and mathematical proofs concerning their properties the reader is referred to Møller (1994). Okabe, Boots and Sugihara (1992) takes a more application oriented approach.

14.3.3 Locally dependent Markov random fields

Consider then the joint distribution of K random variables $\eta_1, \ldots, \eta_K$. In general, the term *Markov random field* (MRF) only means that the joint density $p(\eta_1, \ldots, \eta_K)$ is specified though its full conditionals $p(\eta_k \,|\, \eta_j, j \neq k)$. To define a *locally dependent* MRF a symmetric neighbourhood relation $\sim$ between the sites $1, \ldots, K$ is specified, and it is assumed that the full conditionals depend only on the neighbouring values:

$$p(\eta_k \,|\, \eta_j, j \neq k) = p(\eta_k \,|\, \eta_j, j \sim k).$$

Unobvious consistency conditions are required from the conditional distributions in order to specify a well defined joint distribution by the MRF approach. The general conditions were identified by the famous unpublished Hammersley-Clifford theorem (see Besag, 1974). Sufficient conditions for Gaussian random fields can be found in Besag and Kooperberg (1995), for example.

14.4 Prototype prior

Each realisation θ of the suggested prior distribution is a piecewise constant function on a partition of E, but the number and the boundaries of the subregions are random. Random partitions are obtained as Voronoi tessellations (Section 14.3.2) of random *generating point patterns* $\boldsymbol{\xi}$. Step function θ can then be parametrised by locations $\xi_1, \ldots, \xi_K$ and corresponding function values $\eta_1, \ldots, \eta_K$, letting θ take value η_k on tile $E_k(\boldsymbol{\xi})$

$$\theta(s) = \sum_{k=1}^{K} \eta_k \mathbf{1}\{s \in E_k(\boldsymbol{\xi})\}.$$

In other words, each realisation is parametrised by a marked point pattern

$$(\xi_1, \eta_1), \ldots, (\xi_K, \eta_K).$$

The prior distribution of functions θ is then naturally determined by specifying the joint prior of $\boldsymbol{\xi}$ and $\boldsymbol{\eta} = (\eta_1, \ldots, \eta_K)$. This, in turn, is built via the chain rule decomposition

$$p(\boldsymbol{\xi}, \boldsymbol{\eta}) = p(\boldsymbol{\xi})p(\boldsymbol{\eta} \mid \boldsymbol{\xi}).$$

The prior for the generating points $\boldsymbol{\xi}$ is a homogeneous Poisson process (Section 14.3.1) constrained to have at least two points. It has one parameter, the intensity $\lambda > 0$, and its density (for fixed λ) is specified by

$$p(\boldsymbol{\xi}) \propto \lambda^K \mathbf{1}(K > 1).$$

The choice of λ determines the average tile size in the individual partitions. In other respects, the homogeneous Poisson process serves as a non-informative prior for point patterns.

The prior distribution of marks $\boldsymbol{\eta}$ is developed conditionally on the locations $\boldsymbol{\xi}$ of the generating points, and it contains an assumption of smoothness in the sense that the differences $|\eta_k - \eta_j|$ between function values on nearby tiles are expected to be small. Natural means of formulating this assumption is provided by locally dependent Markov random fields (Section 14.3.3). If the parameter surface may in principle obtain any real values, then a simple model is specified by assuming Gaussian full conditionals with expected values given by the averages of the neighbouring values:

$$\mathrm{E}(\eta_k \mid \boldsymbol{\eta}_{-k}; \boldsymbol{\xi}) = \frac{1}{N_k} \sum_{j \sim k} \eta_j. \tag{3}$$

For such conditional specifications to be consistent, the precision parameters must be set as τN_k for some $\tau > 0$

$$p(\eta_k \mid \boldsymbol{\eta}_{-k}, \boldsymbol{\xi}, \tau) = (\tau N_k / 2\pi)^{1/2} \exp\left[-\frac{\tau N_k}{2} \left(\eta_k - \frac{1}{N_k} \sum_{j \sim k} \eta_j \right)^2 \right].$$

The implied joint distribution is a Gaussian pairwise difference prior

$$p(\boldsymbol{\eta} \mid \boldsymbol{\xi}, \tau) \propto (\tau / 2\pi)^{K/2} \, |W|^{1/2} \exp\left[-\frac{\tau}{2} \sum_{j \sim k} (\eta_k - \eta_j)^2 \right],$$

where $W = W(\boldsymbol{\xi})$ is the positive semi-definite $K \times K$ matrix with elements $W_{kk} = N_k$, $W_{kj} = 1$ for $j \sim k$, and $W_{kj} = 0$ otherwise, and $|W|$ denotes the product of the non-zero eigenvalues of W (Besag and Higdon, 1997). The prior is improper, because the density is invariant under shifts of every coordinate by the same amount, but the posterior is proper in the presence of any informative data.

There are two hyperparameters, λ and τ, in the complete prior distribution $p(\theta) = p(\boldsymbol{\xi}, \boldsymbol{\eta})$. In the algorithm specified in the Appendix the hyperparameters are assumed to be given. It is not technically difficult, however, to extend the algorithm so that one or both of them are treated as unknown (see Heikkinen and Arjas, 1997). Section 14.7 contains further discussion on this topic.

14.5 Posterior inference

The general idea of *Markov chain Monte Carlo* methods in Bayesian inference is to construct a Markov chain that has the posterior distribution of interest as its equilibrium distribution. A realisation obtained by simulating this chain can be regarded as a dependent, approximate sample from the posterior, and various posterior inferences can be drawn from empirical data analysis of this sample. For example, posterior means can be approximated by corresponding averages and posterior probabilities by corresponding proportions in the MCMC sample.

In the simulation of an appropriate Markov chain the *reversible jump Markov chain Monte Carlo* method of Green (1995) follows the general Metropolis-Hastings recipe (Metropolis *et al.*, 1953; Hastings, 1970) by iterating the two-stage basic update step:

- Given the current state θ (defined by $\boldsymbol{\xi}$ and $\boldsymbol{\eta}$ in the present case), propose a move of (a randomly or systematically chosen) type m by drawing the proposed new state θ' (defined by $\boldsymbol{\xi}'$ and $\boldsymbol{\eta}'$) from the proposal distribution corresponding to move type m and conditional on θ.

- With probability $\alpha_m(\theta' \mid \theta)$, chosen so as to maintain detailed balance within each move type, accept θ' as the new state of the sampler chain; otherwise retain the current state θ.

The novelty in Green (1995) was to provide a general approach for handling pairs θ and θ' of possibly differing dimension (different numbers of generating points in our case). It should be mentioned here that such an algorithm was developed already in Geyer and Møller (1994) in the context of spatial point processes, and that Tierney (1995) formulated an even more general framework than that of Green.

To define an irreducible prototype sampler for our posterior it is sufficient to consider the following proposal types:

1. Change the function value to η'_k in a randomly chosen tile E_k.
2. Add a new randomly located (marked) generating point (ξ'_k, η'_k), $k = K+1$.
3. Remove a randomly chosen generating point (ξ_k, η_k).

A detailed description of the suggested sampler is given in the Appendix. This section concludes with some computational considerations related to the evaluation of the posterior ratio

$$\frac{p(\theta' \mid \mathbf{x})}{p(\theta \mid \mathbf{x})} = \frac{p(\theta')}{p(\theta)} \frac{p(\mathbf{x} \mid \theta')}{p(\mathbf{x} \mid \theta)},$$

required in computing the appropriate acceptance probabilities.

For a proposal of type 1 the prior ratio,

$$\frac{p(\theta')}{p(\theta)} = \frac{p(\eta'_k \mid \boldsymbol{\eta}_{-k}, \boldsymbol{\xi})}{p(\eta_k \mid \boldsymbol{\eta}_{-k}, \boldsymbol{\xi})}$$

is immediately available from the Markov random field specification of the function value prior. Owing to the local dependence these full conditionals are also easy to evaluate. Obviously, this is one reason for the popularity of locally dependent MRF priors in Bayesian image analysis.

Move types 2 and 3 are more problematic, since the matrix $W = W(\boldsymbol{\xi})$ in $p(\boldsymbol{\eta} \mid \boldsymbol{\xi})$ is different from $W(\boldsymbol{\xi}')$. The 'determinant' $|W|$ does not factorise, and it would be practically infeasible to evaluate it in each update. Heikkinen and Arjas (1998) suggested a local approximation: Replace W by a smaller matrix corresponding to a reduced generating point pattern that consists of ξ_k (or ξ'_k, as appropriate) and of its neighbours. This means acting as if E were the union of the corresponding tiles only. Then the computations stay local, which is an essential requirement in designing a successful Markov chain Monte Carlo sampler in a high-dimensional parameter space.

The data enter into the computation through the likelihood ratio

$$\frac{p(\mathbf{x} \mid \theta')}{p(\mathbf{x} \mid \theta)},$$

where θ is the current and θ' the proposed parameter function. For the models of Section 14.2 this is easy to evaluate. Consider, for example, likelihood (1) for density estimation, and let $A \subset E$ be the set where θ' differs from θ. For move type 1 A is just E_k, and for move types 2 and 3 A consists of $E_k(\boldsymbol{\xi})$ (or $E_k(\boldsymbol{\xi}')$) and its neighbouring tiles. Then the likelihood ratio

$$\frac{p(\mathbf{x} \mid \theta')}{p(\mathbf{x} \mid \theta)} = \prod_{i : x_i \in A} \frac{\theta'(x_i)}{\theta(x_i)}$$

only involves the restriction of $\mathbf{x}$, θ and θ' to A, i.e. it can be locally evaluated.

In this case, as well as in the other examples of Section 14.2, the essential reason for easy evaluation of the likelihood ratio is the conditional independence of data items x_i given θ. Usually it is only dependencies (not contained in θ) between the data items that can cause serious problems in evaluating the ratio. A concrete example of such case is given in Heikkinen and Penttinen (1995), where θ is the pair potential function of the Gibbs point process model for dependent spatial locations.

14.6 Example in intensity estimation

To illustrate a simple application in a surface estimation context, this section briefly reviews an example from Heikkinen and Arjas (1998), where more details can be found. The top right display of Figure 2 shows a point pattern simulated from a non-homogeneous planar Poisson process (Section 14.3.1) on a unit square with the intensity function shown on the top left. In the bottom row are a simple kernel density estimate (left) and a posterior mean estimate obtained by the method proposed here (right). The latter seems to exhibit better adaptivity: smoothness in the 'flat' regions, but also abrupt changes of intensity level on the steep slopes of the 'ridge'. Heikkinen and Arjas (1998) also contains some quantitative comparisions indicating that the proposed method indeed restores the original intensity surface better than simple kernel density estimates.

It should be remembered that the output of our algorithm is an (approximate) sample from the entire posterior distribution, and that the posterior mean estimate is only its most basic summary. Various other aspects of the posterior can be easily studied from the MCMC sample. As just one example, Figure 3 shows the marginal posterior densities of the intensity $\lambda(s)$ at three locations, $s = (0.2, 0.7), (0.4, 0.4), (0.8, 0.8)$.

14.7 Discussion

The approach described here is not limited to step function approximations. The prior is specified for the marked generating points, and the interpolation between them can be, in principle, carried out in any other way, as long as it is local. In curve estimation, for example, one could add fixed generating point locations at the ends of the interval E, and apply linear interpolation between the generating points. This change would only affect the likelihood computations. On the plane one could obtain piecewise linear functions with the help of the Delaunay triangulation (see Section 14.3.2): The marked generating points determine a unique function that is linear

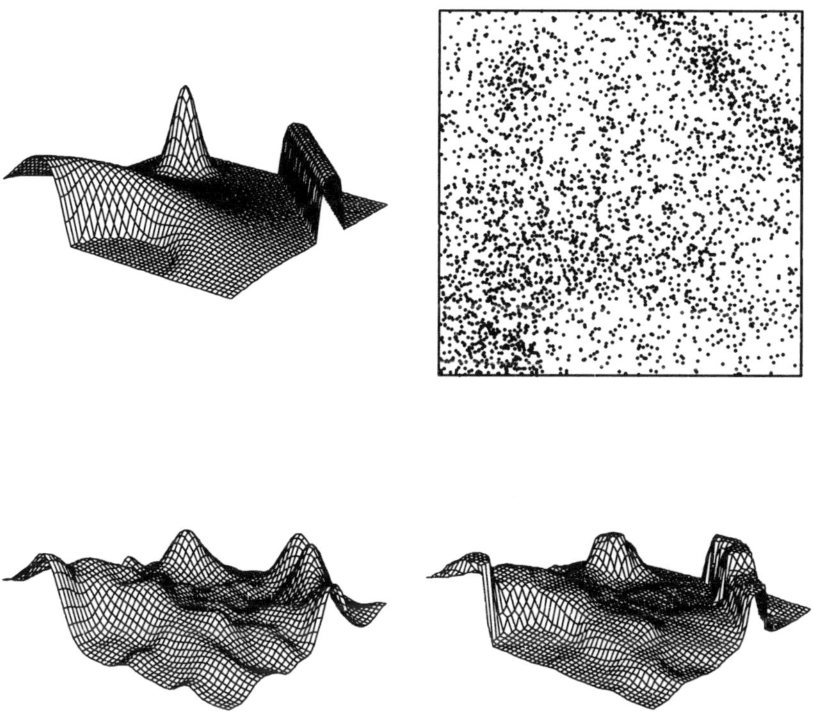

FIGURE 2. Point pattern (top left) simulated using the intensity surface on top right, kernel estimate (bottom left), and posterior mean estimate by the proposed method (bottom right).

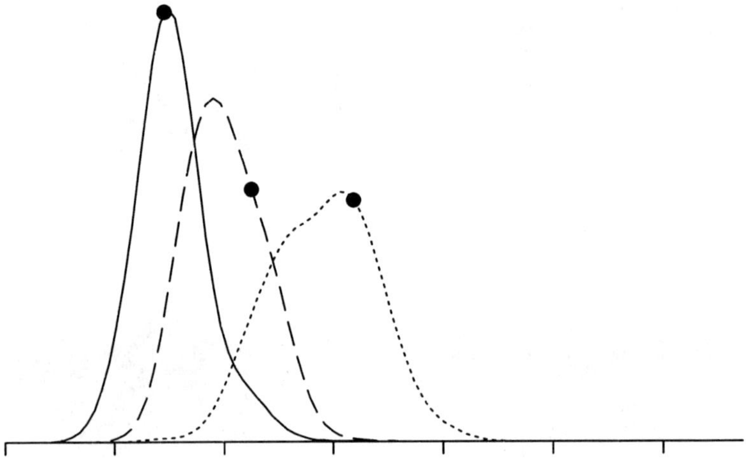

FIGURE 3. Marginal posterior densities of the intensity values at points $s = (0.2, 0.7)$ (solid line), $s = (0.4, 0.4)$ (dashed line), and $s = (0.8, 0.8)$ (dotted line). The dots on the estimated density curves are horizontally located at the corresponding true intensity values.

within each Delaunay triangle. Figure 1 clearly reveals one problem with this idea: How to handle the boundary areas of E? In Nicholls (1998) this is solved by defining another process of generating points along the boundary. Further work is currently being pursued towards implementing this kind of extensions to the method.

The choice of pairwise difference prior in Section 14.4 is convenient for two reasons. First, it has only one parameter, as opposed to the corresponding proper multivariate normal prior used in Heikkinen and Arjas (1998), where three parameters are required. Secondly, the realisations can be re-scaled without changing the prior density value. This is useful in density estimation, and in models where identifiability constraints are needed (see Heikkinen and Arjas, 1997). For positive valued functions, the log-transform can naturally be applied to enable the use of a Gaussian prior. In image classification, for example, the surface of interest is truly piecewise constant and its values are restricted to a (small) finite set of labels. In that case one could use a discrete valued MRF, such as the Potts model, to encourage segments of complex shape than the convex Voronoi polygons.

To specify the conditional expectations (3) we have experimented with different weighted averages

$$\sum_{j \sim k} w_{kj} \eta_j / w_{k+},$$

where $w_{k+} = \sum_{j \sim k} w_{kj}$. One choice is to assign the weight as a function of the distance between the corresponding generating points (Heikkinen and Arjas, 1997). With surfaces there are other natural alternatives such as the length of the common boundary segment (Heikkinen and Arjas, 1998). We have not systematically studied the effects of different weighting schemes, but it seems that they influence the adaptivity to rapid changes (Arjas and Heikkinen, 1997). It should be remembered that in Gaussian schemes the variances $1/\tau w_{k+}$ of the full conditionals are also determined up to scale by the choice of the weights.

In Heikkinen and Arjas (1997) we treat parameter τ as unknown. We believe, however, that it is generally not a good idea to let everything loose. It seems natural to use λ as a control variable to adjust the degree of smoothing, as its choice also has a considerable direct effect on the computational burden through the typical number of generating points.

The role of the normalising 'determinant' $|W|$ is to ensure that the marginal prior of the generating points is indeed a Poisson distribution. Using the wrong normalisation has no effect on the conditional prior of the function values, but it may change the prior distribution of the number and locations of generating points. This far we have not recognized any undesirable implications of using simple approximations. Intended future work includes a more systematic study of this issue.

The prototype sampler specified in Appendix has been designed to be simple rather than efficient. Green (1995) suggests that it is a good idea to preserve the integral of θ in the dimension changing moves by appropriately modifying the function values in the neighbourhood of the added or deleted generating point. The performance of the current sampler can be controlled by tuning the sampler parameters. Parameters δ and C control the magnitude of proposed changes and thereby the proportions of accepted proposals (small changes are typically more often accepted). It has been suggested that proportions around 0.4, or even less, should indicate a reasonably well mixing sampler.

Acknowledgments: This chapter is mainly based on joint work with Elja Arjas. Fruitful discussions with Jesper Møller and Peter Green have had a considerable effect on the views expressed here. I am also grateful to the anonymous referee for valuable comments.

Appendix: Details of the posterior sampler

The reversible jump Markov chain Monte Carlo algorithm described here is nearly the simplest possible for sampling from the posterior distribution $p(\theta \mid \mathbf{x})$ derived from the prior $p(\theta)$ of Section 14.4 and an arbitrary likelihood $p(\mathbf{x} \mid \theta)$. It is intented to serve as a prototype algorithm that contains the vital components and should do the job, but that can certainly be improved for better mixing and extented to cope with more complicated problems.

The procedure is started by choosing an arbitrary initial pattern $\boldsymbol{\xi}^{(0)}$ of at least two generating points located within E and associated function values $\boldsymbol{\eta}^{(0)}$ to define an initial piecewise constant parameter function $\theta^{(0)}$. An iterative sequence $\theta^{(1)}, \theta^{(2)}, \ldots$ of functions is then produced by proposing changes to the current state and accepting or rejecting them according to a probabilistic rule. Denoting by θ' the proposed function, acceptance in the tth iteration implies that $\theta^{(t)}$ is chosen to be θ', and in case of rejection the current state is retained: $\theta^{(t)} = \theta^{(t-1)}$.

To specify the details of one iteration, let θ denote the current state, $\boldsymbol{\xi} = (\xi_1, \ldots, \xi_K)$ its generating points, $\mathcal{E} = \{E_1, \ldots, E_K\}$ their Voronoi tessellation and $\boldsymbol{\eta} = (\eta_1, \ldots, \eta_K)$ the associated function values. The proposal is created by applying one of the following moves.

1. *Change of one function value:* Sample an index k from the uniform distribution on $\{1, \ldots, K\}$ and the new function value η'_k for the associated tile from the uniform distribution on $[\eta_k - \delta, \eta_k + \delta]$, where $\delta > 0$ is a parameter of the sampler. Let $K' = K$, $\boldsymbol{\xi}' = \boldsymbol{\xi}$ and $\boldsymbol{\eta}'_{-k} = \boldsymbol{\eta}_{-k}$.

2. *Birth of a new marked generating point:* Sample a new location ξ' from the uniform distribution on E and the associated function value

η' as described below. Let $K' = K+1$, $\boldsymbol{\xi}' = (\xi'_1, \ldots, \xi'_{K'})$ with $\boldsymbol{\xi}'_{-K'} = \boldsymbol{\xi}$ and $\xi'_{K'} = \xi'$, and define $\boldsymbol{\eta}'$ analogously.

3. *Death of a generating point:* Sample an index k from the uniform distribution on $\{1, \ldots, K\}$. Re-index $\boldsymbol{\xi}$ and $\boldsymbol{\eta}$ so that kth and Kth components are switched. Let $K' = K - 1$, $\boldsymbol{\xi}' = (\xi'_1, \ldots, \xi'_{K'}) = \boldsymbol{\xi}_{-K}$, and define $\boldsymbol{\eta}'$ analogously.

The actual proposal θ' is the piecewise constant function defined by pattern $\boldsymbol{\xi}'$, its tessellation $\mathcal{E}' = \{E'_1, \ldots, E'_{K'}\}$, and function values $\boldsymbol{\eta}'$.

In each iteration one of the proposal types 1–3 is chosen randomly with probabilities h_K, b_K and d_K, respectively, depending on the current number of generating points. Following Green (1995) they are defined as

$$b_K = \begin{cases} c & \text{if } K \leq \lambda\,|E| - 1, \\ c\frac{\lambda |E|}{K+1} & \text{if } K > \lambda\,|E| - 1, \end{cases}$$

$$d_K = \begin{cases} 0 & \text{if } K = 2, \\ c\frac{K}{\lambda|E|} & \text{if } 2 < K \leq \lambda\,|E|, \\ c & \text{if } K > \lambda\,|E|, \end{cases}$$

and

$$h_K = 1 - b_K - d_K,$$

where $|E|$ is the area of E, and sampler parameter $c \in (0, \frac{1}{2})$ controls the rate at which dimension changing moves are proposed.

In type 2 move the function value proposed for the new tile is defined as $\eta'_{K'} = \tilde{\eta} + \varepsilon$, where

$$\tilde{\eta} = \sum_{k \in \partial_{K'}(\boldsymbol{\xi}')} \frac{|E_k| - |E'_k|}{|E'_{K'}|} \eta_k$$

is a weighted average of the current neighbouring values, and perturbation $\varepsilon \in \mathbf{R}$ is drawn from the density $g(\varepsilon) = Ce^{C\varepsilon}/(1 + e^{C\varepsilon})^2$. Here $C > 0$ is yet another sampler parameter.

Type 1 move is a usual dimension preserving Metropolis update, and accordingly it is accepted with probability

$$\alpha(\theta' \,|\, \theta) = \min\left\{1, \frac{p(\theta' \,|\, \mathbf{x})}{p(\theta \,|\, \mathbf{x})}\right\}.$$

After some rather straightforward algebra it is found that

$$\frac{p(\theta' \,|\, \mathbf{x})}{p(\theta \,|\, \mathbf{x})} = \exp\left[-\tau(\eta'_k - \eta_k)\left(N_k \frac{\eta'_k + \eta_k}{2} - \sum_{j \sim k} \eta_j\right)\right] \frac{p(\mathbf{x} \,|\, \theta')}{p(\mathbf{x} \,|\, \theta)}.$$

To work out the appropriate acceptance probabilities for proposal types 2 and 3, let us now forget what used to be move type 1, and re-label the dimension changing moves by index $m = 2, 3, \ldots$. In a move of type m in the new labelling either the current number of generating points is $K = m$ and a birth of a new one is proposed, or $K = m + 1$ and a death is proposed. To ensure that this is a pair of reversible jumps, it was necessary in the death move to give positive proposal probabilities to the deletion of any generating point, and in the birth move to have the proposal density for the location of the new point positive all over E, as well as the proposal density for the new function value positive in the entire real axis (by defining g to be non-zero everywhere). To put it a little more formally let $P(d\theta)$ be the posterior probability of $d\theta$ and $Q_m(d\theta' \mid \theta)$ the probability to propose a move of type m and obtain a proposal in $d\theta'$ given the current state θ. Then a symmetric measure can be constructed, such that the equilibrium joint proposal distribution $R_m(d\theta, d\theta') = P(d\theta)Q_m(d\theta' \mid \theta)$ has density $r_m(\theta, \theta') = p(\theta \mid \mathbf{x})q_m(\theta' \mid \theta)$ with respect to that measure. Here

$$q_m(\theta' \mid \theta) = \begin{cases} b_K g(\eta'_{K'} - \tilde{\eta}) / |E|, & \text{if } K = m \text{ and } K' = m+1, \\ d_K / K, & \text{if } K = m+1 \text{ and } K' = m, \\ 0 & \text{otherwise,} \end{cases}$$

and $r_m(\theta, \theta')$ is positive if and only if $r_m(\theta', \theta)$ is. Hence the dimension matching criterion of Green (1995) is satisfied.

Accordingly, a birth proposal from θ with $K = m$ to θ' with $K' = m+1$ should be accepted with probability

$$\alpha(\theta' \mid \theta) = \min\left\{1, \frac{p(\theta' \mid \mathbf{x}) q_m(\theta \mid \theta')}{p(\theta \mid \mathbf{x}) q_m(\theta' \mid \theta)}\right\},$$

since the Jacobian $\left|\frac{\partial \theta'(\theta, \varepsilon)}{\partial \theta \partial \varepsilon}\right|$ is always equal to 1. The posterior ratio turns out to be

$$\frac{p(\theta' \mid \mathbf{x})}{p(\theta \mid \mathbf{x})} = \lambda \left(\frac{\tau}{2\pi} \frac{|W(\boldsymbol{\xi}')|}{|W(\boldsymbol{\xi})|} \right)^{1/2}$$
$$\exp\left\{ -\frac{\tau}{2} \sum_{k \in \partial_{K'}(\boldsymbol{\xi})} \left[(\eta'_{K'} - \eta_k)^2 - \frac{1}{2} \sum_{j \in \partial_k(\boldsymbol{\xi}) \setminus \partial_k(\boldsymbol{\xi}')} (\eta_k - \eta_j)^2 \right] \right\} \frac{p(\mathbf{x} \mid \theta')}{p(\mathbf{x} \mid \theta)}.$$

For the reverse proposal, the death of ξ_K, $K = m + 1$, the acceptance probability is obtained similarly with

$$\frac{p(\theta' \mid \mathbf{x})}{p(\theta \mid \mathbf{x})} = \lambda^{-1} \left(\frac{2\pi}{\tau} \frac{|W(\boldsymbol{\xi}')|}{|W(\boldsymbol{\xi})|} \right)^{1/2}$$
$$\exp\left\{ -\frac{\tau}{2} \sum_{k \in \partial} \left[\frac{1}{2} \sum_{j \in \partial_k(\boldsymbol{\xi}') \setminus \partial_k(\boldsymbol{\xi})} (\eta_k - \eta_j)^2 - (\eta_K - \eta_k)^2 \right] \right\} \frac{p(\mathbf{x} \mid \theta')}{p(\mathbf{x} \mid \theta)}.$$

Except for the ratios $|W(\boldsymbol{\xi}')| \,/\, |W(\boldsymbol{\xi})|$, all computations required can be handled locally: Only the mark of the generating point to be added or deleted, and those of its neighbours are needed. For the ratio of the 'determinants', an approximation, similar to that in Heikkinen and Arjas (1998), is suggested. To be specific, consider a death move where $\boldsymbol{\xi}' = \boldsymbol{\xi}_{-K}$. Let $\tilde{\boldsymbol{\xi}} = (\tilde{\xi}_1, \ldots, \tilde{\xi}_{N_K+1})$ be the subpattern of $\boldsymbol{\xi}$, where $\tilde{\xi}_{N_K+1} = \xi_K$ and $\tilde{\xi}_1, \ldots, \tilde{\xi}_{N_K}$ are the neighbouring points ξ_k, $k \in \partial_K(\boldsymbol{\xi})$, indexed arbitrarily. Then the approximation consists of replacing $|W(\boldsymbol{\xi})|$ by a $N_K + 1 \times N_K + 1$ matrix $\tilde{W}$ with elements $\tilde{W}_{kk} = \#\{j \in \partial_K \cup \{K\} : \xi_j \underset{\boldsymbol{\xi}}{\sim} \xi_k\}$, $\tilde{W}_{kj} = 1$ for $\tilde{\xi}_j \underset{\boldsymbol{\xi}}{\sim} \tilde{\xi}_k$, and $\tilde{W}_{kj} = 0$ otherwise. In other words, $\tilde{W}$ is built in the same way as the original W, only everything outside the union of E_K and its neighbours is neglected. Matrix $W(\boldsymbol{\xi}')$ is replaced by a $N_K \times N_K$ matrix $\tilde{W}'$ built from pattern $\tilde{\boldsymbol{\xi}}' = \tilde{\boldsymbol{\xi}}_{-(N_K+1)}$ in a similar manner using the neighbourhood relation $\underset{\boldsymbol{\xi}'}{\sim}$.

References

Arjas, E. (1996), Discussion of paper by Hartigan, in *Bayesian Statistics 5* (eds: J. M. Bernardo, J. O. Berger, A. P. Dawid and A. F. M. Smith), Oxford University Press, pp. 221–222.

Arjas, E. and Andreev, A. (1996), "A note on histogram approximation in Bayesian density estimation," in *Bayesian Statistics 5* (eds: J. M. Bernardo, J. O. Berger, A. P. Dawid and A. F. M. Smith), Oxford University Press, pp. 487–490.

Arjas, E. and Gasbarra, D. (1994), "Nonparametric Bayesian inference from right censored survival data, using the Gibbs sampler," *Statistica Sinica*, 4, 505–524.

Arjas, E. and Heikkinen, J. (1997), "An algorithm for nonparametric Bayesian estimation of a Poisson intensity," *Computational Statistics*, 12, 385–402.

Besag, J. (1974), "Spatial interaction and the statistical analysis of lattice systems (with discussion)," *Journal of the Royal Statistical Society, Series B*, 36, 192–236.

Besag, J. and Higdon, D. (1997), "Bayesian inference for agricultural field experiments," technical report, Department of Statistics, University of Washington.

Besag, J. and Kooperberg, C. (1995), "On conditional and intrinsic autoregressions," *Biometrika*, 81, 733–746.

Cressie, N. A. C. (1993), *Statistics for Spatial Data*, revised edition, Wiley, New York.

Denison, D. G. T., Mallick, B. K. and Smith, A. F. M. (1998), "Automatic Bayesian curve fitting," *Journal of the Royal Statistical Society, Series B*, to appear.

Geyer, C. J. and Møller, J. (1994), "Simulation procedures and likelihood inference for spatial point processes," *Scandinavian Journal of Statistics*, 21, 359–373.

Green, P. J. (1995), "Reversible jump MCMC and Bayesian model determination," *Biometrika*, 82, 711–732.

Hastings, W. K. (1970), "Monte Carlo sampling methods using Markov chains and their applications," *Biometrika*, 57, 97–109.

Heikkinen, J. and Arjas, E. (1997), "Nonparametric Bayesian estimation of a spatial Poisson intensity using a partially observed concomitant variable," manuscript.

Heikkinen, J. and Arjas, E. (1998), "Nonparametric Bayesian estimation of a spatial Poisson intensity," *Scandinavian Journal of Statistics*, to appear.

Heikkinen, J. and Penttinen, A. (1995), "Bayesian smoothing in estimation of the pair potential function of Gibbs point processes," preprint 17, Department of Statistics, University of Jyväskylä.

Karr, A. F. (1991), *Point Processes and their Statistical Inference*, 2nd edition, Dekker, New York.

Metropolis, N., Rosenbluth, A. W., Rosenbluth, M. N., Teller, A. H. and Teller, E. (1953), "Equation of state calculations by fast computing machines," *the Journal of Chemical Physics*, 21, 1087–1092.

Møller, J. (1994), *Lectures on Random Voronoi Tessellations*, number 87 in "Lecture Notes in Statistics," Springer-Verlag, New York.

Nicholls, G. (1998), "Bayesian image analysis with Markov chain Monte Carlo and colored continuum triangulation models," *Journal of the Royal Statistical Society, Series B*, to appear.

Okabe, A., Boots, B. and Sugihara, K. (1992), *Spatial Tessellations. Concepts and Applications of Voronoi Diagrams*, Wiley, Chichester.

Stoyan, D., Kendall, W. S. and Mecke, J. (1995), *Stochastic Geometry and its Applications*, 2nd edition, Wiley, New York.

Tierney, L. (1995), "A note on Metropolis Hastings kernels for general state spaces," technical report, School of Statistics, University of Minnesota.

15

Prior Elicitation for Semiparametric Bayesian Survival Analysis

Joseph G. Ibrahim
Debajyoti Sinha

ABSTRACT The quantification of prior information is a very important problem in a Bayesian analysis. In many situations, especially in clinical trials, the investigator has historical data from past studies which are similar to the current study. In this chapter, we discuss a class of informative prior distributions for Cox's proportional hazards model. A novel construction of the prior is developed for this semiparametric model based on the notion of the availability of historical data. The prior specifications focus on the observables in that the elicitation is based on a prior prediction y_0 for the response vector and a quantity a_0 quantifying the uncertainty in y_0. Then, y_0 and a_0 are used to specify a prior for the regression coefficients in a semi-automatic fashion. One of the main applications of our proposed priors is for model selection. Efficient computational methods are proposed for sampling from the posterior distribution and computing posterior model probabilities. A real data set is used to demonstrate our methodology.

15.1 Introduction

Historical data are often available in applied research settings where the investigator has access to previous studies measuring the same response and covariates as the current study. For example, in many cancer and AIDS clinical trials, current studies often use treatments that are very similar or slight modifications of treatments used in previous studies. In carcinogenicity studies, large historical databases exist for the control animals from previous experiments. It is natural to incorporate the historical data into the current study by quantifying it with a suitable prior distribution on the model parameters.

From a Bayesian perspective, historical data from past similar studies can be very helpful in interpreting the results of the current study. However, very few methods exist in Bayesian survival analysis for the formal incorporation of previous data to construct the prior distribution. There is some literature addressing this issue for the linear model and generalized linear models. See for example, Chen, Ibrahim, and Yiannoutsos (1998), Bedrick, Christensen, and Johnson (1996), and the references therein.

In this chapter we discuss elicitation of prior processes and prior distributions for semiparametric Bayesian models in survival analysis. We focus our discussion on Cox's proportional hazards model. Sinha and Dey (1997) give a nice overview of Bayesian semiparametric survival models. Semiparametric Bayesian analyses of complex survival models are becoming computationally feasible due to modern technology and recent advances in computing techniques such as the Gibbs sampler and other Markov chain Monte Carlo (MCMC) methods. See also the contribution from Sinha and Dey in this volume for a current review. Elicitating the hyperparameters associated with the prior distributions of the parameters and prior processes of the unknown functions is an important step for semiparametric Bayesian analysis. It often is not an easy task to elicit the hyperparameters, especially when the entertained models are very complex. There are very few articles in the literature that address prior elicitation for semiparametric Bayesian analysis. Use of reference priors and diffuse prior processes may lead to some difficulties, such as ambiguous posterior probabilities, and may require problem-specific modifications.

Ibrahim, Chen, and MacEachern (1996) propose a class of semi-parametric informative priors for the Cox model and examine the problems of prior elicitation and variable subset selection. Ibrahim and Chen (1997) propose a more general class of priors that can be viewed as a generalization to those of Ibrahim, Chen, and MacEachern (1996). This chapter gives an overview of these two papers. For the proportional hazards model, we specify a nonparametric prior for the baseline hazard rate and a parametric prior for the regression coefficients. The priors are best suited for situations where historical data is available. The potential advantage of using Bayesian methods to jointly model the baseline hazard and the regression coefficients is that one can accurately compute posterior quantities of interest using MCMC simulation techniques. Our prior specification focuses on the observables in that the elicitation is primarily based on a prior prediction y_0 for the response vector and a quantity a_0 quantifying the uncertainty in y_0. Then, y_0 and a_0 are used to specify the prior in a semi-automatic fashion. This type of elicitation scheme is quite attractive in variable subset selection, for example, since all that is needed is a one time specification of the prior prediction y_0 and a_0. This proves to be a useful approach for this problem since specifying meaningful prior distributions for the parameters in each model is a difficult task requiring contextual interpretations of a large number of parameters. We demonstrate our priors in the model selection context in Section 5.

The remainder of the chapter is organized as follows. In Section 2.1, we describe our semiparametric survival model and the necessary notation. In Section 2.2, we discuss nonparametric priors for the baseline hazard rate. In Section 2.3, we give the approximate likelihood based on the semiparametric model. In Sections 2.4-2.5, we discuss the prior distributions for the model parameters, and in Section 2.6, we discuss elicitation of the prior parameters. In Section 3, we discuss methods for sampling from the posterior distribution and in Sections 4.1 and 4.2, we discuss methods for computing prior and posterior model probabilities for variable subset selection. We demonstrate the priors and computational methods with a real example in Section 5. We conclude the chapter with a brief discussion section.

15.2 The Method

15.2.1 Model and Notation

A proportional hazards Cox (Cox, 1972) model is defined by a hazard function of the form

$$h(t,x) = h_0(t)\exp(x'\beta) , \tag{1}$$

where $h_0(t)$ denotes the baseline hazard function at time t, x denotes the $p \times 1$ covariate vector for an arbitrary individual in the population, and β denotes a $p \times 1$ vector of regression coefficients. The likelihood function for a set of right censored data on n individuals for the current study under a Cox model of (1) is given by

$$L\left(\beta, h_0(t)\right) = \prod_{i=1}^{n} \left[h_0(t_i)\eta_i\right]^{\nu_i} \left(S_0(t_i)^{\eta_i}\right), \tag{2}$$

where $\eta_i = \exp(x_i'\beta)$, t_i is an observed event time or censoring time for the ith individual and ν_i is the indicator variable taking on the value 1 if t_i is an event time, and 0 if it is a censoring time. Moreover, x_i is a $p \times 1$ vector of covariates for the ith individual, and X denotes the $n \times p$ covariate matrix of rank p. Further, $F_0(\cdot)$ denotes the baseline cumulative distribution function, and $S_0(\cdot) = 1 - F_0(\cdot)$ is the baseline survivor function, which, since we consider continuous survival distributions, is related to $h_0(\cdot)$ by $S_0(t) = \exp\left(-\int_0^t h_0(u)\, du\right)$.

15.2.2 The Approximate Likelihood Function

We consider a nonparametric prior for the baseline hazard and a parametric prior for β. Specifying a nonparametric prior for the baseline hazard rate allows for a large and flexible class of hazards. Moreover, it also facilitates comparison with Cox's partial likelihood (Cox, 1975). To define an approximation to the prior process on $h_0(t)$, we first construct a finite partition of the time axis. Let $0 = s_0 < s_1 < \ldots < s_M$ denote this finite partition with $s_M > t_i$ for all $i = 1, \ldots, M$. Further, let

$$\delta_i = h_0(s_i) - h_0(s_{i-1})$$

denote the increment in the baseline hazard in the interval $(s_{i-1}, s_i]$, $i = 1, \ldots, M$. Further, let $\Delta = (\delta_1, \ldots, \delta_M)$. We follow Ibrahim, Chen, and MacEachern (1996) for constructing the approximate likelihood function of (β, Δ). For an arbitrary individual in the population, the survival function for the Cox model at time s is given by

$$S(s \mid x) = \exp\left\{-\eta \int_0^s h_0(t)\, dt\right\} \simeq \exp\left\{-\eta \left(\sum_{i=1}^{M} \delta_i (s - s_{i-1})^+\right)\right\}, \tag{3}$$

where $h_0(0) = 0$, $(t)^+ = t$ if $t > 0$, 0 otherwise, and $\eta = \exp(x'\beta)$. This first approximation arises since the specification of Δ does not specify the entire hazard rate, but only the δ_i. For purposes of approximation, we take the increment in the hazard rate, δ_i, to occur immediately after s_{i-1}. Let p_i denote the probability of a failure in the interval $(s_{i-1}, s_i], i = 1, \ldots, M$. Using (3), we have

$$\begin{aligned} p_i &= S(s_{i-1}) - S(s_i) \\ &\simeq \exp\left\{-\eta \sum_{j=1}^{i-1} \delta_j (s_{i-1} - s_{j-1})\right\} \left[1 - \exp\left\{-\eta (s_i - s_{i-1}) \sum_{j=1}^{i} \delta_j\right\}\right]. \end{aligned}$$

Thus, in the ith interval $(s_{i-1}, s_i]$, the contribution to the likelihood function for a failure is p_i and $S(s_i)$ for a right censored observation. For $i = 1, \ldots, M$, let d_i be the number of failures, $\mathcal{D}_i$ be the set of subjects failing, c_i be the number of right censored observations and $\mathcal{C}_i$ be the set of subjects getting censored in the ith interval. Define

$$\begin{aligned} a_i &= \sum_{j=i+1}^{M} \sum_{k \in \mathcal{D}_j} \eta_k (s_{j-1} - s_{i-1}), \\ b_i &= \sum_{j=i}^{M} \sum_{k \in \mathcal{C}_j} \eta_k (s_j - s_{i-1}), \\ T_i(\Delta) &= (s_i - s_{i-1}) \sum_{j=1}^{i} \delta_j. \end{aligned} \tag{4}$$

Let $D = (n, t, X, \nu)$ denote the data for the current study. The approximated likelihood function given the data D for the current study is given by

$$L(\beta, \Delta \mid D) = \prod_{i=1}^{M} \left\{ e^{\{-\delta_i(a_i+b_i)\}} \prod_{k \in \mathcal{D}_i} \left[1 - e^{\{-\eta_k T_i\}} \right] \right\} . \tag{5}$$

We note that this likelihood involves a second approximation. Instead of conditioning on exact event times, we condition on the set of failures and set of right censored events in each interval, and thus we approximate continuous right censored data by grouped data.

15.2.3 The Prior Distributions

For ease of exposition, we assume that we have one previous study, as the extension to multiple previous studies is straightforward. Let $D_0 = (n_0, y_0, X_0, \nu_0)$ denote the data from the previous study, where n_0 denotes the sample size of the previous study, y_0 denotes a right censored vector of survival times with censoring indicators ν_0, and X_0 denotes the $n \times p$ matrix of covariates. We shall refer to D_0 as the historical data throughout.

In general, for most problems, there are no firm guidelines on the method of prior elicitation. Typically, one tries to balance sound theoretical ideas with practical and computationally feasible ones. The issue of how to use D_0 study into the current study has no obvious solution since it depends in large part of how similar the two studies are. In most clinical trials, for example, no two studies will ever be identical. In many cancer clinical trials, the patient populations typically differ from study to study even when the same regimen is used to treat the same cancer. In addition, other factors may make the two studies heterogeneous. These include conducting the studies at different institutions or geographical locations, using different physicians, using different measurement instruments and so forth. Due to these differences, an analysis which combines the data from both studies may not be desirable. In this case, it may be more appropriate to "weight" the data from the previous study so as to control its impact on the current study. Thus, it is desirable for the investigators to have a prior distribution that summarizes the prior data in an efficient and useful manner and allows them to tune or weight D_0 as they see fit in order to control its impact on the current study.

Towards this goal, let $\pi_0(\beta, \Delta)$ denote the joint prior distribution for (β, Δ) used on the previous study. Thus, $\pi_0(\beta, \Delta)$ can be viewed as the "original prior" for the parameters before D_0 was observed. For the current study, we consider a joint prior distribution for (β, Δ) of the form

$$\pi(\beta, \Delta \mid D_0, a_0) \propto \{L(\beta, \Delta \mid D_0)\}^{a_0} \, \pi_0(\beta, \Delta) , \tag{6}$$

where $L(\beta, \Delta \mid D_0)$ is the likelihood function of (β, Δ) based on the data from previous study (i.e., D_0), and thus, $L(\beta, \Delta \mid D_0)$ is (5) with D replaced by $D_0 = (n_0, y_0, X_0, \nu_0)$. The parameter a_0 is a scalar prior parameter that controls the influence of the prior data in the prior distribution and can be interpreted as a dispersion parameter for the prior data. It is reasonable to take $0 \leq a_0 \leq 1$. Setting $a_0 = 1$ in (6) corresponds to the formal update of $\pi_0(\beta, \Delta)$ using Bayes theorem. That is, with $a_0 = 1$, (6) corresponds to the posterior distribution of (β, Δ) from the previous study. When $a_0 = 0$, no historical data is incorporated into the prior, and in this case, $\pi(\beta, \Delta \mid D_0, a_0 = 0) = \pi_0(\beta, \Delta)$. The prior in (6) can be viewed as a generalization of the usual Bayesian update of $\pi_0(\beta, \Delta)$. The parameter a_0 allows the investigator to control the influence of the historical data on the current study. As mentioned earlier, such control is desirable in situations where there is heterogeneity between the previous and current study, or the sample sizes of the two studies are far apart. In these cases, one would not want to equally weight the historical data and the current data.

To simplify the prior specification, we take

$$\pi_0(\beta, \Delta) = \pi_0(\beta \mid c_0)\ \pi_0(\Delta \mid \theta_0)\ ,$$

where c_0 and θ_0 are fixed hyperparameters. Specifically, we take $\pi_0(\beta \mid c_0)$ to be a p dimensional multivariate normal density, $N(0, c_0 W_0)$, with mean 0 and covariance matrix $c_0 W_0$, where c_0 is a specified scalar and W_0 is a $p \times p$ diagonal matrix. We take $\pi_0(\Delta \mid \theta_0)$ to be a product of M independent gamma densities of the form, each with mean f_{0i}/g_{0i} and variance f_{0i}/g_{0i}^2, $i = 1, \ldots, M$. So, we get

$$\pi_0(\Delta \mid \theta_0) \propto \prod_{i=1}^{M} \delta_i^{f_{0i}-1} \ \exp\{-\delta_i g_{0i}\}\ , \tag{7}$$

where $\theta_0 = (f_{01}, g_{01}, \ldots, f_{0M}, g_{0M})$.

The motivation for (7) is that (7) is a discrete approximation to an underlying gamma process prior for $h_0(t)$. Ibrahim, Chen, and MacEachern (1996) discuss the gamma process priors for the baseline hazard rate in detail. In addition, they demonstrate the advantages and disadvantages of the gamma process prior in model selection contexts and show numerically that the prior is quite robust in many situations. It is important to note that our prior elicitation method can be also adapted to other popular discretized prior processes such as discretized gamma process for the cumulative baseline hazard (Sinha et al., 1997), discretized beta processes for the baseline hazard rate (Sinha, 1998) and the discretized correlated process on $h_0(t)$ (Arjas and Gasbarra, 1994). The modification required for these other processes would be an appropriate selection of $\pi_0(\Delta \mid \theta_0)$. We note that in (6), (β, Δ) are not independent, and also the components of Δ are not independent a priori.

The prior specification is completed by specifying a prior for a_0. Since $0 \leq a_0 \leq 1$, a beta prior is reasonable for a_0, so that

$$\pi(a_0 \mid \alpha_0, \lambda_0) \propto a_0^{\alpha_0 - 1}(1 - a_0)^{\lambda_0 - 1}. \tag{8}$$

Thus we propose a joint prior for (β, Δ, a_0) to be of the form

$$\pi(\beta, \Delta, a_0 \mid D_0) \propto L(\beta, \Delta \mid D_0)^{a_0} \, \pi_0(\beta \mid c_0) \, \pi_0(\Delta \mid \theta_0) \, \pi(a_0 \mid \alpha_0, \lambda_0). \tag{9}$$

It can be shown that (9) is proper and is best suited for situations in which a historical data set is available. However, the prior can still be used if a previous experiment does not exist for the current study. In this case, D_0 can be elicited from expert opinion or by using case-specific information about each individual in the current study, or perhaps by a theoretical model giving forecasts for the event times. The elicitation scheme is less automated than the one in which a previous study exists. In this situation, there are a number of ways one could elicit y_0, and this elicitation depends on the context of the problem. We do not attempt to give a general elicitation scheme for y_0 in this setting, but rather mention some general possibilities. In any case, the cleanest specification of D_0 is to use data from a previous study for which y_0 would be taken to be the vector of survival times from the previous study and X_0 is taken as the design matrix from the previous study.

15.2.4 An Approximate Prior

One of the potential drawbacks of (9) is that is does not have a closed form and may be computationally intensive. For these reasons, it may be desirable to approximate (9) in a suitable way. We begin by assuming a priori independence between the the baseline hazard rate and the regression coefficients, and thus the joint prior density of (β, Δ) is given by

$$\pi(\beta, \Delta \mid D_0) = \pi(\beta \mid D_0) \, \pi(\Delta \mid D_0) \; . \tag{10}$$

The assumption of prior independence between β and Δ is a sensible specification, since the hazard rate may be a nuisance parameter for many problems. We consider a fully parametric p dimensional multivariate normal prior for β given by

$$\pi(\beta \mid a_0, \mu, T) \sim N_p(\mu, a_0^{-1} \, T^{-1}) \; , \tag{11}$$

where μ is the prior mean, $a_0 T$ is the precision matrix, and a_0 is defined in the previous section. We take μ to be the solution to Cox's partial likelihood (Cox, 1975) equations for β using D_0 as data. Suppose there are r failures and $n_0 - r$ right censored values in y_0. Cox's partial likelihood for β based on D_0 is given by

$$L^*(\beta \mid D_0) = \prod_{i=1}^{r} \left\{ \frac{\exp\{x'_{0i}\beta\}}{\sum_{\ell \in \mathcal{R}_{(y_{0i})}} \exp\{x'_{0\ell}\beta\}} \right\} \; , \tag{12}$$

where x'_{0i} is the ith row of X_0, $(y_{01}, \cdots, y_{0r})$ are the ordered failures and $\mathcal{R}(y_{0i})$ is the set of labels attached to the individuals at risk just prior to y_{0i}. Now we take μ to be the solution to

$$\frac{\partial \log \left(L^*(\beta \mid D_0)\right)}{\partial \beta_j} = 0 , \tag{13}$$

$j = 1, \ldots, p$. The matrix T is taken to be the Fisher information matrix of β based on the partial likelihood in (12). Thus

$$T = \left[\frac{-\partial^2}{\partial \beta_i \partial \beta_j} \log(L^*(\beta \mid D_0)) \right] \Big|_{\beta=\mu} . \tag{14}$$

These proposed priors represent a summary of the prior data D_0 through (μ, T) which are obtained via Cox's partial likelihood. This is a practical and useful summary of the data D_0 as indicated by many authors including Cox (1972, 1975), and Tsiatis (1981).

We emphasize that since our main goal is to make inferences on β, the parameter Δ is viewed as a nuisance parameter, and thus a noninformative prior for $\pi(\Delta \mid a_0, D_0)$ is suitable for it. Otherwise we can take $\pi(\Delta \mid a_0, D_0)$ as a product of M independent gamma densities with mean ϕ_i and variance $a_0^{-1}\gamma_i$ for $i = 1, \cdots, M$. The ϕ_i's and γ_i can be elicited from D_0 by taking $\phi_i = \hat{\delta}_{0i}$ and $\gamma_i = var(\hat{\delta}_{0i})$ where $\hat{\delta}_{0i}$ is the MLE and $var(\hat{\delta}_{0i})$ is the corresponding estimated variance of δ_i calculated from the profile likelihood $L_c(\Delta \mid D_0, \beta = \mu)$. Here $L_c(\Delta, \beta \mid D_0)$ is proprtional the probability of observing the grouped survival data D_0 given (β, Δ).

The prior given by (10) and (11) has several advantages. First, it has a closed form and is easy to interpret. Second, the prior elicitation is straightforward in the sense that (μ, T) and a_0 completely determine the prior for β. Third, (10) and (11) are computationally feasible and relatively simple. Fourth, our prior assumes a priori independence between (β, Δ), which further simplifies interpretations as well as the elicitation scheme. In addition, as discussed in Ibrahim, Chen, and MacEachern (1996), (10) and (11) provide a reasonable asymptotic approximation to the prior for (β, Δ) given in (9).

15.2.5 Choices of Prior Parameters

The choices of the prior parameters play a crucial role in any Bayesian analysis. We first discuss (α_0, λ_0). For the purposes of prior elicitation, it is easier to work with $\mu_0 = \alpha_0/(\alpha_0 + \lambda_0)$ and $\sigma_0^2 = \mu_0(1 - \mu_0)(\alpha_0 + \lambda_0 + 1)^{-1}$. A uniform prior (i.e., $\alpha_0 = \lambda_0 = 1$), which corresponds to $(\mu_0, \sigma_0^2) = (1/2, 1/12)$ may be a suitable noninformative starting point, and facilitates a useful reference analysis for other choices. The investigator may choose μ_0 to be small (say $\mu_0 \leq .1$), if he/she wishes to have low prior weight on the historical data. If a large prior weight is desired, then $\mu_0 \geq .5$ may be desirable. It is reasonable to choose σ_0^2 in the range $\mu_0/1000 \leq \sigma_0^2 \leq \mu_0/10$. In any case, in an actual analysis, we recommend that several choices of (μ_0, σ_0^2) be used, including ones that give small and large weight to the historical data, and several sensitivity analyses conducted. The choices recommended here can be used as starting points from which sensitivity analyses can be based.

It is reasonable to specify a noninformative prior for $\pi_0(\beta \mid c_0)$ since this is the prior for β corresponding to the previous study and contains no information about the historical data D_0. The quantity $c_0 \geq 0$ is a scalar variance parameter which serves to control the impact of $\pi_0(\beta \mid c_0)$ on $\pi(\beta, \Delta, a_0 \mid D_0)$. To make $\pi_0(\beta \mid c_0)$ noninformative, we take large values of c_0 so that $\pi_0(\beta \mid c_0)$ is flat relative to $L(\beta, \Delta \mid D_0)^{a_0}$. Small values of c_0 will let $\pi_0(\beta \mid c_0)$ dominate (9).

The actual size of c_0 used will depend on the structure of the data set and the prior parameters for a_0. From the example in Section 5, reasonable choices of c_0 are $c_0 \geq 3$. In any case, we do not recommend an automatic one time specification for c_0, but rather we emphasize that several sensitivity analyses be conducted with several values of c_0 to examine the impact of $\pi_0(\beta \mid c_0)$ on posterior quantities of interest. The matrix W_0 has a less crucial role than c_0. We take W_0 to be a diagonal matrix consisting of the sample variances of the covariates on the diagonal. The purpose of picking W_0 in this way is to properly adjust for the different scales of the measured covariates. If the covariates are all standardized or are measured on the same scale, then we take $W_0 = I$. In any case, W_0 plays a minimal role when c_0 is taken large.

For $\pi_0(\Delta \mid \theta_0)$, the values of θ_0 should be chosen so that $\pi_0(\Delta \mid \theta_0)$ is flat relative to $L(\beta, \Delta \mid D_0)^{a_0}$. Here, we choose f_{0i} to be proportional to the interval widths, i.e., $f_{0i} \propto s_i - s_{i-1}$, and $g_{0i} \to 0$, for $i = 1, \ldots, M$. Choosing g_{0i} small creates a noninformative gamma prior, and choosing the shape parameters to be proportional to the length of each interval is useful in allowing the variance of each δ_i to depend on the interval length. We emphasize that we always recommend doing several sensitivity analyses using various choices of $(c_0, \theta_0, \mu_0, \sigma_0^2)$. We demonstrate such sensitivity analyses in the numerical example of Section 5.

15.3 Sampling from the Joint Posterior Distribution of (β, Δ, a_0)

In this section, we describe how to sample from the joint posterior distribution of (β, Δ). Samples form the joint posterior of (β, Δ) will enable us to compute any posterior summaries involving β or Δ. For notational convenience, we denote the distribution of a random vector X by $[X]$.

The posterior density of $[\beta, \Delta, a_0 \mid D]$ for the current study is given by

$$p(\beta, \Delta, a_0 \mid D) \propto \left[\prod_{i=1}^{M} \left\{ e^{\{-\delta_i(a_i+b_i)\}} \prod_{k \in \mathcal{D}_i} \{1 - e^{(-\eta_k T_i)}\} \right\} \right] \\ \times \quad L(\beta, \Delta \mid D_0)^{a_0} \, \pi_0(\beta \mid c_0)\pi_0(\Delta \mid \theta_0)\pi(a_0 \mid \alpha_0, \lambda_0), \qquad (1)$$

where η_k, a_i and b_i are given by (4), and $L(\beta, \Delta \mid D_0)$ is the likelihood function given by (5) with D replaced by D_0. In (1), we use the same subintervals $(s_{i-1}, s_i]$ for both $L(\beta, \Delta \mid D)$ and $L(\beta, \Delta \mid D_0)$ so that Δ has the same meaning across both the historical data and the current data. To specify the s_i's, (i) we combine $\{t_i, i = 1, 2, \ldots, n\}$ and $\{y_{0i}, i = 1, 2, \ldots, n_0\}$ together, denoted by $\{t_i^*, i = 1, 2, \ldots, n + n_0\}$; (ii) the subintervals $(s_{i-1}, s_i]$ are chosen to have equal numbers of failure or censored observations for the combined survival or censored time t_i^*'s, i.e., s_i is chosen to be the $(i/M)th$ quantile of the t_i^*'s, where M is the total number of the subintervals. We take M so that in each subinterval $(s_{i-1}, s_i]$, there is at least one failure or censored observation from the y_{0i}'s and the t_i's. In Section 5, we take $M = 28$ for the myeloma data.

To obtain samples from the posterior distribution $[\beta, \Delta, a_0 \mid D]$ given by (1), we now describe a Gibbs sampling strategy for sampling from $[\beta \mid \Delta, a_0, D]$, $[\Delta \mid \beta, a_0, D]$, and $[a_0 \mid \beta, \Delta, D]$.

To sample β from its conditional posterior distribution, we first observe that the density function of $[\beta \mid \Delta, a_0, D]$ is log-concave in each component of β. Therefore, we may directly use the algorithm of Gilks and Wild (1992) to sample from this conditional posterior distribution. To show that the density function of $[\beta \mid \Delta, a_0, D]$ is log-concave in each component of β, it suffices to show that

$$\frac{\partial^2 \log p(\beta \mid \Delta, a_0, D)}{\partial \beta_r^2} \leq 0$$

for all $r = 1, \ldots, p$, where $p(\beta \mid \Delta, a_0, D)$ is the conditional posterior density of $[\beta \mid \Delta, a_0, D]$. We note that

$$\log(p(\beta \mid \Delta, a_0, D)) = \log(L(\beta, \Delta \mid D)) + a_0 \log(L(\beta, \Delta \mid D_0)) + \\ \log(\pi_0(\beta \mid c_0)). \qquad (2)$$

Since $\pi_0(\beta \mid c_0)$ is a normal density, it is clearly log-concave. If $L(\beta, \Delta \mid D)$ is log-concave, then, clearly $L(\beta, \Delta \mid D_0)^{a_0}$ will be log-concave since $a_0 \geq 0$. Thus it suffices to show that $L(\beta, \Delta \mid D)$ is log-concave. To this end, let $A_{ik}(\beta, \Delta) = \eta_k T_i(\Delta)$, $B_{ik}(\beta, \Delta) = 1 - \exp\{-A_{ik}(\beta, \Delta)\}$, and $C_{ik}(\beta, \Delta) =$

$1 - A_{ik}(\beta, \Delta) - \exp\{-A_{ik}(\beta, \Delta)\}$ we get

$$\begin{aligned}
&\frac{\partial^2 \log(L(\beta, \Delta \mid D))}{\partial \beta_r^2} \\
&= \sum_{i=1}^{M} \sum_{k=\in \mathcal{D}_i} \{x_{kr}^2 \; A_{ik} \; B_{ik}^{-2} \; \exp\{-A_{ik}\} \; C_{ik}\} \qquad (3) \\
&- \sum_{i=1}^{M} \sum_{j=i+1}^{M} \sum_{k \in \mathcal{D}_j} \{\delta_i \; x_{kr}^2 \exp\{x_k' \beta\} \; (s_{j-1} - s_{i-1})\} \qquad (4) \\
&- \sum_{i=1}^{M} \sum_{j=i}^{M} \sum_{k \in \mathcal{C}_j} \{\delta_i \; x_{kr}^2 \exp\{x_k' \beta\} \; (s_j - s_{i-1})\}. \qquad (5)
\end{aligned}$$

We see that the summands in (4) and (5) are negative. Thus, to show that $\frac{\partial^2 \log(L(\beta,\Delta|D))}{\partial \beta_r^2} \leq 0$, it suffices to show that $C_{ik}(\beta, \Delta)$ in (3) is negative, since all of the other terms in the summand of (3) are positive. We see that $C_{ik}(\beta, \Delta)$ is of the form $f(x) = 1 - e^x - e^{-e^x}$. Clearly, when $x > 0$, $f(x) < 0$. For $x \in (-\infty, 0)$, we see that $f(x)$ is a monotonic decreasing function and $\lim_{x \to -\infty} f(x) = 0$. Thus $f(x) < 0$ for all $x \in R^1$, and thus $C_{ik}(\beta, \Delta) \leq 0$ for all (β, Δ). Thus $L(\beta, \Delta \mid D)$ is log-concave in each component of β, which implies that $p(\beta \mid \Delta, a_0, D)$ is log-concave in β.

Now, we consider sampling from $[\Delta \mid \beta, a_0, D]$. Let $p(\Delta|\beta, a_0, D)$ denote the conditional posterior density of Δ. As for $p(\beta \mid \Delta, a_0, D)$, it can be shown that $p(\Delta \mid \beta, a_0, D)$ is log-concave in each component of Δ as long as $\pi_0(\Delta \mid \theta_0)$ is log-concave in each component of Δ. Thus, we may use the adaptive rejection algorithm of Gilks and Wild (1992) to sample Δ. Note that sampling Δ is slightly different from the one for β, that is, $\delta_i \geq 0$ while $-\infty < \beta_r < \infty$. To ensure the log-concavity of $\pi(\Delta|\theta_0)$, we need to take $f_{0i} \geq 1$.

Finally, we briefly discuss the generation of a_0. We note that the generation of a_0 does not depend on the data D from the current study. The conditional posterior density of a_0 can be written as

$$p(a_0 \mid \beta, \Delta, D) \;\propto\; L(\beta, \Delta \mid D_0)^{a_0} \; \pi(a_0 \mid \alpha_0, \lambda_0). \qquad (6)$$

In general, the conditional posterior density of a_0 is not log-concave. Therefore, we use an adaptive Metropolis algorithm to sample a_0 as proposed in Chen, Ibrahim, and Yiannoutsos (1998). The algorithm proceeds as follows. Consider

$$a_0 = \frac{\exp(\xi)}{1 + \exp(\xi)}, \qquad (7)$$

then the conditional posterior distribution $[\xi \mid \beta, \Delta, D]$ is

$$p(\xi \mid \beta, \Delta, D) \;\propto\; p(a_0 \mid \beta, \Delta, D) \; \frac{\exp(\xi)}{(1 + \exp(\xi))^2}, \qquad (8)$$

where $p(a_0 \mid \beta, \Delta, D)$ is given by (6) and a_0 is evaluated at $a_0 = \exp(\xi)/(1+\exp(\xi))$. Instead of directly generating a_0 from (6), we first generate ξ from (8) and then use (7) to obtain a_0. To generate ξ, we use a normal proposal $N(\hat{\xi}, \hat{\tau}^2_{\hat{\xi}})$, where $\hat{\xi}$ is a maximizer of the logarithm of the right side of (8), which can be obtained by the Nelder-Mead algorithm implemented by O'Neill (1971). Also, $\hat{\tau}^2_{\hat{\xi}}$ is minus the inverse of the second derivative of $\log p(\xi \mid \beta, \Delta, D)$ evaluated at $\xi = \hat{\xi}$, given by

$$\hat{\tau}^{-2}_{\hat{\xi}} = - \left. \frac{d^2 \log p(\xi \mid \beta, \Delta, D)}{d\xi^2} \right|_{\xi=\hat{\xi}} .$$

The algorithm to generate ξ operates as follows:

Step 1. Let ξ be the current value.

Step 2. Generate a proposal value ξ^* from $N(\hat{\xi}, \hat{\tau}^2_{\hat{\xi}})$.

Step 3. A move from ξ to ξ^* is made with probability

$$\min \left\{ \frac{p(\xi^* \mid \beta, \Delta, D) \phi\left(\frac{\xi - \hat{\xi}}{\hat{\tau}_{\hat{\xi}}}\right)}{p(\xi \mid \beta, \Delta, D) \phi\left(\frac{\xi^* - \hat{\xi}}{\hat{\tau}_{\hat{\xi}}}\right)}, 1 \right\},$$

where ϕ is the standard normal probability density function.

After we obtain ξ, we compute a_0 by using (7).

15.4 Applications to Variable Selection

One of the main applications of our priors is for variable subset selection. In the next two subsections, we develop novel computational methods for the calculation of prior and posterior model probabilities. We first establish the necessary notation. Let p denote the number of covariates for the full model and let $\mathcal{M}$ denote the model space. We enumerate the models in $\mathcal{M}$ by $m = 1, 2, \ldots, \mathcal{K}$, where $\mathcal{K}$ is the dimension of $\mathcal{M}$ and model $\mathcal{K}$ denotes the full model. Also, let $\beta^{(\mathcal{K})} = (\beta_1, \ldots, \beta_p)'$ denote the regression coefficients for the full model and let $\beta^{(m)}$ denote a $k_m \times 1$ vector of regression coefficients for model m. We write $\beta^{(\mathcal{K})} = (\beta^{(m)'}, \beta^{(-m)'})'$, where $\beta^{(-m)}$ is $\beta^{(\mathcal{K})}$ with $\beta^{(m)}$ deleted. Also, let $D_0^{(m)} = (n_0, y_0, X_0^{(m)}, \nu_0)$ denote the historical data under model m, and let $L(\beta^{(m)}, \Delta \mid D_0^{(m)})$ denote the likelihood function of $(\beta^{(m)}, \Delta)$ for model m based on the historical data $D_0^{(m)}$.

15.4.1 Prior on the Model Space

Let

$$p_0^*(\beta^{(m)}, \Delta | D_0^{(m)}) = L(\beta^{(m)}, \Delta \mid D_0^{(m)})\ \pi_0(\beta^{(m)} \mid d_0)\ \pi_0(\Delta \mid \kappa_0) \tag{1}$$

where $\pi_0(\beta^{(m)} \mid d_0)$ is taken to be a $N(0, d_0 W_0)$ density and $\pi_0(\Delta \mid \kappa_0)$ is the same density as that in (7) with θ_0 replaced by κ_0. We propose to take the prior probability of model m as

$$\begin{aligned} p(m) &\equiv p(m \mid D_0^{(m)}) \\ &= \frac{\int \int p_0^*(\beta^{(m)}, \Delta | D_0^{(m)})\ d\beta^{(m)}\ d\Delta}{\sum_{m \in \mathcal{M}} \int \int p_0^*(\beta^{(m)}, \Delta | D_0^{(m)})\ d\beta^{(m)}\ d\Delta} . \end{aligned} \tag{2}$$

Since Δ is viewed as nuisance, we recommend taking $\kappa_0 = \theta_0$ to simplify the elicitation scheme. The prior parameter d_0 controls the impact of $\pi_0(\beta^{(m)} \mid d_0)$ on the prior model probability $p(m)$. This choice for $p(m)$ has several nice interpretations. First, $p(m)$ in (2) corresponds to the posterior probability of model m based on the data $D_0^{(m)}$ using a uniform prior for the previous study. That is, $p_0(m) = 2^{-p}$ for $m \in \mathcal{M}$, where $p_0(m)$ is the prior probability of model m before observing the historical data $D_0^{(m)}$. Therefore, $p(m) \propto p(m \mid D_0^{(m)})$, and thus $p(m)$ corresponds to the usual Bayesian update of $p_0(m)$ using $D_0^{(m)}$ as the data. Second, as $d_0 \to 0$, $p(m)$ reduces to a uniform prior on the model space. Therefore, as $d_0 \to 0$, the historical data $D_0^{(m)}$ have a minimal impact in determining $p(m)$. On the other hand, with a large value of d_0, $\pi_0(\beta^{(m)} \mid d_0)$ plays a minimal role in determining $p(m)$, and in this case, the historical data plays a larger role in determining $p(m)$. Thus as $d_0 \to \infty$, $p(m)$ will be regulated by the historical data. The parameter d_0 plays the same role as c_0 and thus serves a tuning parameter to control the impact of $D_0^{(m)}$ on the prior model probability $p(m)$. It is important to note that we use a scalar parameter c_0 in constructing the prior distribution $\pi(\beta^{(m)}, \Delta, a_0 \mid D_0^{(m)})$ given in (9), while we use a *different* scalar parameter d_0 in determining $p(m)$. This development provides us with great flexibility in specifying the prior distribution for $\beta^{(m)}$ as well as the prior model probabilities $p(m)$.

To compute $p(m)$ in (2), we adopt a Monte Carlo approach similar to Ibrahim, Chen and MacEachern (1996) to estimate all of the prior model probabilities using a single Gibbs sample from the full model. The details of this procedure are given in Chen, Ibrahim, and Yiannoutsos (1998).

Following the Monte Carlo method of Chen and Shao (1997a) the prior probability of model m can be estimated by

$$\hat{p}(m) \equiv \hat{p}(m|D_0^{(m)}) = \frac{\frac{1}{N_0}\sum_{l=1}^{N_0} \frac{p_0^*(\beta_{0(l)}^{(m)},\Delta_{0(l)}|D_0^{(m)})\ w(\beta_{0(l)}^{(-m)}|\beta_{0(l)}^{(m)},\Delta_{0(l)})}{p_0^*(\beta_{0(l)}^{(\mathcal{K})},\Delta_{0(l)}|D_0^{(\mathcal{K})})}}{\frac{1}{N_0}\sum_{j=1}^{\mathcal{K}}\sum_{l=1}^{N_0} \frac{p_0^*(\beta_{0(l)}^{(j)},\Delta_{0(l)}^{(j)}|D_0^{(j)})\ w(\beta_{0(l)}^{(-j)}|\beta_{0(l)}^{(j)},\Delta_{0(l)}^{(j)})}{p_0^*(\beta_{0(l),\Delta_{0(l)}}^{(\mathcal{K})}|D_0^{(\mathcal{K})})}}, \tag{3}$$

where $p_0^*(\beta^{(m)},\Delta|D_0^{(m)})$ is given by (1), $\beta_{0(l)}^{(\mathcal{K})} = (\beta^{(m)}{}'_{0(l)}, \beta^{(-m)}{}'_{0(l)})'$, $l = 1, 2, \ldots, N_0$, are samples from

$$p_0(\beta^{(\mathcal{K})},\Delta|D_0^{(\mathcal{K})}) \ \propto\ p_0^*(\beta^{(\mathcal{K})},\Delta|D_0^{(\mathcal{K})}), \tag{4}$$

and $w(\beta^{(-m)}|\beta^{(m)},\Delta)$ is a *completely* known conditional density whose support is contained in or equal to the support of the conditional density of $\beta^{(-m)}$ given $\beta^{(m)}$ with respect to the full model joint prior distribution (4). The implementation of Markov chain Monte Carlo sampling, the justification for $\hat{p}(m)$, and the procedure to construct a good $w(\beta^{(-m)}|\beta^{(m)},\Delta)$ are given in Chen, Ibrahim, and Yiannoutsos (1998).

There are several advantages of the above Monte Carlo procedure. Firstly, we need only one random draw from $p_0(\beta^{(\mathcal{K})},\Delta \mid D_0^{(\mathcal{K})})$, which greatly eases the computational burden. Secondly, it is more numerically stable since we calculate ratios of the densities in (3). Thirdly, in (3), $p_0(\beta^{(\mathcal{K})},\Delta|D_0^{(\mathcal{K})})$ plays the role of a ratio importance sampling density (see Chen and Shao 1997b) which needs to be known only up to a normalizing constant since this common constant is cancelled out in the calculation.

15.4.2 Computing Posterior Model Probabilities

The posterior probability of model m is given by

$$p(m|D^{(m)}) = \frac{p(D^{(m)} \mid m)\ p(m)}{\sum_{m\in\mathcal{M}} p(D^{(m)} \mid m)\ p(m)}, \tag{5}$$

where

$$p(D^{(m)}|m) = \int\int\int L(\beta^{(m)},\Delta \mid D^{(m)})\ \pi(\beta^{(m)},\Delta,a_0 \mid D_0^{(m)})\ d\beta^{(m)}\ d\Delta\ da_0 \tag{6}$$

denotes the marginal distribution of the data $D^{(m)}$ for the current study under model m, and $p(m)$ denotes the prior probability of model m in (2), which is estimated by (3). The marginal density $p(D^{(m)} \mid m)$ is precisely the normalizing constant of the joint posterior density of $[\beta^{(m)},\Delta,a_0 \mid D]$.

Computing the posterior model probability $p(m|D^{(m)})$ given in (5) requires a different Monte Carlo method other than the one for computing the prior model probability $p(m)$ given in (2). Let $p(\beta^{(m)}, \Delta, a_0 \mid D^{(m)})$ denote the joint posterior density of $[\beta^{(m)}, \Delta, a_0 \mid D]$, and let $\pi(\beta^{(-m)} \mid D^{(\mathcal{K})})$ and $p(\beta^{(-m)} \mid D^{(\mathcal{K})})$ denote the respective marginal prior and posterior densities of $\beta^{(-m)}$ obtained from the full model. Then it can be shown that the posterior probability of model m is given by

$$p(m \mid D^{(m)}) = \frac{\left(p(\beta^{(-m)} = 0 \mid D^{(\mathcal{K})})/\pi(\beta^{(-m)} = 0 \mid D_0^{(\mathcal{K})})\right)\ p(m)}{\sum_{j=1}^{\mathcal{K}} \left(p(\beta^{(-j)} = 0 \mid D^{(\mathcal{K})})/\pi(\beta^{(-j)} = 0 \mid D_0^{(\mathcal{K})})\right)\ p(j)}, \tag{7}$$

$m = 1, \ldots, \mathcal{K}$, where $\pi(\beta^{(-m)} = 0 \mid D_0^{(\mathcal{K})})$ and $p(\beta^{(-m)} = 0 \mid D^{(\mathcal{K})})$ are the marginal prior and posterior densities of $\beta^{(-m)}$ evaluated at $\beta^{(-m)} = 0$. A derivation of the result in (7) is given in Chen, Ibrahim, and Yiannoutsos (1998). In (7), we use (3) to compute the prior model probability $p(m)$. Due to the complexity of the prior and posterior distributions, the analytical forms of $\pi(\beta^{(-m)} \mid D_0^{(\mathcal{K})})$ and $p(\beta^{(-m)} \mid D^{(\mathcal{K})})$ are not available. However, we can adopt the importance-weighted marginal posterior density estimation (IWMDE) method of Chen (1994) to estimate these marginal prior and posterior densities. The IWMDE is a Monte Carlo method developed by Chen (1994), which is particularly suitable for estimating marginal posterior densities when the joint posterior density is known up to a normalizing constant. The IWMDE method requires using only two respective MCMC samples from the prior and posterior distributions for the full model, making the computation of complicated posterior model probabilities feasible. It directly follows from the IWMDE that a simulation consistent estimator of $p(\beta^{(-m)} = 0 \mid D^{(\mathcal{K})})$ is given by

$$\begin{aligned} &\hat{p}(\beta^{(-m)} = 0 \mid D^{(\mathcal{K})}) \\ &= \frac{1}{N} \sum_{l=1}^{N} w\left(\beta_{(l)}^{(-m)} \mid \beta_{(l)}^{(m)}, \Delta_{0(l)}, a_{0(l)}\right) \\ &\times \left\{ \frac{p(\beta_{(l)}^{(m)}, \beta^{(-m)} = 0, \Delta_{0(l)}, a_{0(l)} \mid D^{(\mathcal{K})})}{p(\beta_{(l)}^{(\mathcal{K})}, \Delta_{0(l)}, a_{0(l)} \mid D^{(\mathcal{K})})} \right\}, \end{aligned} \tag{8}$$

where $w(\beta^{(-m)} \mid \beta^{(m)}, \Delta, a_0)$ is a completely known conditional density of $\beta^{(-m)}$ given $\beta^{(m)}$, Δ, and a_0, $\{(\beta_{(l)}^{(\mathcal{K})}, \Delta_{0(l)}, a_{0(l)}), l = 1, 2, \ldots, N\}$ is a sample from the joint posterior distribution $p(\beta^{(\mathcal{K})}, \Delta, a_0 \mid D^{(\mathcal{K})})$ of $[\beta^{(\mathcal{K})}, \Delta, a_0 | D]$. To construct a good $w(\beta^{(-m)} \mid \beta^{(m)}, \Delta, a_0)$, we can use a similar procedure that is used to construct $w(\beta^{(-m)} \mid \beta^{(m)}, \Delta)$ in (3) for calculating the prior model probabilities. Similarly, we can obtain $\hat{\pi}(\beta^{(-m)} = 0 \mid D_0^{(\mathcal{K})})$, an estimate of $\pi(\beta^{(-m)} = 0 \mid D_0^{(\mathcal{K})})$, using a sample from the joint prior distribution $\pi(\beta^{(\mathcal{K})}, \Delta, a_0 \mid D_0^{(\mathcal{K})})$.

15.5 Myeloma Data

We consider two studies in multiple myeloma. Krall, Uthoff & Harley (1975) analyzed data from a study (historical data) on multiple myeloma in which researchers treated $n_0 = 65$ patients with alkylating agents. Of those patients, 48 died during the study and 17 survived. The response variable for these data was survival time in months from diagnosis. Several covariates were measured for these data at diagnosis. These are blood urea nitrogen, hemoglobin, platelet count, age, white blood cell count, fractures, percentage of the plasma cells in bone marrow, proteinuria, and serum calcium. The data from this study serve as the historical data for the analysis. A few years later, another multiple myeloma study (current study) using similar alkylating agents was undertaken by the Eastern Cooperative Oncology Group (ECOG). This study, labeled, E2479, had $n = 479$ patients with the same set of covariates being measured as the historical data. The results of E2479 are available in Kalish (1992). The two studies have similar patient populations, and thus serve as good examples in which to apply the proposed methodology. Here, y_0 consists of the 65 survival times from the historical study and $X_0^{(m)}$ is an $n_0 \times k_m$ matrix of covariates, where k_m denotes the number of covariates under model m.

Our main goal in this example is to illustrate the variable selection technique and prior elicitation. We also examine the sensitivity of the posterior probabilities to the choices of (μ_0, σ_0^2), c_0, and d_0. Thus, a total of $n = 339$ observations were available from the current study for the analysis, with 8 observations being right censored. Our analysis used $p = 8$ covariates. These are blood urea nitrogen (x_1), hemoglobin (x_2), platelet count (x_3) (1 if normal, 0 if abnormal), age (x_4), white blood cell count (x_5), fractures (x_6), percentage of the plasma cells in bone marrow (x_7), and serum calcium (x_8). To ease the computational burden, we standardize all of the variables. The standardization helps the numerical stability in the implementation of the adaptive rejection algorithm (Gilks and Wild, 1992) for sampling the regression coefficients from the posterior distribution.

We conduct sensitivity analyses with respect to i) c_0, ii) d_0, and iii) (μ_0, σ_0^2). These are shown in Tables 1, 2 , and 3, respectively. To compute the prior and posterior model probabilities, 50,000 Gibbs iterations are used to get convergence. We use $M = 28$, with the intervals chosen so that with the combined data sets from the historical and current data, at least one failure or censored observation falls in each interval. This technique for choosing M is reasonable and preserves the consistency in the interpretation of Δ for the two studies. In addition, we take $\theta_0 = \kappa_0$ and use $f_{0i} = s_i - s_{i-1}$ if $s_i - s_{i-1} \geq 1$ and $f_{0i} = 1.1$ if $s_i - s_{i-1} < 1$, and $g_{0i} = .001$. For the last interval, we take $g_{0i} = 10$ for $i = M$ since very little information in the data is available for this last interval. The above choices of f_{0i} and g_{0i} ensures the log-concavity of $\pi_0(\Delta \mid \theta_0)$ as it is required in sampling

Δ from its conditional prior and posterior distributions (see Section 3). A stepwise variable selection procedure in SAS for the current study yields $(x_2, x_3, x_4, x_7, x_8)$ as the top model.

Table 1 gives the model with the largest posterior probability using $(\mu_0, \sigma_0^2) = (.5, .004)$, (i.e., $\delta_0 = \lambda_0 = 30$) for several values of c_0. For each value of c_0 in Table 1, the model, $(x_1, x_2, x_3, x_4, x_5, x_7, x_8)$, obtains the largest posterior probability, and thus model choice is not sensitive to these values. In addition, for $d_0 = 3$ and for any $c_0 \geq 3$, the $(x_1, x_2, x_3, x_4, x_5, x_7, x_8)$ obtains the largest posterior probability. Although not shown in Table 1, values of $c_0 < 3$ do not yield $(x_1, x_2, x_3, x_4, x_5, x_7, x_8)$ as the top model. Thus, model choice may become sensitive to the choice of c_0 when $c_0 < 3$.

From Table 2, we see how the prior model probability is affected as d_0 is changed. In each case, the true model obtains the largest posterior probability. Under the settings of Table 2, the $(x_1, x_2, x_3, x_4, x_5, x_7, x_8)$ model obtains the largest prior probability when $d_0 \geq 3$. With values of $d_0 < 3$, however, model choice model choice may be sensitive to the choice of d_0. For example, when $d_0 = .0001$ and $c_0 = 10$, the top model is $(x_1, x_2, x_4, x_5, x_7, x_8)$ with posterior probability of .42 and the second best model is $(x_1, x_2, x_3, x_4, x_5, x_7, x_8)$ with posterior probability of .31.

Finally, we mention that as both c_0 and d_0 become large, the $(x_1, x_2, x_4, x_5, x_7, x_8)$ model obtains the largest posterior model probability. In addition, Tables 1 and 2 indicate a monotonic decrease in the posterior probability of model $(x_1, x_2, x_4, x_5, x_7, x_8)$ as c_0 and d_0 are increased. This indicates that the there's is a moderate impact of the historical data on model choice.

Table 3 shows a sensitivity analysis with respect to (μ_0, σ_0^2). Under these settings, model choice is not sensitive to the choice of (μ_0, σ_0^2). We see that in each case, $(x_1, x_2, x_4, x_5, x_7, x_8)$ obtains the largest posterior probability. In addition, there is a monotonic increase in the posterior model probability as more weight is given to the historical data.

Although not shown here, several different partitioning schemes were used for the choices of the subintervals $(s_{i-1}, s_i]$, following a procedure similar to Ibrahim, Chen, and MacEachern (1996). The prior and posterior model probabilities do not appear to be too sensitive to the choices of the subintervals $(s_{i-1}, s_i]$. This is a comforting feature of our approach since it allows the investigator some flexibility in choosing the subintervals. As indicated by Ibrahim, Chen, and MacEachern (1996), sensitivity analyses on (f_{0i}, g_{0i}) show that as long as the shape parameter f_{0i} remains fixed, the results are not too sensitive to changes in the scale parameter g_{0i}. However, when we fix the scale parameter and vary the shape parameter, the results can be sensitive.

15.6 Discussion

We have discussed a class of informative semi-parametric priors for proportional hazards models. The priors are based on the availability of historical data and are quite useful in model selection contexts, due to their semi-automatic elicitation scheme. Computational methods have been developed and have been implemented for these priors. Extensions for these priors are being explored for frailty models and multivariate survival models.

Table 1: Posterior Model Probabilities for $(\mu_0, \sigma_0^2) = (.5, .004)$, $d_0 = 3$ and Various Choices of c_0

c_0	m	$p(m)$	$p(D \mid m)$	$p(m\|D)$
3	$(x_1, x_2, x_3, x_4, x_5, x_7, x_8)$	.015	.436	.769
10	$(x_1, x_2, x_3, x_4, x_5, x_7, x_8)$	.015	.310	.679
30	$(x_1, x_2, x_3, x_4, x_5, x_7, x_8)$	.015	.275	.657

Table 2: The Posterior Model Probabilities for $(\mu_0, \sigma_0^2) = (.5, .004)$, $c_0 = 3$ and Various Choices of d_0

d_0	m	$p(m)$	$p(D \mid m)$	$p(m\|D)$
5	$(x_1, x_2, x_3, x_4, x_5, x_7, x_8)$	.011	.436	.750
10	$(x_1, x_2, x_3, x_4, x_5, x_7, x_8)$	.005	.436	.694
30	$(x_1, x_2, x_3, x_4, x_5, x_7, x_8)$	.001	.436	.540

Table 3: The Posterior Model Probabilities for $c_0 = 10$, $d_0 = 10$ and Various Choices of (μ_0, σ_0^2)

(μ_0, σ_0^2)	m	$p(m)$	$p(D \mid m)$	$p(m\|D)$
$(.5, .008)$	$(x_1, x_2, x_3, x_4, x_5, x_7, x_8)$	.005	.274	.504
$(.5, .004)$	$(x_1, x_2, x_3, x_4, x_5, x_7, x_8)$	.005	.310	.558
$(.98, 3.7 \times 10^{-4})$	$(x_1, x_2, x_3, x_4, x_5, x_7, x_8)$	.005	.321	.572

Acknowledgments: Dr. Sinha's research was supported by the grant R29-CA69222 from the NCI.

References

Arjas, E., and Gasbarra, D. (1994), "Nonparametric Bayesian inference from right censored survival data, using the Gibbs sampler". *Statistica Sinica*,**4**, 505-524.

Bedrick, E. J., Christensen, R., and Johnson, W. (1996), "A New Perspective on Priors for Generalized Linear Models." *Journal of the American Statistical Association, 91*, 1450-1460.

Chen, M.-H. (1994), "Importance-weighted Marginal Bayesian Posterior Density Estimation." *Journal of the American Statistical Association, 89*, 818-824.

Chen, M. -H., Ibrahim, J. G., and Yiannoutsos, C, (1998), "Prior Elicitation, Variable Selection, and Bayesian Computation for Logistic Regression Model," *Journal of the Royal Statistical Society, Series B*, to appear.

Chen, M.-H. and Shao, Q.-M. (1997a), "Estimating Ratios of Normalizing Constants for Densities with Different Dimensions," *Statistica Sinica, 7*, 607-630.

Chen, M.-H., and Shao, Q.-M. (1997b), "On Monte Carlo Methods for Estimating Ratios of Normalizing Constants," *The Annals of Statistics, 25*, 1563-1594.

Cox, D. R. (1972). Regression models and life tables, (with discussion). *J. R. Statist. Soc.* **B 34**, 187-220.

Cox, D. R. (1975). Partial likelihood. *Biometrika* **62**, 269-276.

Gilks, W.R., and Wild, P. (1992), "Adaptive Rejection Sampling for Gibbs Sampling," *Applied Statistics, 41*, 337-348.

Ibrahim, J. G., and Chen, M. -H. (1997), "Prior Distributions and Bayesian Computation for Proportional Hazards Models," submitted.

Ibrahim, J. G., Chen, M. -H., and MacEachern, S. N. (1996), "Bayesian Variable Selection for Proportional Hazards Model," submitted.

Kalish, L. A. (1992), "Phase III Multiple Myeloma: Evaluation of Combination Chemotherapy in Previously Untreated Patients," *Technical Report # 726E, Department of Biostatistics, Dana-Farber Cancer Institute.*

Krall, J. M., Uthoff, V. A., and Harley, J. B. (1975), "A Step-up Procedure for Selecting Variables Associated with Survival," *Biometrics, 31*, 49-57.

Laud, P. W., and Ibrahim, J. G. (1995), "Predictive model selection," *Journal of the Royal Statistical Society, Series B, 57*, 247-262.

O'Neill, R. (1971), "Algorithm AS47-Function Minimization Using a Simplex Procedure," *Applied Statistics, 20*, 338-345.

Sinha, D. and Dey, D. K. (1997), "Semiparametric Bayesian Analysis of Survival Data," *Journal of the American Statistical Association, 92*, 1195-1212.

Sinha, D., Chen, M-H., and Ghosh, S. K. (1997), "Bayesian analysis and model selection for interval-censored survival data". *Submitted for publication.*

Sinha, D. (1998), "Time-discrete beta process model for interval censored survival data". *Canadian Journal of Statist., 27*, to appear.

Tsiatis, A. A. (1981). A large sample study of Cox's regression model. *Ann. Statist.* **9**, 93-108.

16

Asymptotic Properties of Nonparametric Bayesian Procedures

Larry Wasserman

ABSTRACT This chapter provides a brief review of some large sample frequentist properties of nonparametric Bayesian procedures. The review is not comprehensive, but rather, is meant to give a simple, heuristic introduction to some of the main concepts. We mainly focus on consistency but we touch on a few other issues as well.

16.1 Introduction

Nonparametric Bayesian procedures present a paradox in the Bayesian paradigm. On the one hand, they are most useful when we don't have precise information. On the other hand, they require huge amounts of prior information because nonparametric procedures involve high dimensional if not infinite dimensional parameter spaces. The usual hope that the data will dominate the prior was dashed by Freedman (1963, 1965) and then Diaconis and Freedman (1986) who showed that putting mass in weak neighborhoods the true distribution does not guarantee that the posterior accumulates in weak neighborhoods.

Interest in properties like consistency derive from our desire that the posterior not be strongly driven by the prior. Examining the frequentist properties of the posterior is perhaps the only reliable way to understand the import of the many implicit assumptions we make when doing nonparametric Bayesian inference.

This chapter provides a non-technical introduction to the frequentist large sample properties of Bayesian nonparametric procedures. Much of the material in this chapter is derived from or influenced by the aforementioned papers of Diaconis and Freedman, as well as Shen (1995), Tierney (1987), Ghoshal, Ghosh and Ramamoorthi (1997a,b), Cox (1993), Barron (1988), Schwartz (1965) and Barron, Schervish and Wasserman (1997). However, we shall not attempt a comprehensive review of this area. Rather, our hope is merely to give a flavor of some of the main ideas. More thorough reviews include Ghoshal, Ghosh and Ramamoorthi (1997a) and Ghosh and Ramamoorthi (1997). Other relevant references include Diaconis and Freedman (1993) and Freedman and Diaconis (1983).

One of the messages of this chapter is that in some cases, posteriors on infinite dimensional spaces exhibit reasonable large sample behavior. But in other cases, they do not and purely frequentist procedures might be better suited to these problems. Bayesian nonparametric methods – like all statistical methods – have their limitations.

16.2 Frequentist or Bayesian Asymptotics?

In Bayesian inference, there are two styles of asymptotics. Frequentist style asymptotics studies the behavior of the posterior with respect to draws from a fixed sampling distribution. Bayesian style asymptotics studies the behavior of the asymptotics with respect to the marginal distribution obtained by integrating the sampling distribution with respect to the prior.

Diaconis and Freedman (1986) made a compelling case for using frequentist asymptotics. Without delving into too much detail, the case for using frequentist asymptotics can be made using a simple example. Suppose $Y_1, \ldots, Y_n \sim N(\theta, 1)$ and let the prior π be a point mass at zero. Clearly, the posterior is also a point mass at zero. Barring the miracle that the true value of θ is zero, this posterior is inconsistent: the posterior mass never accumulates around the true value.

Yet if we let A denote the event that the posterior eventually accumulates around the true value of θ, then Doob's theorem (Doob 1949) shows that A has a probability one. This would seem to contradict the obvious fact that the posterior is inconsistent. The reason it does not is that Doob's theorem uses the Bayesian's own prior when it claims that A has probability one. Indeed, the posterior is consistent if θ happens to be 0 and, according to the prior, this is a sure event. Consistency fails on a null set, but the null set is huge. In this example, we don't need fancy techniques to see that the null set is big. But in infinite dimensional spaces, it is not so easy know whether a null set is "big," hence the need to study frequentist consistency seriously.

16.3 Consistency

"Consistency" means that the posterior concentrates around the true distribution. Often, one or two conditions are needed for consistency. The first is a support condition which requires that the prior put positive mass around the true distribution. The second is a smoothness condition which requires that not too much mass be put on rough densities. Consistency requires a balancing act between trying to spread the mass around as much as possible but not spreading it on nasty distributions.

Let $\mathcal{P}$ be the set of all distributions. We need to introduce some distance measures on $\mathcal{P}$. Let d_W be any distance that metrizes the weak topology i.e. $d_W(P_n, P) \to 0$ if and only if $\int g(x)dP_n(x) \to \int g(x)dP(x)$ for every bounded, continuous function g. There are many such metrics and we shall not need to concern ourselves with the details. Let d_H be Hellinger distance:

$$d_H(P,Q)^2 = \int (\sqrt{f}(y) - \sqrt{g}(y))^2 d\lambda(y)$$

where f and g are the densities of P and Q with respect to the dominating measure λ Also, let d_K be Kullback-Leibler divergence:

$$d_K(P,Q) = \int dP \log dP/dQ.$$

We shall also write the distances with densities as their arguments when convenient. It is also helpful to define total variation distance $d_1(f,g) = \int |f(y) - g(y)| d\lambda(y)$ which is also called the L_1 distance. In what follows, there is little difference between using d_H or d_1.

The "support" of a prior π is

$$\mathcal{S}(\pi) = \{P;\ \text{for every } \epsilon > 0,\ \pi(N_\epsilon(P)) > 0\} \quad (1)$$

where $N_\epsilon(P) = \{Q; d(P,Q) \leq \epsilon\}$. The support depends on the choice of distance so we write $\mathcal{S}_W(\pi)$, $\mathcal{S}_H(\pi)$ and $\mathcal{S}_K(\pi)$ for the three choices. It is easy to show that $\mathcal{S}_K(\pi) \subset \mathcal{S}_H(\pi) \subset \mathcal{S}_W(\pi)$.

Say that π is consistent at P_0 if, for every $\epsilon > 0$, $\pi(N_\epsilon(P_0)|Y^n)$ tends to 1, almost surely with respect to repeated sampling from P_0. Let $\mathcal{C}(\pi)$ be the set of all f_0 such that π is consistent at f_0. We call $\mathcal{C}(\pi)$, the "consistency class" of π. Again, $\mathcal{C}(\pi)$ depends on the choice of distance and we subscript $\mathcal{C}(\pi)$ appropriately as needed. We would like $\mathcal{C}(\pi)$ to be as large as possible.

Now the question is: if we put positive mass in each neighborhood of P_0, does it imply that the posterior of every neighborhood of P_0 tends to 1? To be precise, the question is whether $\mathcal{S}(\pi) \subset \mathcal{C}(\pi)$.

Freedman (1963) and Diaconis and Freedman (1986) showed that $\mathcal{S}_W(\pi)$ is not a subset of $\mathcal{C}_W(\pi)$. In other words, even if you put positive mass in weak neighborhoods of P_0, it does not follow that the posterior concentrates in weak neighborhoods of P_0.

On other hand, Schwartz (1965) and Barron (1986) proved that $\mathcal{S}_K(\pi) \subset \mathcal{C}_W(\pi)$. In words: if the prior puts positive mass in Kullback-Leibler neighborhoods of P_0, then the posterior concentrates in weak neighborhoods of P_0. To state Schwartz's result in more detail, we must first recall the following definition. A sequence of tests $\phi_n(Y_1, \ldots, Y_n)$ is "uniformly consistent" for $H_0 : f = f_0$ versus $H_1 : f \in A^c$ if $E_0(\phi_n) \to 0$ and $\inf_{f \in A^c} E_f \phi_n \to 1$. Here E_0 and E_f refer to expectation under repeated sampling from f_0 or

f. Schwartz showed that if $f_0 \in S_K(\pi)$ and if there exists a uniformly consistent test for $H_0 : f = f_0$ versus $H_1 : f \in A^c$, then $\pi(A|Y^n)$ tends to 1 almost surely. If A is a weak neighborhood of f_0 then it is known that there does exist a uniformly consistent test and hence consistency holds. As pointed out in Ghoshal, Ghosh and Ramamoorthi (1997), the condition that $f_0 \in S_K(\pi)$ is not a necessary condition. They give the following simple example. Let $Y_1, \ldots, Y_n$ be i.i.d. from a Uniform $(0, \theta)$ with $\theta \in (0, 1]$. The condition fails since the Kullback-Leibler distance from Uniform (0,1) to a Uniform $(0, \theta)$ is infinite. But if a prior has full support on [0,1] then consistency obtains when the true distribution is Uniform (0,1).

Weak neighborhoods are large. It would also be nice if we could say that the posterior concentrates in stronger neighborhoods. Barron (1988), Barron, Schervish and Wasserman (1997), Ghoshal, Ghosh and Ramamoorthi (1997a, 1997b) and Shen (1997) discuss this issue. Generally, one can show that $\mathcal{S}_K(\pi) \subset \mathcal{C}_H(\pi)$ if we insist that the prior π satisfy a regularity condition. The outline of why this is true, is in the next section.

16.4 Consistency in Hellinger Distance

Here we blend some ideas of Barron (1988), Barron, Schervish and Wasserman (1997), Ghoshal, Ghosh and Ramamoorthi (1997a, 1997b), Schwartz (1963), Shen (1995) and others to give a heuristic explanation of when the posterior concentrates in Hellinger neighborhoods of the true density. Specifically, we will show that $\mathcal{S}_K(\pi) \subset \mathcal{C}_H(\pi)$ under a smoothness condition. To do so, we need some definitions.

Let $\mathcal{G}$ be a set of densities with respect to some underlying measure λ. A set of pairs of functions $(\ell_1, u_1), \ldots, (\ell_k, u_k)$ is called a an "ϵ-bracketing" of $\mathcal{G}$ if (i) $d_H(\ell_j, u_j) \leq \epsilon$ for $j = 1, \ldots, k$ and (ii) for every $f \in \mathcal{G}$ there is a $j \in \{1, \ldots, k\}$ such that $\ell_j \leq f \leq u_j$ a.e. with respect to a dominating measure. Let $N_{[\,]}(\epsilon, \mathcal{G})$ be the number of brackets in the smallest ϵ bracketing of $\mathcal{G}$ and let $H_{[\,]}(\epsilon, \mathcal{G}) = \log N_{[\,]}(\epsilon, \mathcal{G})$. The number $H_{[\,]}(\epsilon, \mathcal{G})$ is called the bracketing metric entropy of $\mathcal{G}$. See van der Vaart and Wellner (1996) for more detail about bracketing entropy.

For this section we assume that the prior π lives on a set of distributions which all possess densities with respect to some common dominating measure λ. We shall be somewhat loose in some of our terminology. Thus we shall speak of a density f when we should really talk about an equivalence class of densities. And we shall refer to neighborhoods of a density f to mean the same as a neighborhood around the corresponding measure. The true probability measure generating the data is P_0 with density f_0. The n-fold product measure of P_0 is denoted P_0^n. The corresponding density is f_0^n. Statements like "such and such happens almost surely" refer to repeated sampling from P_0. We shall assume all relevant quantities are measurable.

THEOREM. Suppose that $Y_1, Y_2, \ldots, Y_n$ are i.i.d. from f_0 and suppose that $f_0 \in \mathcal{S}_K(\pi)$. Further, suppose that for every $\epsilon > 0$, there exist a sequence of sets of densities $\mathcal{F}_n$ such that
(i) entropy condition: $\int_{\epsilon^2}^{\epsilon} H_{[\,]}^{1/2}(u, \mathcal{F}_n)du \leq \sqrt{n}\epsilon^2$, and
(ii) tail condition: there exists a positive constant $r > 0$ such that

$$\pi(\mathcal{F}_n^c) < e^{-nr}$$

for all large n. Then, for every $\epsilon > 0$, $\pi(N_\epsilon^c|Y^n)$ tends to 0 exponentially quickly, almost surely, where N_ϵ is a Hellinger neighborhood around P_0 of size ϵ. Thus, $\mathcal{S}_K(\pi) \subset \mathcal{C}_H(\pi)$.

PROOF. Let N be an ϵ Hellinger neighborhood of f_0. By Bayes' theorem we have

$$\pi(N^c|Y^n) = \frac{\int_{N^c} \prod_i f(Y_i)d\pi(f)}{\int \prod_i f(Y_i)d\pi(f)} = \frac{\int_{N^c} R_n(f)d\pi(f)}{\int R_n(f)d\pi(f)} \tag{2}$$

where

$$R_n(f) = \prod_i \frac{f(Y_i)}{f_0(Y_i)}.$$

Now,

$$\int_{N^c} R_n(f)d\pi(f) = \int_{N^c \cap \mathcal{F}_n} R_n(f)d\pi(f) + \int_{N^c \cap \mathcal{F}_n^c} R_n(f)d\pi(f) = A + B, \text{ say.}$$

Property (ii) implies that B is very small. This can be seen using the following trick. Let Pr represent probability statements under P_0^n and let E represent expectation under P_0^n. Fix an arbitrary $c > 0$ and, using Markov's inequality and Fubini's theorem:

$$\begin{aligned}
Pr(B > c) &\leq c^{-1}E(B) \\
&= c^{-1} \int_{\mathcal{Y}^n} \int_{N^c \cap \mathcal{F}_n^c} R_n(f)d\pi(f)dP_0^n \\
&= c^{-1} \int_{N^c \cap \mathcal{F}_n^c} \int_{\mathcal{Y}^n} R_n(f)dP_0^n d\pi(f) \\
&= c^{-1} \int_{N^c \cap \mathcal{F}_n^c} \int_{\mathcal{Y}^n} R_n(f) f_0^n(Y^n)d\lambda^n d\pi(f) \\
&= c^{-1} \int_{N^c \cap \mathcal{F}_n^c} \int_{\mathcal{Y}^n} \frac{f^n(Y^n)}{f_0^n(Y^n)} f_0^n(Y^n)d\lambda^n d\pi(f) \\
&= c^{-1} \int_{N^c \cap \mathcal{F}_n^c} \int_{\mathcal{Y}^n} f^n(Y^n)d\lambda^n d\pi(f) \\
&= c^{-1} \int_{N^c \cap \mathcal{F}_n^c} d\pi(f) \\
&= c^{-1}\pi(N^c \cap \mathcal{F}_n^c) \leq c^{-1}\pi(\mathcal{F}_n^c) \leq c^{-1}e^{-nr}
\end{aligned}$$

by condition (ii). Take $c = e^{-nr/2}$. Conclude that $Pr(B > e^{-nr/2}) \leq e^{-nr/2}$. By the first Borel-Cantelli lemma, $B \leq e^{-nr/2}$ almost surely, for all large n.

Now

$$\int_{N^c \cap \mathcal{F}_n} R_n(f) d\pi(f) \leq \sup_{f \in N_c \cap \mathcal{F}_n} R_n(f) \pi(N^c \cap \mathcal{F}_n) \leq \sup_{f \in N_c \cap \mathcal{F}_n} R_n(f).$$

Now we use a large deviation inequality which is implied by property (i). Specifically, from Wong and Shen (1995, Theorem 1), (i) implies that, almost surely, for all large n,

$$\sup_{f \in N_c \cap \mathcal{F}_n} R_n(f) < e^{-nc} \tag{3}$$

for some $c > 0$. We have upper bounded the numerator of (2) by $e^{-nc} + e^{-nr/2}$. Now we proceed to lower bound the denominator.

Let $\delta = (1/2)[(r/2) + c] > 0$ and let K be a Kullback-Leibler neighborhood of size $\delta/2$. Let

$$D_n(f) = \frac{1}{n} \sum_{i=1}^{n} \log \frac{f_0(Y_i)}{f(Y_i)}.$$

Then,

$$e^{n\delta} \int R_n(f) d\pi(f) = e^{n\delta} \int e^{-nD_n(f)} d\pi(f) \geq e^{n\delta} \int_K e^{-nD_n(f)} d\pi(f).$$

For each $f \in K$, by the law of large numbers, $D_n(f)$ converges almost surely to $D(f) \equiv d_K(f_0, f) < \delta$. Thus, $\lim_n n(\delta - D_n(f)) = \infty$ almost surely. This convergence is almost sure on the sample space, for each fixed f. The null set can depend on f. However, by Fubini's theorem, we can choose a single set of probability one on the sample space such that $D_n(f)$ converges to $D(f)$ expect possibly for a set of $f's$ of π measure 0. So, by Fatou's lemma,

$$\text{liminf}_n e^{n\delta} \int R_n(f) d\pi(f) \geq \int_K \text{liminf}_n e^{n[\delta - D_n(f)]} d\pi(f) = \infty$$

as long as $\pi(K) > 0$, which holds by property (i). Thus, for large n, $e^{n\delta} \int e^{-nD_n(f)} \geq 1$, i.e. $\int R_n(f) d\pi(f) \geq e^{-n\delta}$ almost surely for large n.

Having bounded the numerator and denominator, we can now see that

$$\pi(N^c | Y^n) \leq \frac{e^{-nc} + e^{-nr/2}}{e^{-n\delta}} = e^{-(n/2)[c + (r/2)]} \tag{4}$$

almost surely, for large n. ♣

The conditions of the Theorem are stronger than needed. Ghoshal, Ghosh and Ramamoorthi (1997b) have a consistency result that uses L_1 entropy which is weaker than the bracketing entropy condition. Their result is as follows. Suppose that $f_0 \in \mathcal{S}_K(\pi)$ and that for every $\epsilon > 0$ there is a $\delta < \epsilon$, $c_1, c_2 > 0$, $\beta < \epsilon^2/2$ and $\mathcal{F}_n$ such that (i) $\pi(\mathcal{F}_n^c) < c_1 e^{-nc_2}$ and (ii) $J(\delta, \mathcal{F}_n) < n\beta$ where $J(\beta, \mathcal{F}_n)$ is the logarithm of the cardinality of the smallest number m of densities $f_1, \ldots, f_m$ such that $\mathcal{F}_n \subset \cup_{j=1}^m \{f; d_1(f, f_j) < \delta\}$. Then Hellinger (and total variation) consistency obtains.

A more general result is contained in Barron (1986) which is also summarized nicely in Ghoshal, Ghosh and Ramamoorthi (1997a). Suppose again that $f_0 \in \mathcal{S}_K(\pi)$. Then Barron shows that the following two statements are equivalent:

(1) For any measurable set U, there exists $\beta_0 > 0$ such that

$$P_0(\pi(U^c|Y^n) > e^{-n\beta_0}, i.o.) = 0;$$

(2) There exist subsets V_n and W_n, positive numbers $c_1, c_2, \beta_1, \beta_2$ and tests $\{\phi_n\}$ such that (i) $U^c = V_n \cup W_n$, (ii) $\pi(W_n) \le c_1 e^{-n\beta_1}$, (iii) $Pr(\phi_n > 0, i.o.) = 0$ and $\inf_{f \in V_n} E_f(\phi_n) \ge 1 - c_2 e^{-n\beta_2}$.

The next question is: do commonly used methods satisfy the conditions of the theorem? Fortunately, the answer is yes. For example, Barron, Schervish and Wasserman (1997) show that random histograms, Polya trees and infinite dimensional exponential families all satisfy the conditions of the theorem. For example, Polya trees with $Beta(a_k, a_k)$ distributions at the kth level are shown to be consistent if $a_k = 8^k$. Ghoshal, Ghosh and Ramamoorthi (1997) show that Dirichlet process mixtures are consistent. This result requires technical conditions on the tail of the true density as well as some conditions on the base measure of the Dirichlet.

16.5 Other Asymptotic Properties

Consistency is a very weak property. We would like the posterior to possess other properties as well. If we study how the posterior concentrates in neighborhoods N_{ϵ_n} where ϵ_n tends to zero, then we can assess the rate of convergence of the posterior. Shen (1995), for example, has results on rates of convergence. Some of the results are negative: in certain cases the rate of convergence of the posterior is slower than the best obtainable rates by other means.

Zhao (1993, 1997) studied rates of convergence of the Bayes estimator in the following model. Suppose that

$$Y_i = \beta_i + \epsilon_i \tag{5}$$

for $i = 1, 2, \ldots$ where the ϵ_i's are independent, identically distributed random variables with mean 0 and variance $\sigma_n^2 = 1/n$ and $\sum_i \beta_i^2 < \infty$. Take the prior to make the $\beta_i's$ Normal with mean 0 and variance i^{-2p}. Suppose that β lies in $\Omega = \{\beta; \sum_i i^{2q}\beta_i^2 \leq B\}$ where q and B are known. The optimal minimax rate for squared error loss is known to be $n^{2q/(2q+1)}$. In other words

$$0 < \lim_{n\to\infty} \inf_{\hat{\beta}} \sup_{\theta\in\Omega} n^{\frac{2q}{2q+1}} R(\hat{\beta}, \beta) < \infty \tag{6}$$

where $R(\hat{\beta}, \beta) = E(\sum_i (\hat{\beta}_i - \beta_i)^2)$.

Zhao found that the Bayes estimate achieves the optimal rate when $p = (2q+1)/2$. But in this case, the prior and posterior have measure 0 on Ω! And she shows that any independent Gaussian prior with support on Ω cannot attain the optimal rate. On a positive note, she shows the following result. Suppose we take the prior to be $\pi = \sum_j w_j \pi_j$ where $w_j \sim j^{-2}$ and π_j is Gaussian on the first j coordinates and sets all other $\beta_j's$ to 0. Then this prior has positive mass on Ω and the Bayes estimator achieves the optimal rate.

It is also natural to inquire about asymptotic normality of the posterior. The results are few and mixed. Shen (1995) gives conditions which ensure that the posterior of functions of the distribution has, asymptotically, a normal distribution. He also investigates whether the posterior is efficient, i.e. the asymptotic variance is as small as possible. The conditions needed to guarantee to normality are complicated. Diaconis and Freedman (1997) studied the following problem; a related investigation is in Cox (1993). Here we summarize the results of Diaconis and Freedman.

Suppose we again have the model above used by Zhao where now the prior makes the $\beta_i's$ Normal with mean 0 and variance $\tau_i^2 = A/i^\alpha$, $1 < \alpha < \infty$. Let $T_n = \sum_{i=1}^\infty (\beta_i - \hat{\beta}_i)^2$ where $\hat{\beta}_i = E(\beta_i|Y) = \tau_i^2 Y_i/(\sigma_n^2 + \tau_i^2)$. They prove that, with respect to the posterior,

$$T_n \approx n^{-1+\frac{1}{\alpha}} C + n^{-1+\frac{1}{2\alpha}} \sqrt{D} Z_n \tag{7}$$

where C and D are numbers and Z_n converges in law to a $N(0,1)$. On the other hand, with respect to the sampling distribution,

$$T_n \approx n^{-1+\frac{1}{\alpha}} C + n^{-1+\frac{1}{2\alpha}} \sqrt{F} U_n(\beta) + n^{-1+\frac{1}{2\alpha}} \sqrt{G} V_n(\beta). \tag{8}$$

The last equation holds for almost all β drawn from the prior. That is, it is a frequentist result holding for fixed β, and the β's for which it holds are a set of measure 1 with respect to the prior. Moreover, $G < D$, $U_n(\beta)$ and $V_n(\beta)$ both converge in law to $N(0,1)$ random variables and

$$\text{liminf}_n U_n(\beta) = -\infty \quad \text{and} \quad \text{limsup}_n U_n(\beta) = \infty.$$

Thus, the posterior behaves quite differently from the usual Normal limiting behavior we would expect in finite dimensional models. However, the

Bayesian and frequentist limiting behavior can be made equivalent if the prior variances of the $\beta_j's$ are chosen to decay exponentially quickly.

Another desirable property of the posterior is that sets with posterior probability $1 - \alpha$ should have frequentist coverage nearly equal to $1 - \alpha$. There is a long history of such results for parametric models; see Kass and Wasserman (1996). The correspondence breaks down even in complicated finite dimensional models; For example, Wasserman (1998) showed that for certain finite dimensional mixture models, accurate coverage can only be obtained by data dependent priors. Less is known in the nonparametric case. Cox (1993) showed that in certain Gaussian process models, the correspondence between posterior probability and coverage breaks down. The results of Diaconis and Freedman above also show a breakdown of the coverage properties in infinite dimensions. Suppose we choose c_n such that $I_n = \{\beta; T_n \leq c_n\}$ has posterior probability $1 - \alpha$. Then, it is easy to see from their results that there almost certainly will be subsequences along which the coverage of the I_n is arbitrarily small.

16.6 The Robins-Ritov Paradox

Here we summarize an intriguing example, due to Robins and Ritov (1997) which shows the limitation of nonparametric Bayesian methods. Consider independent, identically distributed data $W_1, \ldots, W_n$ where $W_i = (X_i, R_i, Y_i)$. The $Y_i's$ are binary and our goal is simply to estimate $\psi = E(Y_1)$. The random variable X_i lives in $\mathcal{R}^p$ where p is large. R_i is another binary random variable which indicates whether or not we observe Y_i. We are also given a known function $g : \mathcal{R}^p \to [0, 1]$ such that $\inf_x g(x) > 0$.

The data are obtained as follows. First we observe X_i. Then, with probability $\pi_i = g(X_i)$ we get to observe Y_i i.e. $R_i = 1$ and with probability $1 - \pi_i$ we do not observe Y_i i.e. $R_i = 0$. To make the problem easier, we will assume that the marginal distribution of X_1 is known. (This is not essential to the problem.) The likelihood is

$$\begin{aligned}
\mathcal{L} &= \prod_i f_X(X_i) f_{R|X}(R_i|X_i)[f_{Y|X}(Y_i|X_i)]^{R_i} &(9)\\
&= \prod_i f_X(X_i) \pi_i^{R_i} (1 - \pi_i)^{1-R_i} h(X_i)^{R_i Y_i} (1 - h(X_i))^{R_i(1-Y_i)} &(10)\\
&\propto \prod_i h(X_i)^{R_i Y_i} (1 - h(X_i))^{R_i(1-Y_i)} &(11)
\end{aligned}$$

where $h(X) = Pr(Y = 1|X)$. The only unknown "parameter" in this problem is the function $h(\cdot)$ which is a binary regression function over a p-dimensional covariate.

A Bayesian would put a prior on h, then compute the posterior of h. We will assume that the prior is chosen independently of the function g. The posterior of h induces a posterior on $\psi = E(Y) = \int E(Y|X = x) f_X(x) dx = \int h(x) f_X(x) dx$. For the sake of argument, let us suppose $p = 20$, not at all unrealistic. Performing this 20 dimensional nonparametric regression problem is no small feat. Due to the curse of dimensionality, we cannot realistically expect to do well in such a high dimensional problem. Indeed, Robins and Ritov show that the Bayes estimate of ψ cannot be $\sqrt{n}$ consistent and, moreover, cannot be uniformly consistent, unless one imposes severe smoothness restrictions.

This would not be so disturbing except that there is a simple frequentist estimator which behaves quite well. The estimator is the well-known Horvitz-Thompson estimator

$$\hat{\psi} = \frac{1}{n} \sum_i \frac{Y_i R_i}{\pi_i} \tag{12}$$

which is $\sqrt{n}$ consistent regardless of the joint law of (X, R, Y) (assuming only that g is bounded away from 0). In addition to its appealing asymptotic properties, this estimator is notable for its simplicity compared to the Bayesian nonparametric approach. The appearance of the $\pi_i' s$ is disturbing from the Bayesian point of view. The $X_i' s$ are ancillary and the usual Bayesian methods will not depend on the probability of observing the $X_i' s$. Still, the alternative of trying to estimate h nonparametrically is even less appealing. Imposing a parametric model on h is also unappealing since any such model is bound to be wrong. If we do wish to impose such a model, Robins and Ritov show that the Horvitz-Thompson estimator can be modified in such a way that the resulting estimator retains $\sqrt{n}$ consistency if the model is wrong. But if the model is right, then their estimator attains the smallest possible asymptotic variance (among estimators that are $\sqrt{n}$ consistent when the model is wrong).

There are many other examples that compare Bayesian and frequentist methods, some favoring one or the other. The point is simply that we should not assume that Bayesian nonparametric methods are a panacea.

16.7 Conclusion

The frequentist properties of Bayesian methods are well understood for parametric models. This is less so for nonparametric models. Fortunately, this is now an active area and we can expect in the coming years to have a much deeper understanding of the issues. Of course, most of the work pertains to large sample properties. Little seems to be known about the finite sample properties of posteriors.

Nonparametric Bayesian methods are valuable in many situations. But the properties of the posterior in nonparametric problems are only now becoming well understood. There is still much to learn but, from what is now known, we see that there is reason for caution.

Acknowledgments. I would like to thank Steve MacEachern, Peter Mueller R.V. Ramamoorthi, Jamie Robins and a referee for reading a draft of this chapter and providing valuable feedback. Research supported by NIH grant RO1-CA54852 and NSF grants DMS-9303557 and DMS-9357646.

References

Barron, A. (1988). "The exponential convergence of posterior probabilities with implications for Bayes estimators of density functions." Technical Report 7, Department of Statistics, University of Illinois, Champaign, IL.

Barron, A., Schervish, M. and Wasserman, L. (1997). "The consistency of posterior distributions in nonparametric problems." Technical report, Department of Statistics, Carnegie Mellon.

Cox, D. (1993). "An analysis of Bayesian inference for nonparametric regression." *The Annals of Statistics*, **21**, 903-923.

Diaconis, P. and Freedman, D. (1997). "On the Bernstein-von Mises Theorem with Infinite Dimensional Parameters," unpublished manuscript.

Diaconis, P. and Freedman, D. (1986). "On the consistency Bayes estimates." *The Annals of Statistics*, **14**, 1-26.

Diaconis, P. and Freedman, D. (1993). "Nonparametric binary regression: A Bayesian approach." *The Annals of Statistics*, **21**, 2108-2137.

Doob, J.L. (1949). "Application of the theory of martingales." In *Le Calcul des Probabilités et ses Applications*, 23-27. Colloques Internationaux du Centre National de la Recherche Scientifique, Paris.

Freedman, D. (1963), "On the asymptotic behavior of Bayes' estimates in the discrete case." *The Annals of Mathematical Statistics*, **34**, 1386-1403.

Freedman, D. (1965), "On the asymptotic behavior of Bayes' estimates in the discrete case, II." *The Annals of Mathematical Statistics*, **36**, 454-456.

Freedman, D., Diaconis, P. (1983). "On inconsistent Bayes estimates in the discrete case." *The Annals of Statistics*, **11**, 1109-1118.

Ghosh, J.K. and Ramamoorthi, R.V. (1997). *Lecture Notes on Bayesian Asymptotics,* Under preparation.

Ghoshal, S., Ghosh, J.K. and Ramamoorthi, R.V. (1997a). "Consistency issues in Bayesian nonparametrics," unpublished manuscript.

Ghoshal, S., Ghosh, J.K. and Ramamoorthi, R.V. (1997b). "Posterior consistency of Dirichlet mixtures in density estimation," unpublished manuscript.

Kass, R.E. and Wasserman, L. (1996). "The selection of prior distributions by formal rules." *Journal of the American Statistical Association,* **91**, 1343-1370.

Robins, J. and Ritov, Y. (1997). "Toward a curse of dimensionality appropriate asymptotic theory for semiparametric models." To appear: *Statistics and Medicine.*

Schwartz, L. (1960). "Consistency of Bayes procedures." Ph.D. dissertation, University of California.

Schwartz, L. (1965). "On Bayes procedures." *Z. Wahrsch. Verw. Gebiete* **4**, 10-26.

Shen, X. (1995). "On the properties of Bayes procedures in general parameter spaces." Unpublished manuscript.

Tierney, L. (1987). "Asymptotic nonparametric posterior distributions," Technical report 497, Department of Statistics, University of Minnesota.

van der Vaart, A. and Wellner, J. (1996). *Weak Convergence and Empirical Processes.* Springer: New York.

Wasserman, L. (1997). "Asymptotic inference for mixture models using data dependent priors," In preparation.

Wong, W.H. and Shen, X. (1995). "Probability inequalities for likelihood ratios and convergence rates of sieve MLE's." *Annals of Statistics,* **23**, 339-362.

Zhao, L. (1993). "Frequentist and Bayesian aspects of some nonparametric estimation." Ph.D. Thesis, Cornell University.

Zhao, L. (1997). "Bayesian Aspects of Some Nonparametric Problems." Technical report, University of Pennsylvania.

17

Modeling Travel Demand in Portland, Oregon

Katja Ickstadt
Robert L. Wolpert
Xuedong Lu

ABSTRACT An important problem in transportation planning is the modeling of patterns of trip-making—especially trips from home (origin) to work (destination) within a fixed geographic area. The standard tool used to study such OD flows is the so-called "gravity model", a Poisson log-linear regression model for studying the number of trips from origins within one element of a fixed partition area (traffic analysis zone, for example) to destinations within another.

We present an alternative: a gridless Bayesian hierarchical Poisson/gamma random field model, allowing us to incorporate spatial correlation explicitly. The models are fitted to a subset of data from the 1994/95 METRO survey of Portland, Oregon, and are internally validated on a reserved portion of the survey data. Bayesian posterior probability distributions are calculated using a Markov chain Monte Carlo integration scheme based on a novel method for simulating samples from gamma random fields.

Some key words: Bayesian mixture models, gamma process, Markov chain Monte Carlo, simulation.

17.1 Introduction

One of the most important problems in transportation planning is to model the patterns of trip-making from an origin (O) to a destination (D), in order to predict travel demand for such OD flows.

Usually, travel demand studies are based on a system of coarsely partitioned geographical areas, known as traffic analysis zones (TAZs). Interest centers on estimating the expected number of trips T_{ij} from an origin TAZ i to a destination TAZ j. The model most commonly used is the so-called gravity model (Casey (1955), Sen and Smith (1995))

$$T_{ij} = A_i \cdot B_j \cdot \exp\left(- \sum_{n=1}^{N} c_{ij}^n \theta_n \right) ,$$

where the traffic flows T_{ij}, and the costs, travel times or distances c_{ij}^n, $n = 1, ..., N$, are observed or known, and the parameters $A_1, ..., A_I$, $B_1, ..., B_J$, and $\theta_1, ..., \theta_N$ are estimated from the data; for $c_{ij}^n = \log|x_i - x_j|$ and $\theta_n = 1$ the equation reduces to the Newtonian potential, hence the name "gravity model". Methods commonly used for analyzing gravity models include fitting marginals in trip-distribution tables, Poisson log-linear regression, and multinomial logit regression.

New legislation, the Clean Air Act Amendments (CAAA) of 1990 and the Intermodal Surface Transportation Efficiency Act (ISTEA) of 1991, requires higher standards of travel demand modeling, such as finer spatial and temporal resolution and incorporating inherent urban structural patterns into the analysis. Although disaggregated forms of the gravity model such as the continuous Poisson log-linear model (Smith (1987)) and the continuous multinomial logit model (Ben-Akiva and Watanatada (1981)) refine the spatial resolution infinitely and do not force the transportation planner to pick an arbitrary partition of the study area before the analysis, they do assume independence between neighboring areas for both origin zones and destination zones. However, neighboring sites often exhibit similar travel behavior, enabling us to learn about travel demands in nearby, structurally similar regions if we model spatial correlation explicitly.

Although little used in transportation science, Bayesian statistical models for possibly correlated, spatial count data are increasingly used in other areas of application. Most commonly applied are Gaussian random field models in which the count data, following a variance-stabilizing transformation, are treated as if normally distributed with a partially unknown covariance structure (this is closely related to the spatial interpolation method of Kriging (Matheron (1963))), and log-linear Poisson/log-normal models, in which the data are assumed to be Poisson distributed with intensities whose logarithms are modeled with a Gaussian random field. Unfortunately Gaussian models (Cressie (1993)) fail to respect the discrete nature of the count data, while log-linear models (Besag et al. (1991); Clayton and Kaldor (1987)) do not scale properly under spatial refinement of partitions (the logarithmic structure leads to products rather than sums for the Poisson means for unions of neighboring partition elements) excluding a continuous, disaggregated version.

Other methods often applied to spatial point count data include auto-Poisson models (Besag (1974)) and Strauss models (Strauss (1975)), which cannot reflect the positive associations that might be expected for nearby reporting units (all count correlations must be negative in those models), and conjugate Poisson/gamma hierarchical models (Clayton and Kaldor (1987)), under which all counts are independent under both prior and posterior distributions. Recently a flexible class of Bayesian hierarchical Poisson/gamma random field models was introduced by Wolpert and Ickstadt (1998a) for possibly spatially-dependent count data. These models generalize both conjugate Poisson/gamma models and mixture of Dirichlet process models, and are suitable for spatially continuous applications without aggregation.

In this paper, we apply these geographically continuous hierarchical Poisson/gamma models to analyze part of the 1994/95 Portland Metropolitan Area Household Activity and Travel Survey concentrating on trips from home (origin) to work (destination). We treat all home locations and some of the work locations as known, and forecast the missing work locations. At the lowest stage of the hierarchy the home locations will be modeled as a Poisson random field. At the next level of the hierarchy the intensity measure of this Poisson random field will be modeled as a distance-based kernel mixture of a gamma random field describing the work locations.

In previous applications of this model class the gamma random field was unobserved and latent, and the analysis mainly focussed on smoothing the intensity measure of the Poisson random field (e.g. Wolpert and Ickstadt (1998a), Ickstadt and Wolpert (1997)). In this application interest will mainly center on estimating the gamma random field (here partially observed and describing the work locations). Using the given home locations and the estimated work locations we obtain the resulting estimated origin/destination pairs. We also estimate the parameters of the kernel function in order to help us understand the influence of travel distance on the home/work travel pattern.

Our analysis is meant as a first step toward modeling origin/destination patterns in Portland, Oregon, intended eventually to enable the prediction of how changes in zoning or road construction work will affect the OD flows.

In section 2 the data are described, in section 3 the class of hierarchical Bayesian Poisson/gamma models is introduced, and its connection to the mixture of Dirichlet processes is pointed out. The Markov chain Monte Carlo sampling scheme is given in section 4. The results of our analysis are presented in section 5, followed by a discussion in section 6.

17.2 The Data

We use part of the Portland Metropolitan Area Household Activity and Travel Survey conducted in Spring 1994 and Fall/Winter 1994/95 (available on-line at `http://www.webconcepts.com/m2lib/pdxsvy.htm`). The sample includes 4,400 households, selected by random dialing, with about 10,000 people altogether. For each respondent the activity and travel behavior for a 48 hours period was reported. For every change in location the exact (x, y) coordinates and the activity were recorded for the destination. The pairs of days were staggered across the seven days of the week. In addition, household characteristics, such as household size, income, type of dwelling, and number of vehicles, and personal socio-demographics, such as age, gender, employment status, and occupation, were reported.

For our analysis we are particularly interested in the adults' regular travel behavior from home to work. Thus, we take only data entries of employed individuals of age 16 years and older, who made home to work trips on the survey day; we restrict ourselves to the first survey day to avoid correlation; we pool observations from Monday to Friday; and we use only the spring sample.

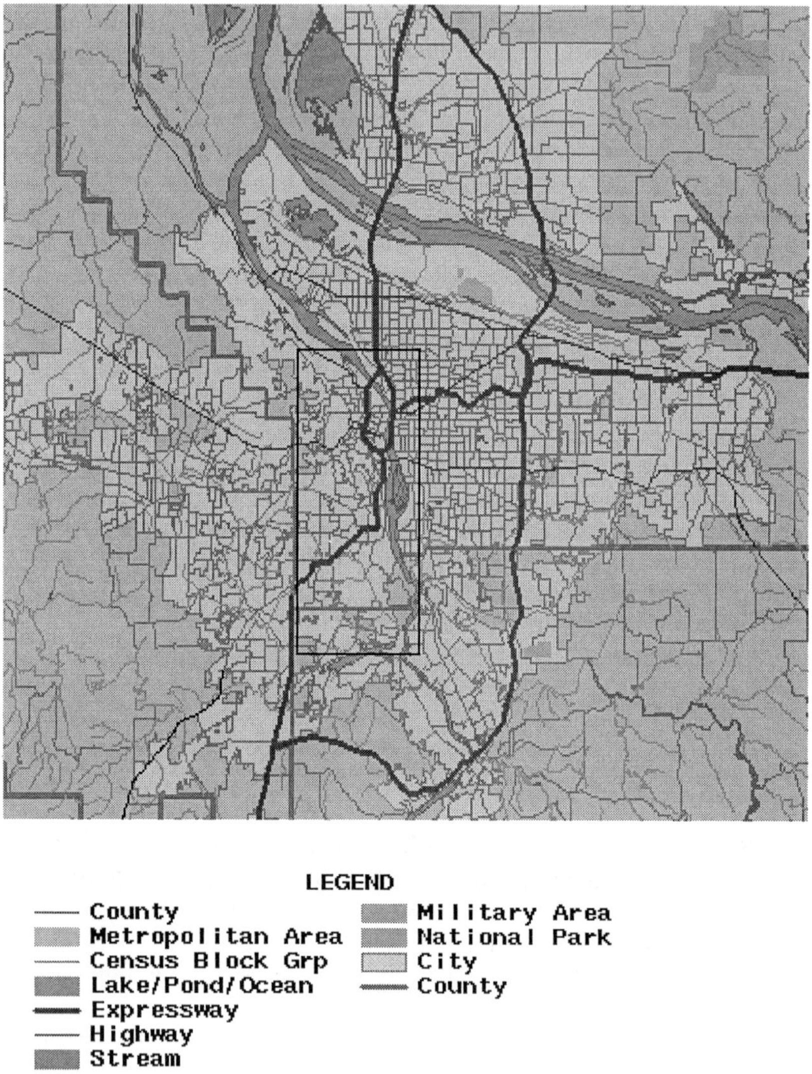

FIGURE 1. Metro and downtown Portland. The box indicates the study region.

The survey data covers the entire Multinomah county, as well as parts of Clackamas county, Washington county, and Clark county. Figure 1, obtained from `http://tiger.census.gov/cgi-bin/mapbrowse-tbl`, shows large parts of Multinomah county, including the Metro Portland and downtown Portland areas. For illustration purposes we base our analysis only on the 290 home to work trips originating and ending in a 6.1*km* by 15.2*km* rectangular study region in downtown Portland, Multinomah County, indicated by a box in figure 1.

Figure 2 displays the 290 OD data pairs in the study region box; overlaid are the TAZs. In figure 2a) the home locations are shown as circles, in figure 2b) the work locations are shown. Here, filled and unfilled circles represent the unknown and known work locations, respectively. In our analysis in section 5, we will treat different fractions of work locations as known.

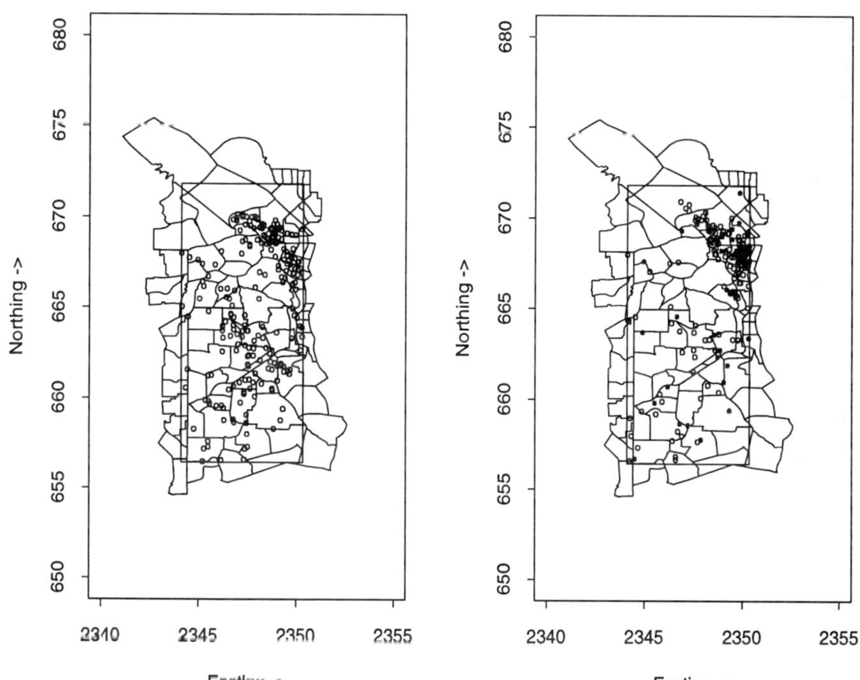

(a) Home (b) Work

FIGURE 2. (a) Home locations and (b) work locations in the study region.

17.3 Poisson/Gamma Random Field Models

In this section we describe the models in somewhat more generality than needed for the present application, both to clarify their structure and to simplify their use in other applications; for example, we maintain the distinction between the spaces $\mathcal{X}$ and $\mathcal{S}$ of trip origins and destinations, even though in our application both represent the study region.

17.3.1 The Model Class and its Interpretation

At the first stage of the hierarchy we will model the point pattern of the origin locations as a Poisson random field in which the number of points $N(A)$ in each set A of the origin space $\mathcal{X}$, part of the plane, is a Poisson random variable with mean $\Lambda(A)$. The representations of the Poisson random field are (almost surely) discrete, consisting of a number of jump locations σ_m, all with unit mass (jump height), so that

$$N(A) = \sum_{\sigma_m \in A} 1 .$$

In our application the jump locations represent the house locations in the study region; equal jump height for all locations means that all households are regarded equally influential for the OD behavior. Since the mean intensity $\Lambda(A)$ of the Poisson random field is unobserved, we regard it as a non-negative random variable and thus use a doubly stochastic Poisson model.

Doubly stochastic models that assign independent "increments" $\Lambda(A_i)$ to disjoint sets A_i, such as the gamma and stable random fields, would necessarily feature independent counts $N(A_i)$ as well, and thus could not directly be used to study possible spatial dependence. Instead we construct the intensity measure at the second stage of the hierarchy as a kernel mixture $\Lambda(A) \equiv \int_{\mathcal{S}} \int_A k(x,s) dx \Gamma(ds)$ for an independent-increment infinitely divisible random measure $\Gamma(ds)$ on an auxiliary space $\mathcal{S}$, usually a subset of the plane or a discrete set.

In our application the random measure $\Gamma(ds)$ describes the destinations, i.e. the work locations. Work places with many employees will be more influential for the OD behavior than places with a small number of employees. Thus, we model the destination locations by an independent-increment infinitely divisible random field that features a range of possible jump heights at different locations, larger jumps representing a larger number of employees, such as the gamma or stable random fields (Wolpert and Ickstadt (1998b)). The entire set of possible workplaces is infinite (and dense) in such a model, but for each bounded subregion $A \subset \mathcal{S}$ and number $\varepsilon > 0$ the number of work places in A with employment intensity greater than ε will be finite, again with a Poisson distribution. In this paper we illustrate the ideas taking $\Gamma(ds)$ to have the gamma random field distribution

$\Gamma(ds) \sim \text{Ga}\big(\alpha(ds), \beta(s)^{-1}\big)$ with shape measure $\alpha(ds)$ and inverse scale function $\beta(s) > 0$.

At the third stage of the hierarchy we introduce a prior probability distribution $\pi(d\theta)$ on a parameter space Θ in order to express uncertainty about the kernel, $k(x,s) = k^\theta(x,s)$, and some aspects of the $\Gamma(ds)$ random field, $\alpha(ds) = \alpha^\theta(ds)$ and $\beta(s) = \beta^\theta(s)$. Thus, our class of doubly stochastic Poisson random field models is given hierarchically by:

$$\begin{aligned}
\theta &\sim \pi(d\theta) \\
\Gamma(ds)\big|_\theta &\sim \text{Ga}(\alpha^\theta(ds), \beta^\theta(s)^{-1}) \\
N(dx)\big|_{\theta,\Gamma} &\sim \text{Po}\big(\Lambda(dx) \equiv \int_S k^\theta(x,s)\, dx\Gamma(ds)\big).
\end{aligned}$$

Figure 3 presents the hierarchical Poisson/gamma random field model in pictures. On the lowest level the home locations (dots) are shown; on the middle level a contour plot of the estimated mean intensity $\Lambda(dx)$ is given; and on the highest level a realization of the gamma random field (jump locations are work locations, jump heights are proportional to the number of employees) is presented. The dashed lines indicate that the gamma random field jumps are averaged (weighted by the kernel) to obtain an estimate for $\Lambda(dx)$ at a certain home location x.

Conditional on the parameter values θ the means and covariances of the Poisson random field are given by:

$$\begin{aligned}
\text{E}^\theta(N(A)) &= \int_S k^\theta(A,s)\beta^\theta(s)^{-1}\alpha^\theta(ds) \\
\text{Cov}^\theta(N(A),N(B)) &= \int_S k^\theta(A\cap B,s)\beta^\theta(s)^{-1}\alpha^\theta(ds) \\
&\quad + \int_S k^\theta(A,s)k^\theta(B,s)\beta^\theta(s)^{-2}\alpha^\theta(ds),
\end{aligned}$$

where $k^\theta(A,s) \equiv \int_A k^\theta(x,s)\, dx$. For further details on this model class see Wolpert and Ickstadt (1998a).

17.3.2 *The Specific Model*

In our application we represent uncertain model features with $\theta \in \mathbb{R}^3$ as follows. We take $\Gamma(ds)$ to be a gamma random field with a uniform shape measure $\alpha^\theta(ds) = (\theta_1/\theta_2)\, ds$, and constant scale $\beta^\theta(s)^{-1} = \theta_2$, hence mean $\beta^\theta(s)^{-1}\alpha(ds) = \theta_1\, ds$.

We choose a gravity model kernel with a single cost function expressing the undesirability of commuting from x to s as squared Euclidean distance, leading to the Gaussian kernel density

$$k^\theta(x,s) \equiv \frac{1}{2\pi\theta_3^2} e^{-|x-s|^2/2\theta_3^2}$$

that allows for a range of distances over which spatial interaction between the home and work locations might extend.

For sets A of large diameter the above parameterization leads to the following approximate means and variances (see Wolpert and Ickstadt (1998a)):

$$\mathrm{E}^{\theta}(N(A)) \approx \theta_1 |A|, \qquad \mathrm{Var}^{\theta}(N(A)) \approx \theta_1 |A|(1+\theta_2).$$

Evidently, θ_1 is the home location density, in counts per unit area (hectare (ha), in our application), and θ_2 is the fraction of the extra-Poisson variation or large-scale overdispersion.

17.3.3 The Connection to Mixture of Dirichlet Processes

If we specifically set $\beta^{\theta}(s) \equiv \beta$ for a constant β, and scale the gamma random field to total mass one, $\mathcal{D}(ds) = \Gamma(ds)/\Gamma(\mathcal{S})$, our model can be viewed as a mixture of Dirichlet processes model (Antoniak (1974), Escobar and West (1998), MacEachern (1998)) for estimating the Poisson intensity measure $\Lambda(dx)$. In our hierarchical Poisson/gamma model class nonconstant functions $\beta^{\theta}(s)$ are allowed reflecting our prior opinion about the space $\mathcal{S}$ to be more informed in some places than in others; the kernel function $k^{\theta}(x, s)$ need not be a Gaussian density; and the gamma random field can be replaced by another Lévy random field (Wolpert and Ickstadt (1998b)) to model a different tail behavior for the distributions on the hierarchical (gamma) level.

17.4 The Computational Scheme

Here we describe the Markov chain Monte Carlo (MCMC) computational scheme needed for the posterior analysis presented in section 5. While the likelihood function for θ from our Poisson/gamma model is unavailable analytically, we can study the intractable posterior distributions of the uncertain quantities θ and the gamma random field $\Gamma = \Gamma(ds)$ by simulating steps (θ^t, Γ^t) from an ergodic Markov chain whose stationary distribution is the posterior distribution of θ and Γ given the observed home locations $N(dx)$ (Gelfand and Smith (1990); Gilks et al. (1996); Tierney (1994)).

The simulation approach requires sampling from the complete conditional distributions of each uncertain quantity $(\theta^t, \Gamma^t, etc.)$ given all the others; see Wolpert and Ickstadt (1998a). The simulation is simplified dramatically by the introduction of data augmentation, in this application selecting an augmented work location $s_n \in \mathcal{S}$ for each of the $N = 290$ survey participants residing at $x_n \in \mathcal{X}$. With this augmentation the data become conjugate for the gamma random field, enabling us to simulate the conditional distribution of Γ given θ and the augmented points $\{s_n\}_{n \leq N}$

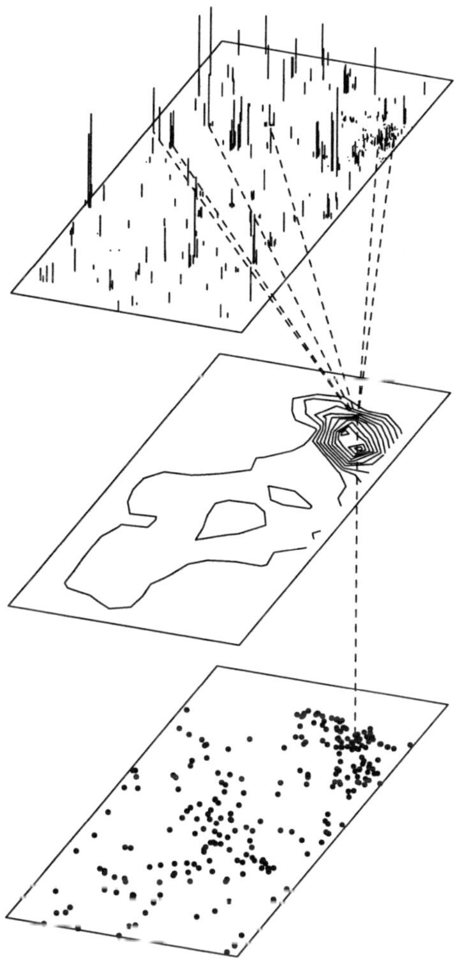

FIGURE 3. Graphical representation of the hierarchical Poisson/gamma random field model.

using the methods of Wolpert and Ickstadt (1998b). The conditional distribution of the augmentation points given θ and Γ is discrete and easy to sample from. The conditional distribution of θ given Γ and the augmentation points $\{s_n\}_{n\le N}$ depends on the arbitrary prior distribution $\pi(d\theta)$. It has a density function if $\pi(d\theta)$ does, and can be sampled from by the Metropolis/Hastings step given below.

The following version of the MCMC algorithm merges steps 1 (updating Γ) and 3 (updating θ) into a single block, improving the efficiency of the algorithm given in Wolpert and Ickstadt (1998a). The algorithm is also more efficient if the sampling distribution $\Pi(ds)$ can be chosen to be nearly proportional to the gamma shape measure $\alpha^\theta(ds)$, so that the density function $\alpha^\theta(s) = \alpha^\theta(ds)/\Pi(ds)$ is nearly constant; in this application we address this by sampling heavily from the downtown region, where most employees work.

For the properties and computational algorithms for evaluating the exponential integral function $E_1(x) \equiv \int_x^\infty u^{-1}e^{-u}\,du$ see (Abramowitz and Stegun, 1964, p. 228). We denote the integral of the kernel by $k^\theta(\mathcal{X}, s) \equiv \int_{\mathcal{X}} k^\theta(x, s)\,dx$. Note that gamma random field realizations are almost-surely discrete, so admit a representation $\Gamma(ds) \equiv \sum_{m<\infty} \upsilon_m \delta_{\sigma_m}(ds)$; in the algorithm we can draw only a large finite number M of the mass points, of course.

Given a prior density function $\pi(\theta)$ on the parameter space Θ, a probability measure $\Pi(ds)$ on $\mathcal{S}$, a kernel function $k^\theta(x, s)$, a gamma random field shape density $\alpha^\theta(s)$ and inverse scale function $\beta^\theta(s)$, a Markov transition density $Q(\theta, \theta^*)$ on Θ, an integer $M \gg N$ and initial points $\theta^0 \in \Theta$ and $\{s_n^0\}_{n\le N}$, successive points can be generated by the following hybrid Gibbs/Metropolis scheme starting at iteration $t = 1$.

1. Gibbs step to update the gamma random field:

Given θ^{t-1} and $S^{t-1} = \{s_n^{t-1}\}_{n\le N}$,

a) Set $\sigma_m^t \equiv s_m^{t-1}$, $1 \le m \le N$, and generate independent $\sigma_m^t \sim \Pi(ds)$, $N < m \le M$;

b) Set $\alpha_m^t \equiv \alpha^{\theta^{t-1}}(\sigma_m^t)$, $\beta_m^t \equiv \beta^{\theta^{t-1}}(\sigma_m^t) + k^{\theta^{t-1}}(\mathcal{X}, \sigma_m^t)$, and $i_m^t \equiv 0$ if $s_n^{t-1} = \sigma_m^t$ for some $n < m$, otherwise $i_m^t \equiv 1$;

c) Generate successive jumps $\{\tau_m\}_{m\le M}$ of a standard Poisson process;

d) Set $\upsilon_m^t \equiv (\tau_m - \tau_{m-1})/\beta_m^t$, $1 \le m \le N$, and $\upsilon_m^t \equiv E_1^{-1}\big((\tau_m - \tau_N)/\alpha_m^t\big)$, $N < m \le M$;

e) Set $\Gamma^t(ds) \equiv \sum_{m\le M} \upsilon_m^t \delta_{\sigma_m^t}(ds)$.

2. Gibbs step to update the augmentation points:

Given θ^{t-1} and Γ^t,

a) Generate independent $S^t = \{s_n^t\}_{n \le N}$ with $\mathrm{pr}(S_n^t = \sigma_m^t) \propto v_m^t k^{\theta^{t-1}}(x_n, \sigma_m^t)$.

3. Metropolis/Hastings step to update the parameter θ:
Given θ^{t-1} and $S^t = \{s_n^t\}_{n \le N}$,

a) Set $\theta^- = \theta^{t-1}$ and generate a new candidate $\theta^+ \sim Q(\theta^-, \theta^+)$;

b) Set $k_n^- \equiv k^{\theta^-}(x_n, s_n^t), \quad k_n^+ \equiv k^{\theta^+}(x_n, s_n^t)$;

c) Set $\alpha_n^- \equiv \alpha^{\theta^-}(s_n^t), \quad \alpha_n^+ \equiv \alpha^{\theta^+}(s_n^t)$;

d) Set $\beta_n^- \equiv \beta^{\theta^-}(s_n^t) + k^{\theta^-}(\mathcal{X}, s_n^t), \quad \beta_n^+ \equiv \beta^{\theta^+}(s_n^t) + k^{\theta^+}(\mathcal{X}, s_n^t)$;

e) Calculate the Metropolis/Hastings acceptance probability

$$\begin{aligned} P^t \equiv\ & \{\pi(\theta^+)/\pi(\theta^-)\}\{Q(\theta^+,\theta^-)/Q(\theta^-,\theta^+)\} \prod_{n \le N} (k_n^+/k_n^-) \\ \times\ & \exp\Big(\sum_{n \le N} \big(i_n^t \log(\alpha_n^+/\alpha_n^-) - v_n^t(\beta_n^+/\beta_n^-)\big)\Big) \\ \times\ & \exp\Big(-\int_S \log(1 + k^{\theta^+}(\mathcal{X}, s)/\beta^{\theta^+}(s))\, \alpha^{\theta^+}(ds) \\ & + \int_S \log(1 + k^{\theta^-}(\mathcal{X}, s)/\beta^{\theta^-}(s))\, \alpha^{\theta^-}(ds)\Big); \end{aligned}$$

f) Set $\theta^t \equiv \begin{cases} \theta^+ & \text{with probability} \min(1, P^t) \\ \theta^- = \theta^{t-1} & \text{otherwise.} \end{cases}$

4. Increment $t \leftarrow t + 1$ and return to step **1**.

Software (in ISO C) is available from the authors on request.

17.5 Posterior Analysis

Underlying our analysis is the specific Poisson/gamma random field model introduced in section 3.2. For the three parameters we use vague but proper independent exponential prior distributions.

In our survey data set home and work locations are both known. Home locations for every Portland resident are available from census data, but the corresponding work locations are only known for respondents of the survey. Motivated by this practical problem we treat a portion of the survey work locations as if unknown. In order to see how well the model is able to replicate these "unknown" work locations, we omit a range (10%, 33%, 50%, 66%, and 90%) of the work locations.

For our analysis 6,000 iterations of the MCMC scheme introduced in section 4 (after a burn-in period of 2,000 steps) turned out to be more than adequate for convergence. For the different work location percentages treated as unknown, all three Markov chains passed various statistical tests such as Raftery and Lewis, Heidelberger and Welch, and Gelman and Rubin as implemented in CODA (Best et al. (1995)). We used a symmetric Gaussian random walk for the Metropolis step (3a), with appropriate step sizes to ensure acceptance probabilities of about 40%.

An important goal of our analysis is to obtain estimates for all work locations, both those we treat as known and those we treat as unknown. Figure 4 shows contour plots of these mean work location densities versus the left-out work locations (given as dots), for the different proportions of work locations assumed to be known. It can be seen that the model replicates the work location density quite well; with only 10% of the data treated as known, the high destination density area of downtown Portland (in the upper right corner) is identified. Assuming higher percentages of the work places to be known leads to less smoothed estimates.

From the posterior estimates for all work locations and the given home locations, we can derive the trip OD tables, the traditional objects of central interest for the transportation planning engineer. Since our analysis is independent of any arbitrary grid structure, we can decide a posteriori how to group the OD flows in order to display them. In our example we partitioned the study region into a regular 3 by 3 grid (numbered from 1 to 9, starting at the south-west corner of the study region, assigning numbers 1, 2, 3 moving north in the most western column, 4, 5, 6 moving north in the middle column, and 7, 8, 9 moving north in the most eastern column) and aggregated home locations and mean work location estimates on that grid. Figure 5 shows the resulting OD tables for the different proportions of work locations assumed to be known. It can be seen that destination grid zone 9 is most popular, followed by zones 6 and 8. The diagonal entries include people working at their home locations.

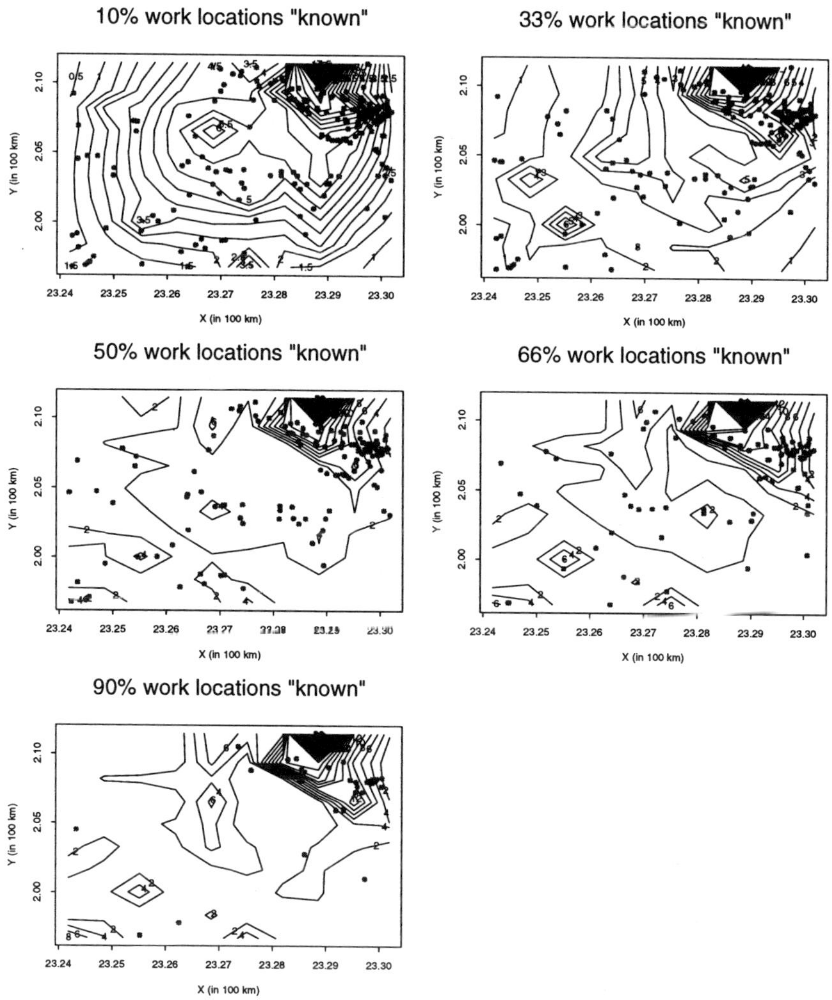

FIGURE 4. Estimated mean posterior work locations.

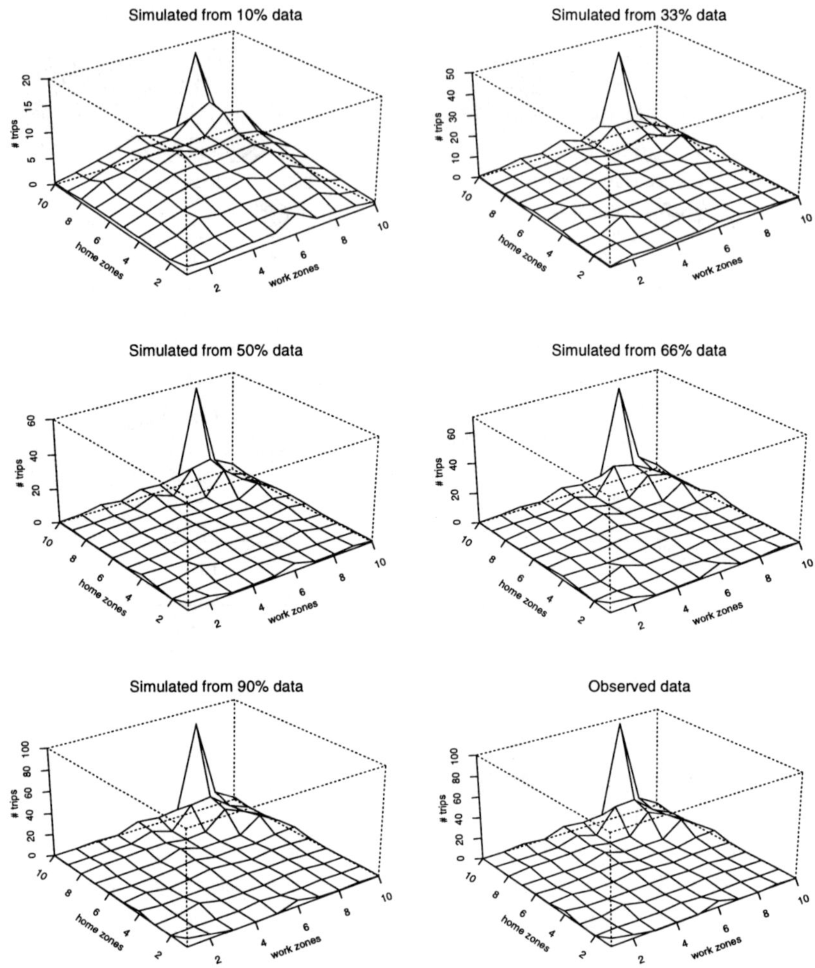

FIGURE 5. Estimated trip OD tables.

Figure 6 shows the prior (curve) and the estimated posterior (histogram) density functions for the home location density θ_1 (in trips/ha), the large-scale extra-Poisson variation θ_2, and the interaction distance θ_3 (in km) for 50% of the work locations assumed to be unknown. The posterior means and standard deviations are summarized in table 1.

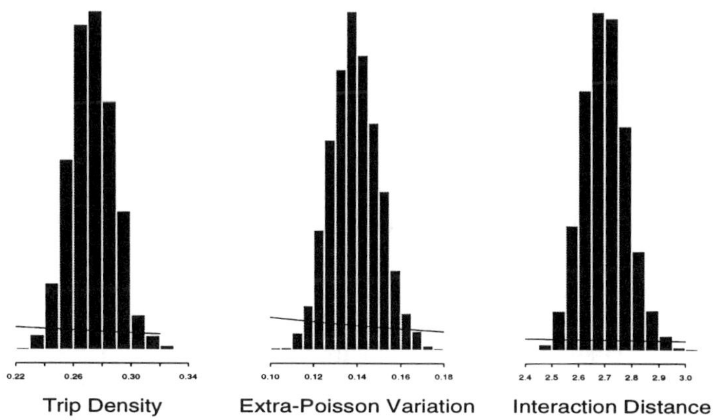

FIGURE 6. Prior densities and posterior histograms.

	Units	Par.	Mean	Std. Dev.
Daily Trip Density	trips/ha	θ_1	0.274	0.015
Extra-Poisson Variation		θ_2	0.139	0.011
Interaction Distance	km	θ_3	2.70	0.083

TABLE 1: Posterior summary statistics.

The interaction (i.e. commuting) distance has a posterior mean of 2.7 km, suggesting that employment opportunities at distances of up to a few times this distance (five or six km) have a strong effect on the choice of home location.

Results similar to those of table 1 are obtained for 10%, 33%, 66%, and 90% of the work locations treated as known. The posterior means of θ_1 and θ_2 remain unchanged, while the mean of the interaction distance θ_3 increases with increasing percentages of data points treated as known, and turns out to be 1.84 km, 2.26 km, 2.80 km, 2.91 km, respectively.

17.6 Discussion

The hierarchical Poisson gamma random field models underlying the present analysis are a useful tool for estimating a continuous work location surface, enabling us in turn to look at OD flows on any arbitrary discretization of both origin and destination space. We model possible spatial correlation explicitly, and allow for incorporating observed data on our hierarchical gamma level.

The model used here can be extended to include additional covariate information. Individual covariates (e.g. age, gender) can be incorporated using a marked Poisson random field on the lowest level of the hierarchy; spatial covariate information (e.g. employment density per TAZ, land use, GIS information about rivers and parks) can be reflected in a spatially dependent measure $\alpha(ds)$ for the gamma random field and might lead to less smoothing for the work location densities. We are also exploring kernels based on different cost functions, e.g. the travel time between home location $x \in \mathcal{X}$ and work location $s \in \mathcal{S}$.

The class of hierarchical Poisson/gamma models introduced in section 3 is applicable in a wide range of problems featuring correlated count data including disease mapping, bioabundance and forest growth studies, image analysis and survival analysis.

Acknowledgments: This work was supported by U.S. N.S.F. grant DMS-9313013 to the National Institute of Statistical Sciences, U.S. E.P.A. grant CR822047-01-0, U.S. N.S.F. grant DMS-9626829, and the Deutsche Forschungsgemeinschaft. The authors would like to thank Keith Lawton and his staff at METRO Portland for providing the data, and the U.S. Census Bureau for offering the TIGER Mapping Service. This work was inspired by and would have been impossible without the ideas, advice and encouragement of Eric Pas, who passed away while the work was in progress. We will all miss him deeply.

References

Abramowitz, M. and Stegun, I. A., eds., (1964), *Handbook of Mathematical Functions With Formulas, Graphs, and Mathematical Tables*, volume 55 of *Applied Mathematics Series*, Washington, D.C.: National Bureau of Standards.

Antoniak, C. E., (1974), "Mixtures of Dirichlet Processes with Applications to Bayesian Nonparametric Problems," *Annals of Statistics*, 2, 1152–1174.

Ben-Akiva, M. and Watanatada, T., (1981), "Application of a Continuous Spatial Choice Logit Model," in *Structural Analysis of Discrete Data with Econometric Applications* (eds: C. F. Manski and D. McFadden), 320–343, Cambridge, Mass.: MIT Press.

Besag, J., (1974), "Spatial interaction and the statistical analysis of lattice systems," *Journal of the Royal Statistical Society, Series B (Methodological)*, 35, 192–236, (With discussion).

Besag, J., York, J., and Mollié, A., (1991), "Bayesian image restoration, with two applications in spatial statistics," *Annals of the Institute of Statistical Mathematics*, 43, 1–59, (With discussion).

Best, N. G., Cowles, M. K., and Vines, S. K., (1995), *Convergence diagnosis and output analysis software for Gibbs sampling output, Version 0.3*, Cambridge, UK: MRC Biostatistics Unit.

Casey, H. J., (1955), "Applications to traffic engineering of the law of retail gravitation," *Traffic Quarterly*, 100, 23–35.

Clayton, D. G. and Kaldor, J., (1987), "Empirical Bayes estimates of age–standardized relative risks for use in disease mapping," *Biometrics*, 43, 671–681.

Cressie, N. A. C., (1993), *Statistics for Spatial Data*, New York, NY, USA: John Wiley & Sons.

Escobar, M. D. and West, M., (1998), "Computing nonparametric hierarchical models," in *Practical Nonparametric and Semiparametric Bayesian Statistics* (eds: Dey, D. and Müller, P. and Sinha, D.), New York: Springer-Verlag.

Gelfand, A. E. and Smith, A. F. M., (1990), "Sampling-based approaches to calculating marginal densities," *Journal of the American Statistical Association*, 85, 398 409.

Gilks, W. R., Richardson, S., and Spiegelhalter, D. J., eds., (1996), *Markov Chain Monte Carlo in Practice*, New York, NY, USA: Chapman & Hall.

Ickstadt, K. and Wolpert, R. L., (1997), "Multiresolution Assessment of Forest Inhomogeneity," in *Case Studies in Bayesian Statistics, Volume III* (eds: C. Gatsonis, J. S. Hodges, R. E. Kass, R. E. McCulloch, P. Rossi, and N. D. Singpurwalla), volume 121 of *Lecture Notes in Statistics*, 371–386, New York: Springer-Verlag.

MacEachern, S. (1998). Computations for Mixture of Dirichlet Process models, in *Practical Nonparametric and Semiparametric Bayesian Statistics* (eds: Dey, D. and Müller, P. and Sinha, D.), New York: Springer-Verlag.

Matheron, G., (1963), "Principles of geostatistics," *Economic Geology*, 58, 1246–1266.

Sen, A. and Smith, T. E., (1995), *Gravity Models of Spatial Interaction Behavior*, Berlin: Springer-Verlag.

Smith, T. E., (1987), "Poisson Gravity Models of Spatial Flows," *Journal of Regional Science*, 27, 315–340.

Strauss, D. J., (1975), "A Model for Clustering," *Biometrika*, 62, 467–476.

Tierney, L., (1994), "Markov chains for exploring posterior distributions," *Annals of Statistics*, 22, 1701–1762, (With discussion).

Wolpert, R. L. and Ickstadt, K., (1998a), "Poisson/gamma random field models for spatial statistics," *Biometrika*, to appear.

Wolpert, R.L. and Ickstadt, K., (1998b), "Simulation of Lévy Random Fields," in *Practical Nonparametric and Semiparametric Bayesian Statistics* (eds: Dey, D. and Müller, P. and Sinha, D.), New York: Springer-Verlag.

18

Semiparametric PK/PD Models

Peter Müller
Gary Rosner

ABSTRACT We consider Bayesian analysis of hematology profile data. The discussion is based on a pharmacodynamic study involving longitudinal data consisting of hematologic profiles (i.e., blood counts measured over time) of cancer patients undergoing chemotherapy. Specifically, we analyze data from two studies carried out by the Cancer and Leukemia Group B (CALGB).

Patient populations are known to be heterogeneous. For a thorough consideration of uncertainties it is important to account for this. Specifically, in modeling the distribution of patient-specific random effects, we need to consider models that allow for heterogeneity, outliers, skewness, etc. We review semi-parametric longitudinal data models which use a flexible mixture of normal representation for the random effects distribution. We define a probability model on the mixture by assuming that the mixing measure is generated by a Dirichlet process.

Building on these models we develop a hierarchical meta-model to combine data from multiple studies. Using this model to combine data from two studies, we achieve joint inference over different levels of the anticancer agent cyclophosphamide (CTX) administered in one study (CALGB 8881), and the different levels of the drug amifostine, purported to reduce side effects, administered in another study (CALGB 9160).

18.1 Introduction

People respond to different pharmaceutic agents differently. Some cancer patients respond to chemotherapy while others do not; some experience more toxicity than others. Understanding the reasons behind this heterogeneity should contribute to safer and more effective chemotherapy for each patient. Population pharmacokinetic (PK) and pharmacodynamic (PD) studies seek to identify within a large sample of patients those characteristics that significantly alter a drug's disposition and a patient's response to the drug. Population PK studies often presuppose some sort of model relating drug concentration to time and evaluate relationships between patient characteristics and the parameters in the model. PD studies of treated patients examine relationships of clinical outcomes, such as toxicity or response, to characteristics of the drug's pharmacokinetic profile in each patient. These PD analyses will also investigate relationships with patient characteristics that may themselves affect the clinical outcome directly or indirectly by altering the drug's PK.

In this chapter we discuss the use of non-parametric Bayesian inference for PK/PD modeling, with primary focus on the analysis of two studies carried out by the Cancer and Leukemia Group B (CALGB) (Lichtman et al. 1993). CALGB 8881 was a phase I study that sought the highest dose of the anticancer agent cyclophosphamide one could give cancer patients every two weeks. Patients also received the drug GM-CSF to help reduce the ill effects of cyclophosphamide on the patients marrow. The other study, CALGB 9160, built upon the experience gained in 8881. The drug amifostine had been shown in some studies to reduce some of the toxic side effects of anticancer agents, such as cyclophosphamide and radiation therapy (Spencer and Goa, 1995). The objective of CALGB 9160 was to determine if adding amifostine would reduce the hematologic side effects of aggressive chemotherapy with cyclophosphamide and GM-CSF. CALGB 9160 randomized patients to receive amifostine or not, along with cyclophosphamide (3 grams per square meter of body surface area) and GM-CSF (5 micrograms per kilogram of body weight). The main study question in CALGB 9160 concerned the effect of amifostine on various measures of hematologic toxicity, such as nadir (i.e., minimum) granulocyte counts or days of granulocytopenia. Since only 46 patients entered the randomized trial, we wished to use data already gathered in the earlier study to help make inference in CALGB 9160 more precise.

In both studies, the measured outcome was white blood cell count (WBC) and absolute granulocyte counts (AGC) for each patient over time. In study 8881, we have data on $I_1 = 52$ patients. The other study includes data on $I_2 = 46$ patients. We will use y_{ij} to denote the j-th blood count measurement on the i-th patient. Because of the approach we will take to accommodate the two types of blood count (WBC and AGC) and the two studies (CALGB 8881 and 9160), we will not need subscripts corresponding to response and study. Instead y_{ij} will generically denote measurements in one of the two studies on one of the two responses. In CALGB 8881 we had a total of 674 observations on WBC, and 497 observations on AGC, with the number of observations for one patient varying between 2 and 19. Figure 1 shows responses for some typical patients.

The statistical models which we will use for the analysis of these data fall within the wider ambit of statistical models collectively known as population models (PK/PD models, longitudinal data models, mixed-effects models, nonlinear models for repeated measurements), characterized by the hierarchical structure

$$p(y_{ij}|\theta_i, t_{ij}), \quad p(\theta_i|\eta, x_i), \quad p(\eta), \tag{1}$$

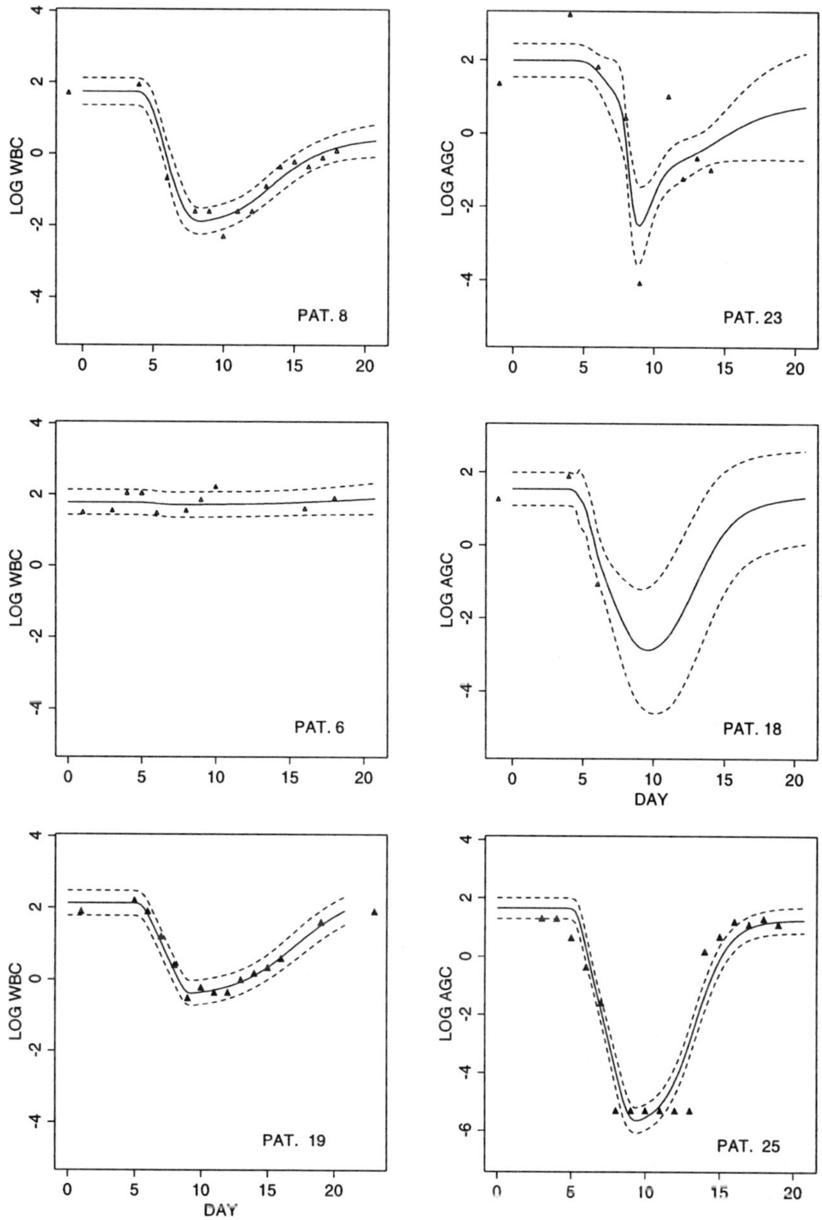

FIGURE 1. Fitted curves for two "typical" patients from study 8881 (patients $i = 8$ and 18), two outliers (patients 23 and 6), and two patients from study 0160 (patients 19 and 25). For patients 8, 6 and 19 the panels show fitted lines for the model with WBC as response, for patients 23, 18 and 25 the panels show fitted profiles for AGC. The points are the actual observations. The solid line plots the fitted curve $\hat{f}_i(t) = E(f(\theta_i, t)|y)$. The dotted lines show plus/minus one posterior standard deviation $SD(f(\theta_i, t)|y)$. WBC and AGC are in 1000s per μl.

$i = 1, \ldots, I$, $j = 1, \ldots, n_i$. In our application, y_{ij} is the j-th measurement on the i-th individual (we will discuss later how we deal with the two additional levels in the hierarchy corresponding to the two responses and the two studies). The top level likelihood $p(y_{ij}|\theta_i, t_{ij})$ explains the intra-individual variation. Here θ_i are parameters explaining the joint distribution of $y_i = (y_{ij}, j = 1, \ldots, n_i)$, and t_{ij} are occasion-specific covariates, in our case time after initiation of the treatment. Typically, $p(y_{ij}|\ldots)$ takes the form of a non-linear regression. Below, in Section 18.2, we will discuss a specific model for the longitudinal blood count data. The first-level prior $p(\theta_i|\eta, x_i)$ explains the inter-individual variation and represents the distribution of PK or PD parameters θ_i across the population. This prior model often includes a regression on treatment indicators x_i. In our analysis x_i is a vector including dosage of cyclophosphamide (CTX), dose of GM-CSF, and an indicator of whether or not the patient received amifostine. In other applications, x_i could be any other subject-specific covariates x_i, including treatment, age, indicators of family history, etc., or, in a PD model, parameters derived from a related PK model. Commonly used assumptions for the population distribution $p(\theta_i|\eta, x_i)$ are multivariate normal or Student t distributions. The second prior level describes the distribution of the hyperparameters η.

Bayesian models similar to (1) have been considered in Zeger and Karim (1991) for generalized linear mixed models, in Wakefield, Smith, Racine-Poon and Gelfand (1994) for the general population model assuming a multivariate normal population distribution, in Dellaportas and Smith (1993), and in Wakefield (1996) with multivariate t priors. See Wakefield et al. (1998) for a recent review of related literature. Alternative maximum likelihood and least-squares-type methods are summarized in Davidian and Giltinan (1995).

In maximum likelihood-based inference, semi-parametric models with non-parametric assumptions on the population distribution have been very successful, mainly because they allow for common critical deviations from more strictly parametric forms, including heterogeneity, outliers, over-dispersion, multimodality, etc. Popular approaches are NPML (Mallet 1986), with no restrictions on the distribution of the random-effects in the model and yielding a discrete estimate of this distribution; and SNP (Davidian and Gallant 1992, 1993), a method that assumes the model's random effects have a "smooth" density and produces estimates from such a class of densities. In the Bayesian literature, similar approaches using semi-parametric random effects models are discussed in Rosner and Müller (1997) and Müller and Rosner (1997). In these papers we discuss non-parametric models for the random effects distribution $p(\theta_i|x_i)$, based on Dirichlet process prior models. The same model is used in Kleinman and Ibrahim (1996). Ibrahim (1998), earlier in this volume, introduces the model in a general framework of mixed models with non-parametric random effects distributions. Hirano (1998) uses Dirichlet process prior mixture models in an econometric application. A similar model, using a Dirichlet process prior for the random effects distribution, without the additional mixture, is proposed by Walker and Wakefield (1994).

In Section 18.2 we review the non-parametric Dirichlet process mixture model for the random effects distribution, and analyze data from CALGB 8881. In Section 18.3 we extend the model to allow meta- analysis in several related studies and extend the earlier results to include data from CALGB 9160. Section 18.4 concludes the discussion.

18.2 A semiparametric population model

We now describe the specific population model (1) used for the analysis of CALGB 8881. The model is a special case of the general mixed effects model with random effects introduced by Ibrahim (1998) earlier in this volume.

Also, we will estimate the model described in this section for each of the two responses separately, i.e., we will estimate two separate models, one with WBC as response, and another model with AGC as response. Alternatively, we could have considered one model with a bivariate response $y_{ij} = (WBC, AGC)$. But for the sake of simplicity, and since the relationship of WBC and AGC is not the central focus of this paper, we deal with the two responses by considering two separate models.

We start by considering a nonlinear regression model for the top level likelihood $p(y_i|\theta_i, t_{ij})$ in model (1). From the data in Figure 1, some common patterns can be seen. At the start of chemotherapy, the initial base line count drops off sharply, reaches a minimum a few days later, and then slowly recovers in an S-shaped profile. To model this common pattern we choose a piecewise linear and logistic regression with patient-specific shift, scale and change point parameters for the mean response y_{ij} at time t_{ij}. Let $\theta_i = (z_{1i}, z_{2i}, z_{3i}, \tau_{1i}, \tau_{2i}, \beta_{0i}, \beta_{1i})$ denote a patient-specific parameter vector, let $r = (\tau_{2i} - t_{ij})/(\tau_{2i} - \tau_{1i})$ and let $g(\theta_i, t) = z_{2i} + z_{3i} \cdot [1 + \exp\{-\beta_{0i} - \beta_{1i}(t - \tau_{2i})\}]^{-1}$ denote a rescaled and shifted logisitic.

$$y_{ij} = f(\theta_i, t_{ij}) = \begin{cases} z_{1i} + \epsilon_{ij} & \text{if } t_{ij} < \tau_{1i}, \\ r \cdot z_{1i} + (1 - r)g(\theta_i, \tau_{2i}) + \epsilon_{ij} & \text{if } \tau_{1i} \leq t_{ij} < \tau_{2i}, \\ g(\theta_i, t_{ij}) + \epsilon_{ij} & \text{if } t_{ij} \geq \tau_{2i}. \end{cases} \tag{2}$$

The response is modeled as a horizontal line up to time τ_{1i}, a logistic recovery after the nadir τ_{2i}, and a straight line connecting the two. All parameters are patient specific, i.e., each patient has possibly different change points τ_{1i}, τ_{2i}, logistic regression parameters β_{0i}, β_{1i} and scale and shift parameters z_{1i}, z_{2i}, z_{3i}. Figure 2a shows a plot of $f(\theta_i, t_{ij})$ for typical values of the parameters θ_i.

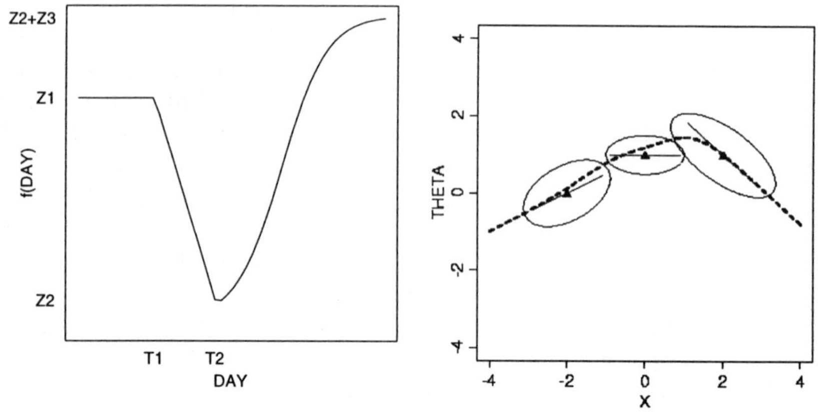

(a) Linear logistic mean curve (b) A mixture of normals.

FIGURE 2. Panel (a): The piecewise linear/logistic mean curve $f(\theta_i, t_{ij})$ defined in (2) as a function of t_{ij}. Panel (b): A mixture of normal model $p(\theta_i, x_i|\eta)$ defines non-linear regression $E(\theta_i|x_i, \eta)$ via the conditional mean process which takes the form of a locally-weighted mixture of straight lines corresponding to each of the normal kernels. The ellipses indicate the terms of a mixture of three normals. The short lines are the conditional mean lines for each of the normal kernels. A locally-weighted mixture of these lines gives $E(\theta_i|x_i, \eta)$ (dashed curve).

Model (2) defines a non-linear regression model for each patient i, allowing a different vector of random effects θ_i for each patient. In the next level of the hierarchy (1) we model the joint distrubution $p(\theta_i|\eta, x_i)$ of these patient specific random effects in the population, using hyperparameters η, and possibly including a regression on covariates x_i. We will start the discussion with a model $p(\theta_i|\eta)$ without a regression. It will later easily be extended to include a regression on x_i. Specifying a model for $p(\theta_i|\eta)$ we face two competing aims. On one hand, we need to include some structure to allow prediction. The technically most convenient solution is to assume some simple parametric form, such as a multivariate normal random effects model. And, in fact, this is the most common choice in the literature, possibly generalized to multivariate Student t distributions. On the other hand, the model needs to allow for outliers like patients 25 and 6 shown in Figure 1. We find a compromise between these two competing goals by choosing a mixture of normal model with a random mixture measure. Using a prior probability model which favors mixtures with few dominant terms amounts to *a priori* assuming a simple model with one or few multivariate normals, and adding *a posteriori* additional structure only if and as dictated by the data. Also, technically the mixture of normal model is only slightly more complex than the single normal model. Specifically, we choose a discrete mixture over multivariate normal kernels.

$$\theta_i \sim \underbrace{\sum_h w_h N(\mu_h, S)}_{=\int N(\mu,S)dG(\mu)}, \qquad (G,S) \sim p(G)p(S). \tag{3}$$

Here (G, S) play the role of η in the general model (1). Let $\delta_x(\cdot)$ denote a point mass at x. The mixing measure G is the discrete measure $G(\mu) = \sum w_h \delta_{\mu_h}(\mu)$. We will discuss below a specific choice for the prior model $p(G)$ on the mixing measure G.

We replace $p(\theta_i|G)$ in (3) by a mixture model jointly on (θ_i, x_i) to add a regression on x_i to the model. Doing this we achieve a non-parametric regression of the subject-specific effects on pretreatment covariates x_i in the following sense. The mixture of normals for (θ_i, x_i) implies a locally-weighted mixture of normal linear regressions for $E(\theta_i|x_i)$, with each term $N(\mu_h, S)$ in the mixture corresponding to a local linear regression model. This is illustrated in Figure 2b. A similar approach in maximum likelihood-based inference was proposed by Mallet et al. (1988) who consider unconstrained maximum likelihood estimation for the joint distribution of (θ_i, x_i) with the desired non-parametric regression given by the implied conditional distribution of θ_i given x_i.

The mixture of normal model is akin to traditional kernel density estimation, albeit with a random number of terms, weights and orientations (if the mixture is extended over S as well) of the individual kernels. Data-driven choice of these parameters is formalized by full posterior inference in the model. As a prior distribution for the mixing measure G, we choose a Dirichlet process (DP) model

$$G \sim DP(a\, G_0), \quad G_0(\mu) = N(m, B), \tag{4}$$

where G_0 is the base measure of the DP and a is the total mass parameter. We assume a multivariate normal base measure G_0. The DP model is the most commonly used approach for putting a prior model on an unknown probability measure in non-parametric Bayesian methods. See the discussion in Escobar and West (1998) and MacEachern (1998) earlier in this volume. The DP generates random probability distributions G with an expected value $E(G) = G_0$. The total mass parameter a is a precision parameter, something akin to a prior sample size. The higher a, the closer G is to G_0.

One of the critical advantages of the DP prior assumption over other alternatives is the straightforward implementation of posterior inference by Markov chain Monte Carlo simulation. See earlier chapters in this volume for a discussion of posterior simulations in DP mixture models in general (Escobar and West, 1998; MacEachern 1998), and Müller and Rosner (1997) for a detailed discussion of the posterior simulation specifically in model (2), (3) and (4).

Figure 1a-d shows fitted profiles for four patients from study CALGB 8881. Note how the hierarchical model "borrows strength" across patients to fit profiles for patients with sparse data, like patient 18. At the same time the structure is sufficiently flexible to fit outliers like patient 6.

Figures 3 and 4 show summaries of posterior predictive inference on a hypothetical future patient. Figure 3 plots posterior predictive hematologic profiles for different levels of cyclophosphamide. All inference is based on Markov chain Monte Carlo posterior simulation. Therefore, posterior predictive summaries of any desired predictive statistic can be obtained. For example, Figure 4 shows the predictive inference for the probability $Pr(\text{WBC} > 1000 \text{ on day } 14)$ for a future patient as a function of dosage. Depending on the desired assurance one can now directly read off the highest dose of cyclophosphamide which can be given if the patient should after 14 days be ready for the next cycle of chemotherapy. Posterior predictive inference allows formal comparison of AGC and WBC recovery profiles. From plots like those shown in Figure 3 it appears that AGC lags around 1.5 days behind WBC recovery. A more extensive discussion of posterior inference in CALGB 8881 under the described model appears in Rosner and Müller (1997) and Müller and Rosner (1997).

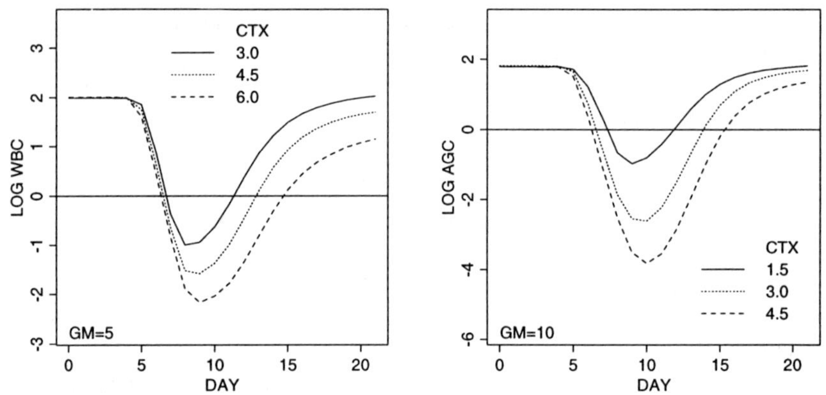

(a) WBC: $E(y^W_{I+1,t}|x_{I+1}, y)$ (b) AGC: $E(y^A_{I+1,t}|x_{I+1}, y)$

FIGURE 3. Predicted response curve for hypothetical future patients by dose of cyclophosphamide and GM-CSF. All patients are assumed to start with WBC=7700/μl and AGC= 6000/μl at day 0 (7700 and 6000 are arbitrarily chosen "typical" values).

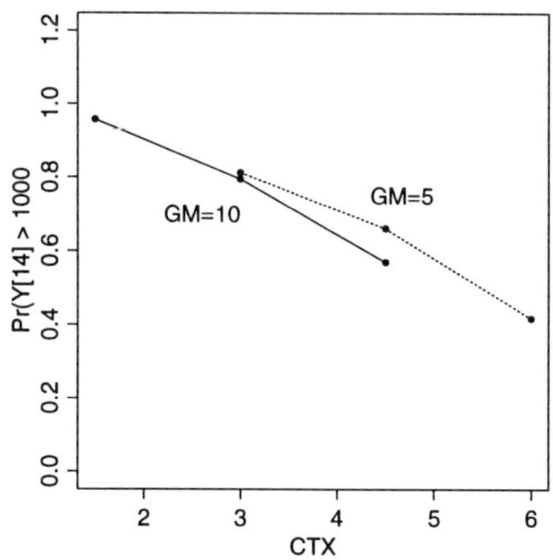

FIGURE 4. Predicted probability $Pr\{WBC > 1000 \text{ on day } t = 14\}$ for a future patient by dose of cyclophosphamide and GM-CSF. The estimation error for the plotted predictive probabilities is neglible (around 0.01).

18.3 Meta-analysis over related studies

We now consider an extension of model (2), (3) and (4) to allow joint analysis of both studies: CALGB 8881 and CALGB 9160. A full scale meta-analysis over both studies would consider a hierarchical extension of the model by adding to the model an additional level corresponding to the two studies. In this analysis we choose instead to use an informal implementation of the complete meta-analysis. Instead of the large hierarchical model, we will consider two separate implementations of the model described in Section 18.2. We use the posterior distribution of the model hyperparameters conditional on the data from CALGB 8881 to define an informative prior for the analysis of CALGB 9160 and, thereby, pool information across the two studies. For reference, we restate the full model (2), (3) and (4)

$$
\begin{aligned}
y_{ij} &= f(\theta_i, t_{ij}) + \epsilon_{ij}, \quad \epsilon_{ij} \sim N(0, S) \\
\theta_i &\sim \int N(\mu, S) dG(\mu) \\
G &\sim DP(a\, G_0), \quad G_0(\mu) = N(m, B),
\end{aligned}
$$

extended with hyperprior distributions on parameters m, B, a and S:

$$S^{-1} \sim W(q, [qR]^{-1}),\ m \sim N(m_0, A),\ B^{-1} \sim W\left[c, (cC)^{-1}\right],\ a \sim G(a_0, b_0).$$

Here $W(\cdot,\cdot)$ and $G(\cdot,\cdot)$ denotes the Wishart and Gamma distribution, respectively. The first parameter of the Wishart distribution is the scalar parameter, the second parameter is the matrix parameter.

For the analysis of the CALGB 9160 data we choose hyperprior parameters based on the posterior distributions under CALGB 8881 by fixing hyperparameters at the posterior means conditional on the data from the earlier study CALGB 8881. Thus we fix R, C and m_0 at the posterior means of S, B and a, respectively, and A at the posterior covariance matrix for a, where these posterior moments are conditional on the CALGB 8881 data. It is important to use a full tri-variate treatment vector x_i corresponding to cyclophosphamide , GM-CSF and amifostine treatment for both studies, even though no patients in CALGB 8881 received amifostine treatment and all CALGB 9160 patients received the same dose of cyclophosphamide (3 gm/m^2) and GM-CSF (5 μg/kg). Posterior inference on elements related to amifostine, conditional on the CALGB 8881 data, will only echo the original prior assumptions. Estimation of the model for the CALGB 9160 data, however, will include information on all three covariates.

Figure 5 shows posterior predictive inference conditional on the date from CALGB 9160. By using the posterior from the earlier analysis as prior this posterior inference already shows updated beliefs conditional on all data, including CALGB 8881 and 9160. Posterior predictive inference for hypothetical future patients treated at levels of cyclophosphamide and amifostine not represented in the data is subject to the usual caveat of extrapolation beyond the range of the data. Figure 6 shows the significantly increased posterior predictive uncertainty for a future patient treated at a combination of cyclophosphamide and amifostine beyond the range of the data.

An important measure of hematologic toxicity is the duration of myelosuppression. Figure 7a shows posterior predictive inference on the number of days below some critical threshold (1000/μl). Panel (b) of the same figure shows the probability of a future patient recovering within 14 days beyond $WBC = 1000/\mu l$. Both plots report the predictive inference as a function of dosage cyclophosphamide and amifostine. This inference would not be possible using the data from either study alone, CALGB 8881 or CALGB 9160. Another important statistic is the predicted nadir WBC. Figure 8 reports inference by dose of cyclophosphamide , with and without amifostine treatment. The implementation is based on posterior MCMC simulation. Thus full probabilistic descriptions of uncertainties in the reported posterior predictive inference are available. For example, Figure 9 plots the posterior predictive distribution for the day of nadir WBC and nadir AGC for a future patient receiving GM-CSF at 5 μg/kg and cyclophosphamide at 3 gm/m^2. Note how the nadir AGC count trails the nadir WBC by around two days. Like in all other predictive inference, the effect of amifostine is minimal.

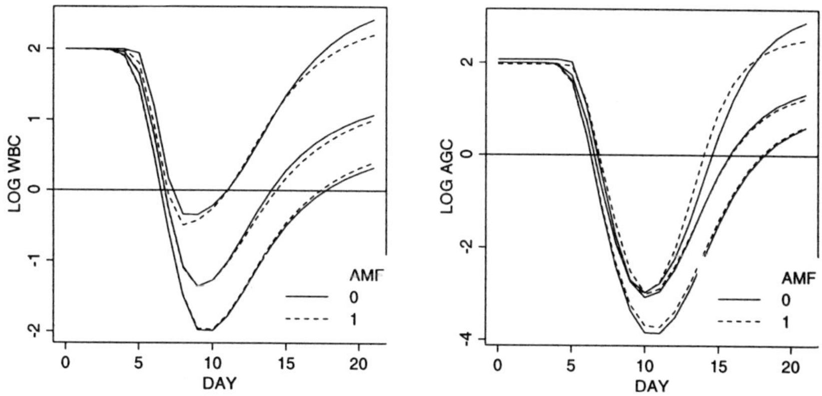

FIGURE 5. Predicted WBC (left panel) and AGC (right panel) response curve for hypothetical future patients by dose of cyclophosphamide and amifostine. The dashed curves are for patients receiving amifostine treatment. The solid curves correspond to no amifostine treatment. The three dashed and solid curves correspond to, from top to bottom, cyclophosphamide = 3.0, 4.5 and 6.0 gm/m^2, respectively. All patients are assumed to start with WBC=7700/μl at day 0 (arbitrarily chosen "typical" value).

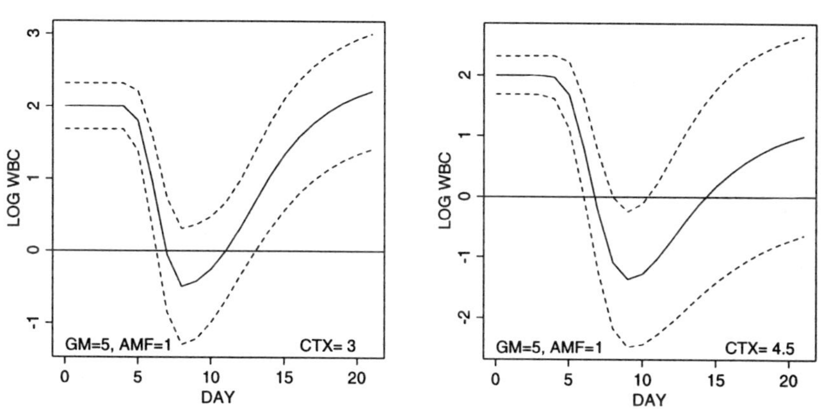

FIGURE 6. Predicted WBC response curve for two hypothetical future patients. The dashed lines indicate one posterior predictive standard deviation margins. The right panel corresponds to a future patient treated at a combination of cyclophosphamide and amifostine not represented in the data. Notice the significantly larger posterior predictive uncertainty.

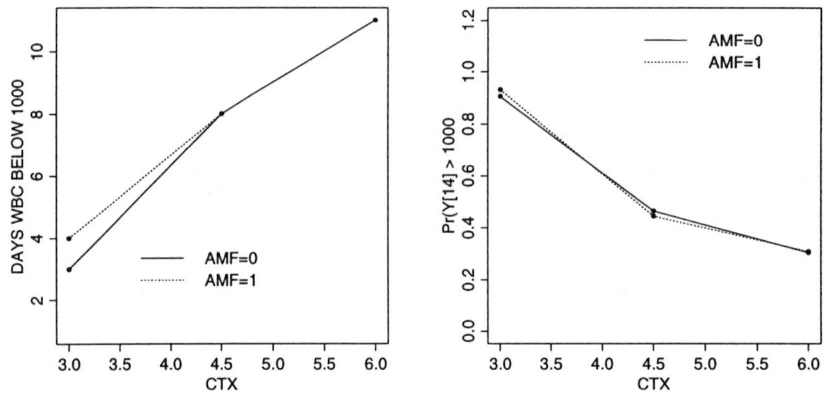

(a) Days below $WBC = 1000$ (b) $Pr\{WBC > 1000 \text{ on day } t = 14\}$

FIGURE 7. Panel (a) shows median predicted number of days WBC $\leq 1,000$ as a function of cyclophosphamide and amifostine. Panel (b) shows the predicted probability $Pr\{WBC > 1000 \text{ on day } t = 14\}$ for a future patient by dose of cyclophosphamide and amifostine. The estimation error for the plotted predictive probabilities is neglible (around 0.01).

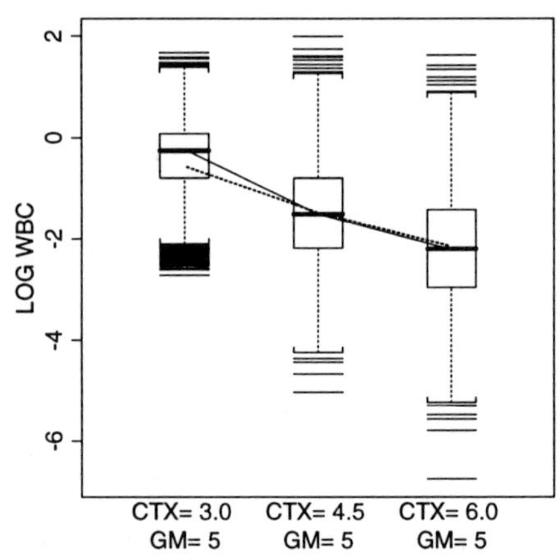

FIGURE 8. Nadir WBC by levels of cyclophosphamide . The boxplots show the posterior predictive distribution of the nadir log WBC for a hypothetical future patient by dose of cyclophosphamide and without amifostine treatment. The solid line connects the posterior predictive means. For comparison, the dashed line connects the corresponding posterior predictive means for hypothetical future patients with amifostine treatment.

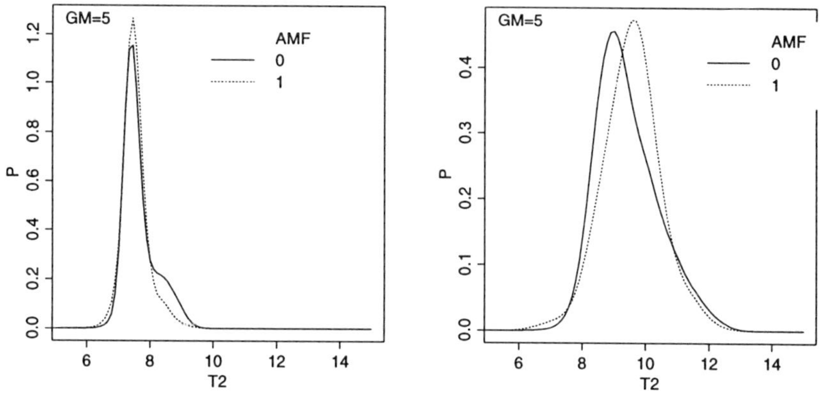

(a) WBC: $p(\tau_{2,I+1}|x_{I+1}, y)$ (b) AGC: $p(\tau_{2,I+1}|x_{I+1}, y)$

FIGURE 9. Posterior predictive distribution of day of recovery τ_2 for a hypothetical future patient by dose of cyclophosphamide and amifostine.

18.4 Discussion

We discussed an extension of commonly used hierarchical models for longitudinal data with random effects to allow for non-parametric prior models on the unknown random effects distribution. Choosing a random mixture of standard parametric models allows for efficient implementation. Conditional on imputed latent mixture indicators, posterior simulation remains the same as under the corresponding fully parametric model. A simple extension to the mixture model for the random effects allows a non-parametric regression on subject-specific covariates.

We proposed an informal approach to meta-analysis over related studies by using the posterior conditional on data from the first study to define an informative prior distribution for the analysis of the second study.

In future work we will pursue a formal implementation of simultaneous inference in related studies by considering an encompassing hierarchical model which takes the form of a hierarchical extension of (1).

Acknowledgment

Research supported by NSF under grant DMS-9404151, and in part by a grant to the Duke University Comprehensive Cancer Center from the American Cancer Society. The authors also thank the Cancer and Leukemia Group B (CALGB) for use of the data.

References

Davidian, M. and Gallant, A.R. (1992), "Smooth nonparametric maximum likelihood estimation for population pharmacokinetics, with application to Quinidine." *Journal of Pharmacokinetics and Biopharmaceutics* **20**, 529-556.

Davidian, M., and Gallant, A. R. (1993), "The nonlinear mixed effects model with a smooth random effects density." *Biometrika*, **80**, 475-488.

Davidian, M., and Giltinan, D. M. (1995), "Nonlinear Models for Repeated Measurement Data". London: Chapman and Hall.

Dellaportas, P. and Smith, A.M.F. (1993), "Bayesian analysis for generalized linear models via Gibbs sampling." *Applied Statistics* **42**, 443-459.

Escobar, M.D. and West, M. (1998), "Computing Bayesian nonparametric hierarchical models," in *Practical Nonparametric and Semiparametric Bayesian Statistics* (eds: Dey, D. and Müller, P. and Sinha, D.), New York: Springer-Verlag.

Hirano, K. (1998). "Nonparametric Bayes models for longitudinal earnings data," in *Practical Nonparametric and Semiparametric Bayesian Statistics* (eds: Dey, D. and Müller, P. and Sinha, D.), New York: Springer-Verlag.

Ibrahim, J. (1998). "Mixed effect population models with mixture of Dirichlet," in *Practical Nonparametric and Semiparametric Bayesian Statistics* (eds: Dey, D. and Müller, P. and Sinha, D.), New York: Springer-Verlag.

Kleinman, K. and Ibrahim, J. (1996), "A semi-parametric Bayesian approach to GLIM," technical report, University of Michigan.

Lichtman, S. M., Ratain, M. J., Echo, D. A. V., Rosner, G., Egorin, M. J., Budman, D. R., Vogelzang, N. J., Norton, L., and Schilsky, R. L. (1993), "Phase I trial of granulocyte-macrophage colony-stimulating factor plus high-dose cyclophosphamide given every 2 weeks: a Cancer and Leukemia Group B study," *Journal of the National Cancer Institute*, **85**, 1319-1326.

MacEachern, S. (1998). Computations for Mixture of Dirichlet Process models,

Mallet, A.(1986), "A maximum likelihood estimation method for random coefficient regression models" *Biometrika* **73**, 645-656.

Mallet, A., Mentré, F., Gilles, J., Kelman, A. W., Thomson, A. H., Bryson, S. M., and Whiting, B. (1988), "Handling covariates in population pharmacokinetics, with an application to Gentamicin," *Biomedical Measurement and Information Control*, **2**, 138-146.

Müller, P. and Rosner, G. (1997). "A semiparametric Bayesian population model with hierarchical mixture priors," *Journal of the American Statistical Association*, **92**, to appear.

Rosner, G. and Müller, P. (1997). "Bayesian population pharmacokinetics and pharmacodynamic analyses using mixture models," *Journal of Pharmacokinetics and Biopharmaceutics*, **25** (2), 209-233.

Spencer, C.M. and Goa, K.L. (1995). "Amifostine. A review of its pharmacodynamic and pharmacokinetic properties, and therapeutic potential as a radioprotector and cytotoxic chemoprotector." *Drugs*, **50**, 1001-1031.

Wakefield, J.C. (1996), "The Bayesian analysis of population pharmacokinetic models," *Journal of the American Statistical Association*, **91**, 62-75.

Wakefield, J.C., Smith, A.F.M., Racine-Poon, A., and Gelfand, A. (1994), "Bayesian analysis of linear and nonlinear population models using the Gibbs sample," *Applied Statistics*, **43**, 201-221.

Wakefield, J.C., Aarons, L. and Racine-Poon, A. (1998), Population Pharmacokinetic Modeling in Drug Development, in *Case Studies in Bayesian Statistics 4*, (R. Kass, ed.), Springer-Verlag, to appear.

Walker, S. (1996). An EM algorithm for nonlinear random effects models, *Biometrics*, **52**, 934-944.

Walker, S. and Wakefield, J. (1994). Population Models with a Nonparametric Random Coefficient Distribution. Technical Report, Imperial College London.

Zeger, S.L. and Karim, M.R. (1991), "Generalized linear models with random effects," *Journal of the American Statistical Association*, **86**, 79-86.

19

A Bayesian Model for Fatigue Crack Growth

Liping Liu
Elja Arjas

ABSTRACT In this chapter we apply hierarchical Bayesian modeling to describe fatigue crack growth. Due to the special pattern of crack growth data, we propose a nonparametric model which is based on a minimum of probabilistic assumptions. We put a (prior) probability model on an unknown growth curve by considering a grid of time points and assuming that the increments between such grid points are independent exponential random variables, with expectations proportional to the length of the grid interval. The nonparametric structure enables us to arrive at a model which is both flexible enough to be fitted to different sets of growth data and, at the same time, enables us to interpret different parts of the model in an intuitively plausible manner. A Markov chain Monte Carlo method is adopted for the numerical computation. Numerical results show that our model behaves well in both being fitted to the data and in predicting future observables.

KEY WORDS: Exchangeability; Hierarchical modeling; Predictive distribution; Growth curve; Markov chain Monte Carlo.

19.1 Introduction

Physical devices or systems in use are typically subject to monotonically increasing wear or degradation. The failure or death of a system is then conveniently defined to be the event in which degradation reaches a certain threshold level. This point of view becomes particularly relevant in cases where the level of degradation can be measured to some degree of accuracy, and where the goal of the statistical modeling is to provide a systematic method for predicting the time to failure from such follow-up measurements. Utilization of system specific degradation readings can be expected to lead to a much smaller prediction error variance than methods which only use the failure time data (see Lu and Meeker, 1993).

The connection between the study of degradation or wear and that of growth curves is obvious. In the former "time" can be simply interpreted as mileage of driving, number of cycles in the trial, etc. There is extensive literature concerning statistical growth curve models. Most papers assume linear or polynomial structure, and traditional and modern regression techniques were often adopted (see e.g. von Rosen, 1991, Fearn, 1975, Lindley and Smith, 1972, Tian, Shukla and Buncher, 1994, and Rao, 1987). For specific studies concerning cumulative damage, Bogdanoff and Kozin (1985) used the ideas of finite Markov chains, and formulated the so-called B-models. Lu and Meeker (1993) established a parametric nonlinear mixed-effects model to describe degradation measurements.

Due to the convenience of expressing uncertainties probabilistically in terms of posterior and predictive distributions, the Bayesian approach looks quite attractive to us. Combined with hierarchical structure and nonparametric modeling, we can arrive at a model which is flexible enough to be fitted to different sets of growth data. Furthermore, it is usually very easy to assign a "flat" prior probability to the unobservables, and to work with very weak structural and probabilistic assumptions.

The set-up considered in this paper is as follows. Suppose there are n physical devices under observation. For the kth device, measurements $0 < Y_{k,1} < Y_{k,2} < \cdots < Y_{k,p_k}$ are made at times $0 < t_{k,1} < t_{k,2} < \cdots < t_{k,p_k} \leq T_{max}$, respectively, where T_{max} is the largest possible observational time (across all devices) and p_k is the number of observations available on device k.

In a designed experiment, the $t_{k,i}$'s above are often equally spaced. However, some observations might be missing, and some devices can fail before it has been possible to make as many measurements as were originally planned. In our model, we do not ask that the number of observations for different devices be the same, nor the times at which these observations were made. In this way, the data can be of a very general form and hence there is no need to consider separately a "missing data" or "incomplete observations" problem.

Our ultimate purpose is to predict the growth of device after a last recorded observation. This device can be a "new" one (previously unobserved), or one about which there are already some observations in the data set. We are especially interested in the time at which this device reaches a certain target level, Y^*. If Y^* is taken to be the threshold level described above, our task becomes naturally the prediction of the failure time, as we have to do in survival analysis.

In the next section we will present our model in detail. Section 3 describes a Metropolis-Hastings algorithm associated with the model. In section 4 an example is analyzed, and our results on prediction are compared with those of Lu and Meeker (1993). The paper is concluded with a brief discussion in section 5.

19.2 The model

The model is based on the assumption that the growth curves of all devices have the same shape. This is, in fact, our only structural assumption. The differences between devices are thought to arise from the fact that they grow at different but constant speeds. In other words, their "operational ages" are linear transformations of calendar time. More precisely, we assume that a measurement of the kth device at time t, $Y_k(t)$, can be expressed as

$$Y_k(t) = g_0(\theta_k \cdot t) + \varepsilon_{k,t}, \tag{1}$$

where θ_k is a positive random variable representing the relative speed of growth of device k, g_0 is an unknown function which is continuous and strictly increasing, and $\varepsilon_{k,t}$ is a measurement error.

As is apparent, the function g_0 can be interpreted as the "common" growth curve of the population under investigation. The coefficient θ_k is the relative speed of growth of device k with respect to calendar time t. In the present Bayesian model both g_0 and θ_k are random. The random function

$$g_k(t) = g_0(\theta_k \cdot t) \tag{2}$$

can be simply interpreted as the growth curve associated with the device indexed by k. Notice that model (1) is not identifiable, since for any positive constant c,

$$g_0(\theta_k \cdot t) = g_0(c\theta_k \cdot \frac{t}{c}) = \tilde{g}_0((c\theta_k) \cdot t),$$

where $\tilde{g}_0(u) = g_0(u/c)$. But assuming that the functions $g_1, \ldots, g_n$, as well as the parameters $\theta_1, \ldots, \theta_n$, are exchangeable (see e.g. Bernardo and Smith, 1994), we now somewhat arbitrarily use the growth curve of device 1 as a baseline, writing then (2) in the form

$$g_k(t) = g_1(\xi_k \cdot t) = g(\xi_k \cdot t), \tag{3}$$

where $\xi_k = \theta_k/\theta_1$ and $g = g_1$. Now, since $\xi_1 \equiv 1$, model (3) is identifiable.

Suppose now that $\theta_1, \theta_2, \ldots, \theta_n$ are i.i.d. and have prior distribution Gamma(α_θ, β_θ), where α_θ is the shape and β_θ the scale parameter. Then, it is easy to verify that given θ_1, the derived parameters ξ_k will be i.i.d. and will follow the (prior) distribution Gamma(α_θ, $\theta_1\beta_\theta$). This confirms the fact that the particular choice of device 1 as a baseline is irrelevant since it only influences the value of the scale parameter. A reasonable assumption for the prior distribution of ξ_k would be therefore: for $k = 2, 3, \ldots, n$, ξ_k follows a Gamma distribution Gamma(α, β), where α and β are two independent random parameters with prior distributions Exp(λ) and Exp(ρ), respectively, with λ and ρ being given hyperparameters. We will generally choose the hyperparameters λ and ρ to be the same, which corresponds to the exchangeability idea that we don't in our prior assume that devices indexed from 2 to n grow faster or slower than device 1. Intuitively, a large value of α indicates that all devices will grow with a speed close to that of device 1, whereas a small value of α reflects the fact that the speeds can be very different. For this reason we also let α be random, so that the model can be updated to correspond to the heterogeneity between devices in the data. Apparently a large value of ξ_k indicates that device k grows faster than device 1, and if most devices in the data actually grow faster than device 1, we will have a large posterior probability that β is smaller than α.

In principle, g is thought of as a nonparametrically defined function. In practice, we assume here for simplicity that g is piecewise linear. The points at which the slope of g changes its value belong to a fixed grid of points (see Arjas and Gasbarra, 1994). So what we need to specify for the model is a positive, strictly increasing sequence $g(t_0) < g(t_1) < g(t_2) < \cdots < g(t_N)$, where $0 = t_0 < t_1 < t_2 < \cdots < t_N = CT_{max}$ are equally-spaced points on $[0, CT_{max}]$. The constant C must be greater than 1, so that in case $\xi_k > 1$, $g(\xi_k t)$ is still well-defined. The prior distribution of g is here specified by assuming that, for $i = 1, 2, \ldots, N$, each increment $\Delta g(t_i) = g(t_i) - g(t_{i-1})$ of g follows an exponential distribution $\text{Exp}((t_i - t_{i-1})^{-1}\mu)$. Here μ is a given hyperparameter. Note that this is somewhat different from considering the increments of a gamma process on the grid intervals.

This prior of g guarantees that g is increasing. The prior mean of $\Delta g(t_i)$ is $(t_i - t_{i-1})/\mu$, which is proportional to the length $\Delta t_i = t_i - t_{i-1}$. This means that the speed of growth has no trend according to the prior, and the posterior distribution of g will acquire additional information from the data so as to be able to describe the behaviour of the "true" growth curve of device 1. If one knows more about g, one can add more specifications or restrictions to the prior of g. Later in this paper, we will consider a convexity condition on g.

It is quite common to assume that the measurement errors follow a normal distribution with mean 0. Here the variables $\varepsilon_{k,t}$ are assumed to be mutually independent and follow a normal distribution $\mathcal{N}(0, \sigma^2)$, with σ^2 having an inverse Gamma distribution $IG(a, b)$, where a and b are hyperparameters. Therefore conjugacy can be realized here (see Roberts and Smith, 1993).

Now the hierarchical structure can be summarized in Figure 1 below. Hyperparameters and data are surrounded by squares, and random parameters are surrounded by circles. Arrows are used to indicate their dependence. The function $g(\cdot)$ in the figure corresponds to a set of parameters as described above.

19.3 A Markov Chain Monte Carlo method

A Markov chain Monte Carlo (MCMC) method is used for the numerical computation. As evidenced by recent literature, MCMC is a very powerful tool in dealing with hierarchical Bayesian models. For the general theory of MCMC, the reader is referred to, for example, Besag et al. (1995). In the following we present the main steps of the algorithm handling our crack growth model.

Denote by $x = (\alpha,\ \xi_2,\ \xi_3,\ \ldots\ ,\ \xi_n,\ g(\cdot),\ \sigma^2)$ the vector of all the model

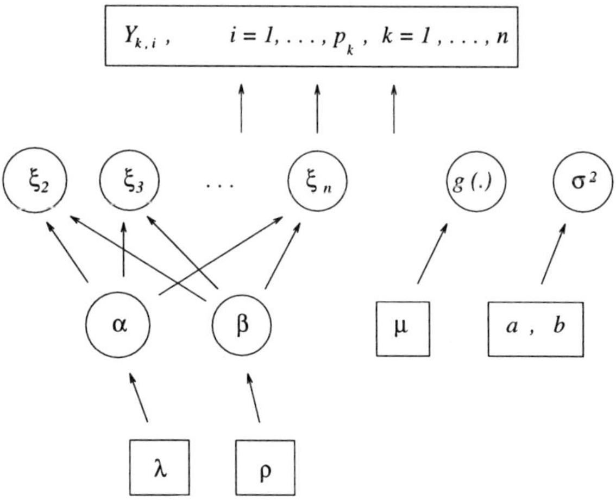

FIGURE 1. Hierarchical structure of the model.

parameters, and by $\pi(dx)$ and $\pi(dx|Y_{k,i} \quad i = 1, \ldots, p_k, \; k = 1, \ldots, n) = \pi(dx|data)$ the prior and posterior distributions of x, respectively. The goal is to construct an ergodic Markov chain $x^{(j)} = (\alpha^{(j)}, \xi_2^{(j)}, \xi_3^{(j)}, \ldots, \xi_n^{(j)}, g^{(j)}(\cdot), (\sigma^2)^{(j)})$, which has the unique stationary distribution given by $\pi(dx|data)$ (see Roberts and Smith, 1994). We first generate an initial value $x^{(0)}$, for example, from the prior distribution $\pi(dx)$. This is then updated from $x^{(j)}$ to $x^{(j+1)}$, always one coordinate at a time, and keeping all other coordinates fixed.

Given all other coordinates of x, the conditional density of α is proportional (in α) to

$$\frac{\beta^{(n-1)\alpha}}{(\Gamma(\alpha))^{n-1}} \Big(\prod_{k=2}^{n} \xi_k\Big)^{\alpha-1} e^{-\lambda\alpha},$$

so the Metropolis-Hastings algorithm can be used here. We generate a proposal α' of $\alpha^{(j+1)}$ from $\mathrm{Exp}(\lambda)$, then calculate the Hastings ratio

$$r_\alpha = \left(\frac{\Gamma(\alpha^{(j)})}{\Gamma(\alpha')}\right)^{n-1} \cdot \beta^{(n-1)(\alpha'-\alpha^{(j)})} \Big(\prod_{k=2}^{n} \xi_k^{(j)}\Big)^{\alpha'-\alpha^{(j)}},$$

and let $\alpha^{(j+1)} = \alpha'$ with probability $\min(1, r_\alpha)$, and $\alpha^{(j+1)} = \alpha^{(j)}$ otherwise.

It is easy to see that the conditional distribution of β, given all other coordinates, is the Gamma distribution $\mathrm{Gamma}((n-1)\alpha + 1, \; (\sum_{k=2}^{n} \xi_k + \rho))$. Therefore β can be updated by draws directly from the complete conditional posterior distribution.

The conditional distribution of ξ_k is proportional to

$$\xi_k^{\alpha-1} \cdot e^{-\beta\xi_k} \prod_{i=1}^{p_k} [\frac{1}{\sqrt{2\pi}\sigma} \exp\{-\frac{1}{2\sigma^2}(Y_{k,i} - g(\xi_k \cdot t_{k,i}))^2\}].$$

Again we can use the Metropolis-Hastings algorithm. We first generate a proposal ξ' of $\xi_k^{(j+1)}$ from Gamma$(\alpha^{(j+1)}, \beta^{(j+1)})$, then calculate the corresponding Hastings ratio

$$r_\xi = \exp\{-\frac{1}{2(\sigma^2)^{(j)}} \sum_{i=1}^{p_k} [(g^{(j)}(\xi' \cdot t_{k,i}))^2 - (g^{(j)}(\xi_k^{(j)} \cdot t_{k,i}))^2$$

$$-2Y_{k,j}(g^{(j)}(\xi' \cdot t_{k,i}) - g^{(j)}(\xi_k^{(j)} \cdot t_{k,i}))]\},$$

and let $\xi_k^{(j+1)} = \xi'$ with probability $\min(1, r_\xi)$.

For the piecewise linear function $g(\cdot)$, we use the increments $\Delta g(t_i) = g(t_i) - g(t_{i-1})$ as our model parameters $(i = 1, 2, \ldots, N)$. When $\Delta g(t_i)$ is updated, all $g(t_j)$'s $(j \geq i)$ will move upwards or downwards by the same amount, according to whether $\Delta g(t_i)$ is made larger or smaller. The updating is realized in a similar way as in the previous step, i.e., by generating a proposal from the prior distribution, then calculating the corresponding Hastings ratio from the likelihood, and finally deciding whether to accept it or not.

The updating of σ^2 is equally simple. Given all other parameters and the data, the conditional distribution of σ^2 is inverse Gamma

$$IG(a + \frac{1}{2}\sum_{k=1}^{n} p_k \quad , \quad b + \frac{1}{2}\sum_{k=1}^{n}\sum_{i=1}^{p_k}(Y_{k,i} - g(\xi_k \cdot t_{k,i}))^2).$$

A new value $(\sigma^2)^{(j+1)}$ can be drawn from this distribution directly to replace $(\sigma^2)^{(j)}$.

After the jth generic step of the algorithm is finished, we can simulate any quantity of interest. From (3), we have

$$g_k^{(j)}(t) = g^{(j)}(\xi_k^{(j)} \cdot t),$$

and a corresponding measurement can be simulated by using the expression

$$Y_k^{(j)}(t) = g_k^{(j)}(t) + \varepsilon_{k,t}^{(j)},$$

where $\varepsilon_{k,t}^{(j)}$ is generated independently from $\mathcal{N}(0, (\sigma^2)^{(j)})$.

If the kth device is still below the target level Y^* at the end of the trial, since the simulated curve $g_k^{(j)}(t)$ is strictly increasing and piecewise linear, we can easily get the simulated crossing time $(t_k^*)^{(j)}$ which satisfies

$$g_k^{(j)}((t_k^*)^{(j)}) = Y^*.$$

It is not difficult to check that the Markov chain we constructed is irreducible and aperiodic; therefore by the ergodic theorem, the predictive distribution of $g_k(t)$ is approximated by

$$P(g_k(t) \in B | Y_{k,i} \;\; i = 1, \ldots, p_k \;, k = 1, \ldots, n)$$

$$= \int_{\mathcal{X}} P(g_k(t) \in B | x) \pi(dx | data) \approx \frac{1}{J} \sum_{j=1}^{J} I(g_k^{(j)}(t) \in B),$$

where B is any Borel set on R^+, $\mathcal{X}$ is the parameter space, and J is a large enough positive integer. A similar formula for t_k^* is also true.

From a computational point of view, it may be more attractive to sample from the algorithm only at times j which are $\tilde{J} > 1$ units apart from each other. Choosing two positive integers, say J_0 and $\tilde{J}$, if J_0 and $\tilde{J}$ are not too small and if the Markov chain is constructed properly, then $\{g_k^{(J_0 + j \cdot \tilde{J})}(t) \;;\;\; j = 1, 2, \ldots\}$ can be regarded approximately as a random sample from the predictive distribution of $g_k(t)$. A similar statement holds again for t_k^*.

19.4 An example: growth of crack lengths

This study was largely motivated by the paper from Lu and Meeker (1993), who analyzed the same crack growth data that we are using. This data set was first used by Hudak et al. (1978), and later by Bogdanoff and Kozin (1985). The measurements are the crack lengths in a compact tension steel test operated in different laboratories. Lu and Meeker (1993) got the data visually from Figure 4.5.2 of Bogdanoff and Kozin (1985), and presented the data set in Table 1 of their own paper. Only the measurements up to the "time" of 0.12 million cycles were reported, and a target level corresponding to failure was defined to be 1.6 inches. The data are shown in a graphical form in Figure 2. It should be mentioned that in the original data set there were also measurements beyond these cut-off points. There were altogether 21 test trials, and in 12 of these trials the length of the crack had reached the target level before the end of observation. In all 21 cases, the initial crack length was chosen to be 0.9 inches. Our goal is to reanalyze this data set, by applying the growth curve model and the inferential method introduced above. From Figure 2, it appears that our structural assumption regarding a common shape for the growth curves is quite reasonable.

From the data it is natural to set the value of T_{max} at 0.12. The constant C was here taken to be 2, which turned out to be large enough in the actual computations performed. Each observational interval (of length 0.01) was further divided into 4 subintervals of equal length, i.e., we let $t_i = 0.01 \times i/4$ (see section 2). Unless stated otherwise, we expressed "vague" prior knowledge in specifying the prior distribution, choosing the values of

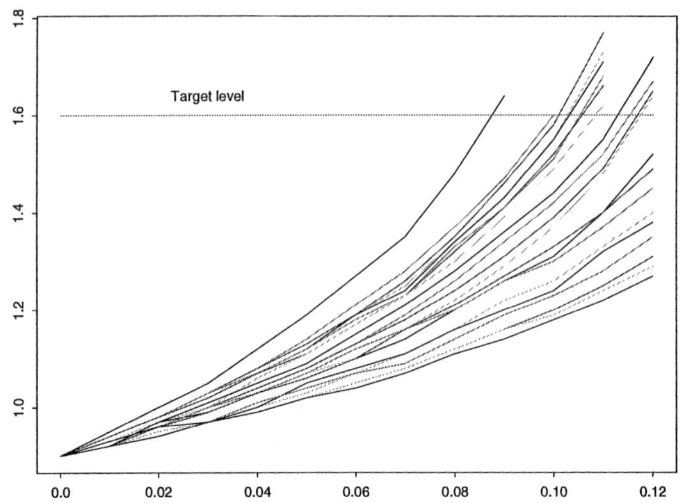

FIGURE 2. Fatigue crack growth data of 21 trials. The figure plots Y_{ki} on the ordinate against t on the abscissa.

the hyperparameters as follows: $\lambda = \rho = 0.05$, $\mu = 20$, $a = 3$ and $b = 0.02$. In this way the prior expectations of α and β are the same (see section 2).

The program was coded in C and run in a Sun Ultra 1 workstation. It ran quite fast, but the burn-in times were sometimes long. In the experiments, we discarded the first $J_0 = 6000$ iterations, picked a sample from every $\tilde{J} = 50$ iterations (see section 3), and did altogether 156000 iterations to get an MCMC sample of 3000 observations. This appeared to be sufficient to lead to stable Monte Carlo estimates.

An obvious concern in a study like this, where there are no strong a priori arguments either in favor or against some particular model, is how to assess the performance of the model we have decided to use. Considering that the primary goal here is prediction, an obvious suggestion is to pretend that one or several readings $Y_{k,i}$ in the data set were actually unknown, then use our model to determine their predictive distributions using the method described in section 3, and finally compare these with the true readings. If $Y_{k,i}$ can really be regarded as a random variable drawn from its (continuous) predictive distribution $F_{Y_{k,i}}$, then $F_{Y_{k,i}}(Y_{k,i})$ is uniformly distributed in (0, 1). So, by varying the readings pretended to be unknown, we should get a "pseudo" (0, 1)-uniform random sample. 21 such simulation trials were performed, one for each growth curve. In the kth trial, we pretended to know all other length readings in the data, except for device k for which only the values up to 0.07 million cycles were assumed known. In this way we obtained 94 "pseudo" (0, 1)-uniform random variables (2 from trial 1, 3 from trial 2, 4 from each trial indexed from 3 to 8, and 5 from each of the remaining trials). The corresponding cumulative empirical distribution is plotted in Figure 3. Note, however, that since we have repeated observations from the same device, these 94 variables are not independent and therefore straightforward goodness-of-fit testing based on this empirical distribution would not be correct. The main message in this figure appears to be that the predictive distributions tend to be somewhat more dispersed than the actual future length readings. At least in part this must be a consequence of having used a "flat" prior for σ^2 with prior mean $E_{prior}(\sigma^2) = 0.01$. Calculating the Kolmogorov-Smirnov statistic gave $d = 1.374$, which for an independent sample would be close to the 5 percent critical value 1.36. Changing the value of hyperparameter b from 0.02 to 0.0006 to correspond to a smaller measurement error variance σ^2 (with prior mean $E_{prior}(\sigma^2) = 0.0003$) improving the fit of the model somewhat. The value of d was then 0.919, which for an independent sample would be below the 5 percent critical value. The prior and posterior means and standard deviations of σ^2 in the above two cases are listed in Table 1.

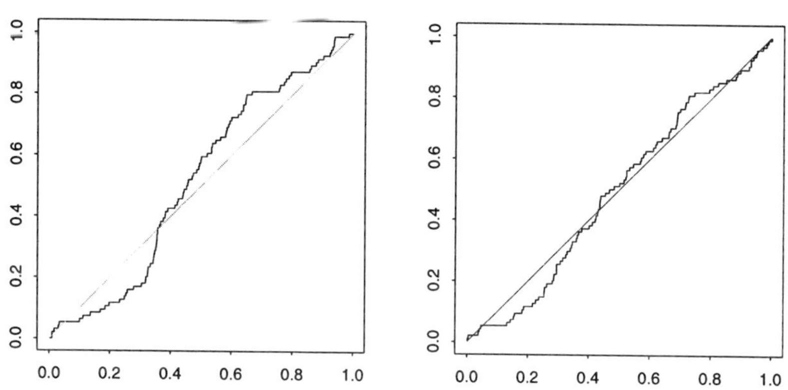

FIGURE 3. Cumulative empirical distributions of 94 "pseudo" (0, 1) uniform random variables. In the left the hyperparameter b was taken to be 0.02, and in the right $b = 0.0006$.

As stated earlier, our main goal was to make prediction. There are two obvious kinds of prediction: one is to predict the true value of $g_k(t)$ at a certain time t, and the other is to predict the crossing (failure) time of an device which has not yet reached the target level Y^* at the end of the trial. Since the first 12 devices in the data had reached the target level of

	Prior		Posterior	
	Mean	Standard error	Mean	Standard error
Experiment 1	0.01	0.01	0.000285	0.0000245
Experiment2	0.0003	0.0003	0.000109	0.0000105

TABLE 1. Prior and posterior means and standard deviations of σ^2 in two goodness-of-fit experiments.

$Y^* = 1.6$ inches before 0.12 million test cycles, in the following, predictions are made only for devices indexed from 13 to 21. Figure 4 shows the predictive distributions of the times at which these 9 crack lengths would reach the target level. As this data set is a part of a larger one (see Bogdanoff and Kozin, 1985), the true crossing times are actually known. The dotted straight lines show their positions, and the corresponding quantiles of their predictive distributions.

As an alternative method of assessment, we can use the mean or median of the predictive distribution as a point estimate, and compare this with the observed true value. This enables us to also compare our results with those obtained by Lu and Meeker (1993). Let us denote by T_k the true crossing time of the device indexed by k, by $\hat{T}_k$ Lu and Meeker's estimate of T_k (calculated from the estimated crack growth curves in their article), and by $\tilde{T}_k$ (resp. $\tilde{T}'_k$) our estimate of T_k by using the mean (resp. median) of the predictive distribution. Define the statistics

$$\Delta\hat{T}_k = \sum_{k=13}^{21} (\hat{T}_k - T_k), \qquad \hat{s} = \sqrt{\frac{1}{8}\sum_{k=13}^{21} (\hat{T}_k - T_k)^2};$$

$$\Delta\tilde{T}_k = \sum_{k=13}^{21} (\tilde{T}_k - T_k), \qquad \tilde{s} = \sqrt{\frac{1}{8}\sum_{k=13}^{21} (\tilde{T}_k - T_k)^2}$$

and

$$\Delta\tilde{T}'_k = \sum_{k=13}^{21} (\tilde{T}'_k - T_k), \qquad \tilde{s}' = \sqrt{\frac{1}{8}\sum_{k=13}^{21} (\tilde{T}'_k - T_k)^2}.$$

Then the numerical values obtained here were $\Delta\hat{T}_k = -0.0090$, $\hat{s} = 4.57 \times 10^{-3}$; $\Delta\tilde{T}_k = -0.0041$, $\tilde{s} = 3.51 \times 10^{-3}$ and $\Delta\tilde{T}'_k = -0.0029$, $\tilde{s}' = 3.89 \times 10^{-3}$. From this it seems that our point estimates are slightly more accurate than those provided by Lu and Meeker. Notice that our model assumptions

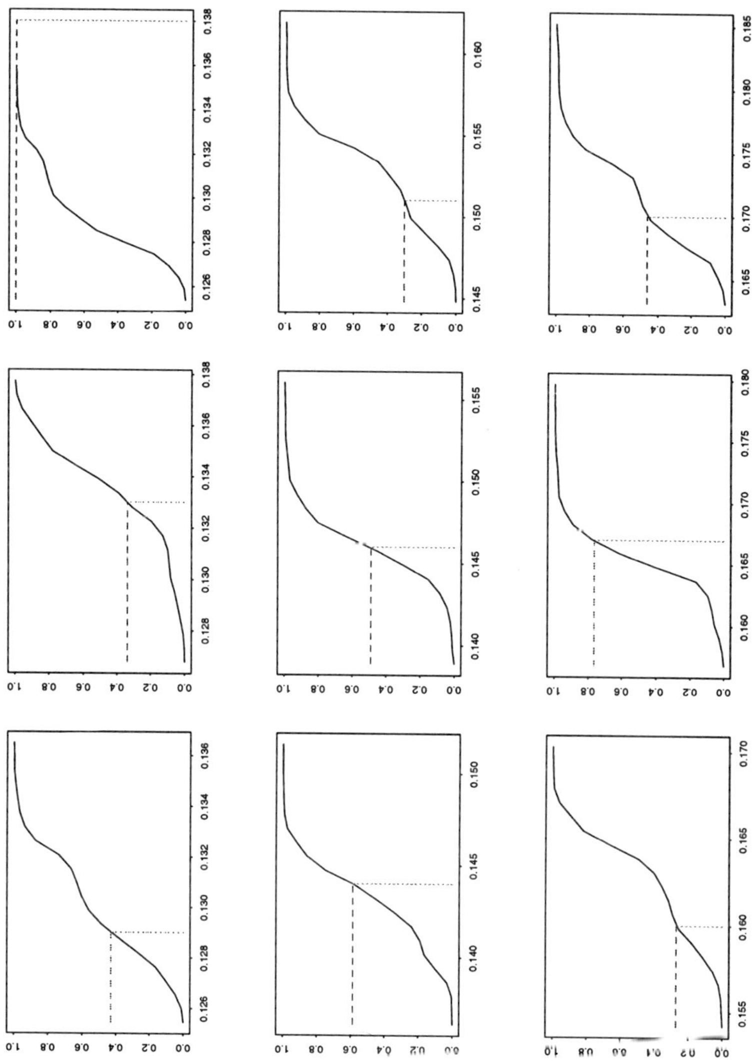

FIGURE 4. Predictive distributions of the crossing times to the target level $Y^* = 1.6$ inches, for trials indexed from 13 to 21. The dotted lines are the true crossing times, and the dashed lines indicate their corresponding quantiles.

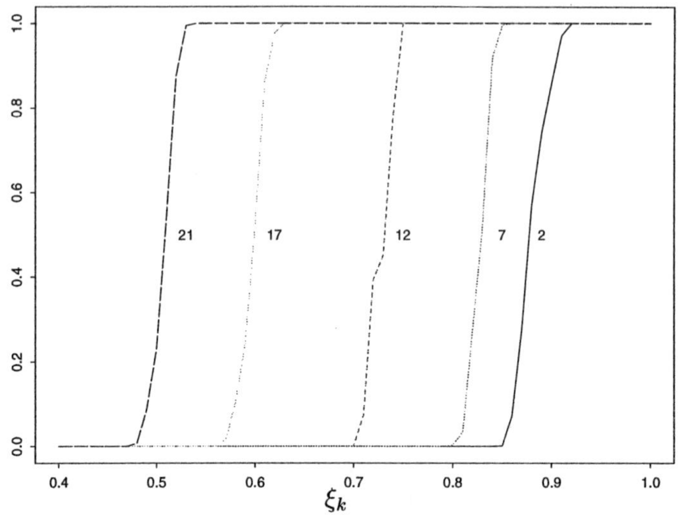

FIGURE 5. Posterior distribution functions for the parameter ξ_k, when $k = 2, 7, 12, 17$ and 21.

were much weaker than theirs. There is one bad prediction, viz. for the crossing time in trial 15. This same value was apparently underestimated also by Lu and Meeker (1993), so that the crossing time of this trial seems to be an "outlier". Some trials with other hyperparameter values were also made, and it seems that different choices of the hyperparameters have little influence on the corresponding results of both kinds of predictions.

Further study about some particular parameters can also be carried out conveniently by using the simulated sample $x^{(j)}$. In Figure 5, we plotted the posterior cumulative distribution functions of the "relative speed" parameters ξ_k for k = 2, 7, 12, 17 and 21. They are all less than 1, which was the assumed value of ξ_1. It is easy to see that a curve on the left side corresponds to slow growth, and on the right to fast growth. The posterior variances of the ξ_k's are small, which partly validates our common shape assumption in all growth curves.

19.5 Discussion

A model's ability to predict future growth beyond the point T_{max} at which all observation is terminated is often doubtful. Here, since the baseline $g(\cdot)$ is estimated when the condition $g(t) \leq Y^*$ is true, we can reasonably use our model to predict future growth for those slowly growing devices which were still below Y^* at time T_{max}. In other words, the information in the data seems to be sufficient for reasonable prediction.

Another advantage of the present approach is that different random quantities in this hierarchical model can be viewed as corresponding to different sources of variation in a crack growth curve, and hence they can be given an intuitive interpretation. The common growth curve g_0 in (1), and later the baseline g in (3), are parameters describing population characteristics, whereas the speed θ_k in (1), and later ξ_k in (3), are individual characteristics representing the (relative) differences between different devices.

We did not estimate what might be described as the time-to-failure distribution of a generic device from the same population as the data, which was a primary goal of the analysis of Lu and Meeker (1993). In our Bayesian setting, this would have corresponded to introducing a hypothetical "new" device about which there was no crack growth data available, and then considering the predictive distribution of its failure time. This could have been done routinely by adding corresponding "dummy" growth curves to the simulation scheme.

We do not claim that our method improves the accuracy of predictions very much. Rather, the emphasis of this paper is on how to establish a hierarchical model under weak assumptions, how to assess the adequacy of the model, and how to make reasonable predictions. However, if some additional knowledge besides the data is available, it could be integrated into the prior distribution, and better predictions would be obtained.

If we add the further structural assumption that the growth curves are all convex, which is indeed strongly suggested by Figure 2, the only thing we need to change is the prior distribution of g, which is then constrained to be in the domain

$$0 < \Delta g(t_1) \leq \Delta g(t_2) \leq \cdots \leq \Delta g(t_N).$$

The corresponding posterior density is simply the original density restricted to the domain above and divided by a normalizing constant. In Figure 6, we have plotted 1000 simulated growth curves for the 13th trial, by using the original model (3), and the model based on this further convexity assumption, respectively. It is not surprising that, from the figure as well as from the numerical results, it makes almost no difference whether the convexity condition is added to the prior or not, since the growth curves sampled from the posterior exhibit an apparent convex shape in any case. The convexity message convoyed by the data seems to be so strong that it is unnecessary for the prior to do the same. The situation becomes of course different if we are actually aiming at predicting beyond the target level Y^*, where there is no data.

In Bogdanoff and Kozin (1985), there is a second crack growth data set, which likewise appears to be consistent with our assumption of a common shape for the growth curves. Although this assumption is weaker than in most growth models considered elsewhere, it still puts a tight restriction on the "operational ages" $\theta_k \cdot t$. Arjas et al. (1996) used stochastic processes to characterize these "operational ages", so that their model is more flexible than the present one in fitting different kinds of empirical data. Here,

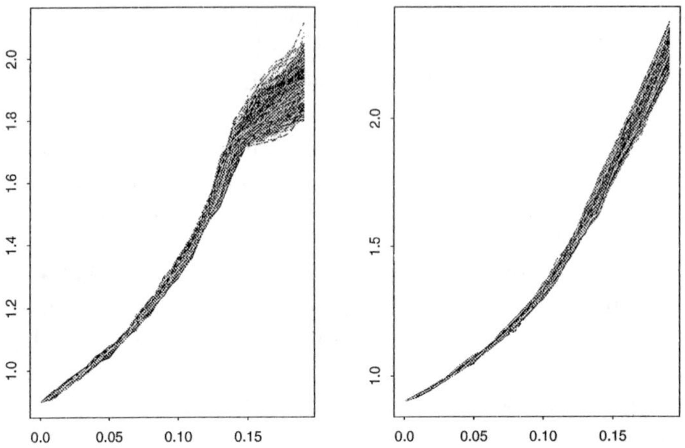

FIGURE 6. Simulated fatigue crack growth curves of size 1000 for trial 13, when the original model and the model with further convexity assumption are used respectively. The figure plots Y_{ki} on the y-axis against t on the abscissa.

however, the much simpler model works quite well, especially in prediction. We believe that, if the data set were less regular, a similar technique as in Arjas et al. (1996) could be used to tackle the problem successfully.

19.6 Acknowledgements

This research was undertaken while the first author was visiting the Department of Mathematical Sciences, University of Oulu, Finland, and was supported by grants from the Academy of Finland.

References

Arjas, E., and Gasbarra, D. (1994), "Nonparametric Bayesian Inference from Right Censored Survival Data, Using the Gibbs Sampler," *Statistica Sinica*, 4, 505-524.

Arjas, E., Liu, L., and Maglaperidze, N. (1996), "Prediction of growth: a Hierarchical Bayesian Approach," *to appear in Biometrical Journal.*

Bernardo, J.M., and Smith, A.F.M. (1994), *Bayesian Theory*, West Sussex: John Wiley & Sons Ltd.

Besag, J., Green, P., Higdon, D., and Mengerson, K. (1995), "Bayesian Computation and Stochastic Systems," *Statistical Science*, 10, 3-66.

Bogdanoff, J.L., and Kozin, F. (1985), *Probabilistic Models of Cumulative Damage*, New York: John Wiley.

Fearn, T. (1975), "A Bayesian Approach to Growth Curves," *Biometrika*, 62, 89-100.

Hudak, S.J., Jr., Saxena, A., Bucci, R.J., and Malcolm, R.C. (1978), "Development of Standard Methods of Testing and Analyzing Fatigue Crack Growth Rate Data," Technical report AFML-TR-78-40, Westinghouse R & D center, Westing house electric corporation, ittsburgh, PA 15235.

Lindley, D.V., and Smith, A.F.M. (1972), "Bayes Estimates for the Linear Model," *Journal of the Royal Statistical Society, Series B*, 34, 1-42.

Lu, C.J., and Meeker, W.Q. (1993), "Using Degradation Measures to Estimate a Time-to-Failure Distribution," *Technometrics*, 35, 161-174.

Rao, C.R. (1987), "Prediction of Future Observations in Growth Curve Models," *Statistical Science*, 2, 434-471.

Roberts, G. O., and Smith, A. F. M. (1994), "Simple Conditions for the Convergence of the Gibbs Sampler and Metropolis-Hastings Algorithm," *Stochastic Processes and their Applications*, 49, 207-216.

Smith, A.F.M., and Roberts, G.O. (1993), "Bayesian Computation via the Gibbs Sampler and Related Markov Chain Monte Carlo Methods," *Journal of the Royal Statistical Society, Series B*, 55, 3-23.

Tian, J.J., Shukla, R., and Buncher, C.R. (1994), "On Prediction of Future Observation in Growth Curve Model," *Statistics in Medicine*, 13, 2205-2217.

von Rosen, D. (1991), "The Growth Curve Model: a Review," *Communications in Statistics, Theory and Methods*, 20, 2791-2822.

20

A Semiparametric Model for Labor Earnings Dynamics

Keisuke Hirano

ABSTRACT We apply semiparametric Bayesian methodology to produce predictive distributions for labor earnings of individuals in the U.S., based on longitudinal data from the Panel Study of Income Dynamics. We generalize conventional models for earnings dynamics, by allowing the model disturbances to have an arbitrary distribution. This distribution is modelled as having a mixture of normals representation, where the mixing distribution is given a Dirichlet process prior. The semiparametric models for earnings result in predictive distributions that are quite different from those produced by conventional parametric models. In particular, the predictive distributions appear to be heavy-tailed and have different persistence properties.

20.1 Introduction

This paper applies semiparametric Bayesian methodology to model labor earnings of individuals in the U.S. Understanding earnings dynamics is crucial for policy analysis, for example in predicting how likely it is that an individual or family below a poverty threshold will continue to be in poverty one year later. In addition, many economic theories of consumption view saving partly as a means to manage earnings risk, so that characterizing this risk is relevant both for understanding individuals' saving behavior, and devising normative guidelines for saving via expected utility calculations. In all of these applications, one needs to be able to provide predictive distributions for an individual's future earnings, conditional on some earnings history, so we will take this to be our primary goal in this analysis.

There is a rich econometric literature on modelling earnings dynamics. Recent work using classical techniques includes MaCurdy (1982), Abowd and Card (1989), Holtz Eakin, Newey, and Rosen (1988), Moffitt and Gottschalk (1995), and Horowitz and Markatou (1996); parametric, fully Bayesian analyses are conducted by Geweke and Keane (1997) and Chamberlain and Hirano (1997). It is often thought that relatively simple, low-order linear models work well, provided that one allow for some source of permanent heterogeneity between individuals. We will focus here on a simple linear autoregression, with individual-specific intercept, though there is clearly scope to extend and improve this model in future work. Let y_{it} denote the logarithm of earnings of individual i in time period t; our basic model has

$$y_{it} = \gamma_i + \rho y_{i,t-1} + \epsilon_{it}.$$

Here the γ_i are interpreted as individual-specific "permanent" components of income, perhaps arising from unmeasured differences in skills. These are treated as random effects.

A convenient starting point is to assume that the model disturbances ϵ_{it} are normally distributed. However, this is often thought to be an unrealistic assumption for earnings data, as there is considerable evidence of nonnormality.[1] As a consequence, much recent work on labor earnings data uses classical semiparametric methods, which would estimate ρ and the variance of ϵ_{it} without specifying the distribution of ϵ_{it}. Some examples of this approach are MaCurdy (1982), Chamberlain (1984), and Holtz–Eakin, Newey, and Rosen (1988). These methods typically do not directly estimate the distribution of ϵ_{it}, so they cannot be used to generate predictive distributions without auxiliary assumptions. Moreover, it is not clear how to deal with parameter uncertainty in the construction of the predictive distributions in a classical framework. For example, the work of Horowitz and Markatou (1996) does estimate the distribution of ϵ_{it} using kernel smoothing methods, but the nonstandard asymptotic theory required in their analysis does not provide interval estimates or other measures of parameter uncertainty that can be incorporated into predictive inference.

The approach taken here, and in Hirano (1997), is to give the distribution of ϵ_{it} a nonparametric prior, using the Dirichlet mixture of normals model of Ferguson (1983), Lo (1984), Escobar and West (1995), and West, Müller, and Escobar (1994). This hierachical model is closely related to the models developed in the chapters by Ibrahim and Kleinman (1998) and Müller and Rosner (1998) in this volume; Taking a nonparametric Bayesian approach is convenient because the prior distribution can be tailored to reflect a priori judgments of smoothness in a flexible way, and because predictive distributions that integrate over the posterior distribution for unknown parameters can be readily simulated. Our model is similar to Geweke and Keane (1997), who use a finite normal mixture with a fixed number of components, but we believe that moving to more flexible models helps to answer concerns in this and many other economic contexts, where prior reasoning provides little guidance on the parametric form of the underlying distributions, and only limited knowledge of how many mixture components are needed to adequately model the underlying distributions.

[1] See for example MaCurdy (1982), who fits normal-based models by maximum likelihood, but then adjusts his standard errors to reflect possible violations of the normality assumption.

20.2 Longitudinal Earnings Data

The data are drawn from the Panel Study of Income Dynamics (PSID), an ongoing longitudinal survey of U.S. households, and records annual labor earnings for males who were heads of households between the ages of 24 and 33, during the period from 1967 to 1991. The PSID was designed to provide information on a broad range of economic, demographic, sociological and psychological aspects of U.S. individuals and households, in an explicitly dynamic context. It is publicly available at `http://www.isr.umich.edu/src/psid`; Hill (1992) provides further information on the survey. Only males with complete data on race, education, and earnings, who were heads of household with positive earnings at least one year prior to age 24, and with positive earnings for the ten consecutive years, were extracted and used in the analysis here. This resulted in 10 years of earnings data for each of 516 individuals. The construction of this sample is similar to the "young men" sample of Geweke and Keane (1996), and is designed to catch individuals early in their careers, when there could be considerable uncertainty about future earnings. Note that our "time" index corresponds to age: $t = 1$ means the year the individual was 24, and this can correspond to any calendar year betwen 1967 and 1982 (since an individual who was 24 in 1982 would be 33 in 1991, when our sample ends). However, we do retain and use calendar year information in the "first-stage" regressions described next.

Define z_{it} to be the logarithm of earnings for individual $i = 1, \dots, n$ in year $t = 1, \dots, T$, adjusted to 1991 dollars using the U.S. Consumer Price Index (CPI). We calculate ordinary least-squares regressions of z_{it} on a constant, an indicator for race, indicators for education equal to some high school, high school, some college, college, and beyond college, and calendar year. Separate regressions are run for each age t, and the residuals from the least–squares regressions are kept and used as our data. We denote these residuals by y_{it}. Working with residuals simplifies the analysis and lets us focus attention on earnings dynamics, at the cost of ignoring the parameter uncertainty in the preliminary regressions; it would be possible to extend the models developed below to directly incorporate the covariates.

We draw three subsamples from this initial sample, by selecting only white heads of household who were not part of the Survey of Economic Opportunity (SEO) sample (which oversampled poor households relative to U.S. population) and grouping by the highest level of education obtained. We are focusing on this very narrow group—white male heads of household—only because reliable and adequately large longitudinal samples for other equally interesting subpopulations are considerably more difficult to obtain. There are 37 high school dropouts, 100 high school graduates, and 122 college graduates who meet this further sample restriction.[2]

We wish to develop a flexible model for the earnings residuals. One possibility would be to build a completely general, nonparametric model for the vector of earnings residuals $(y_{i1}, \dots, y_{iT})$. However, we want to be able to directly extend the model to future periods $T+1, \dots, T+H$, in order to generate predictive distributions for $(y_{i,T+1}, \dots, y_{i,T+H})$ conditional on an earnings history $(y_{i1}, \dots, y_{iT})$. Without imposing some additional structure, it seems difficult to obtain useful predictive distributions of this kind. We will therefore consider placing restrictions on the joint distribution of $y_{i1}, \dots, y_{i,T+H}$, while allowing the distribution of the model disturbances to be quite general, so that our model is *semiparametric.*

20.3 A Parametric Random Effects Model

We begin by estimating a parametric version of the random effects model. Preliminary examination of the empirical autocovariances suggests that a relatively simple, low-order linear model might work adequately, provided there is some form of "permanent" heterogeneity between individuals. However, there seem to be differences in earnings dynamics by educational attainment, so we will treat each of the three educational subsamples separately in the following analysis. Since high school dropouts tend to hold very different kinds of jobs than college graduates, with possibly different forms of fluctuation over time, it seems sensible to allow their earnings dynamics to differ, even though the education information was used in the preliminary regressions discussed in the previous section. Our model conditions on earnings in the first period, y_{i1}, and specifies that for $i = 1, \dots, n$, $t = 2, \dots, T$,

$$y_{it} = \gamma_i + \rho y_{i,t-1} + \epsilon_{it}, \tag{1}$$

where both the innovation term ϵ_{it} and the random effect γ_i are assumed to be normal:

$$\epsilon_{it} \sim \mathcal{N}(0, \sigma^2), \qquad \gamma_i \sim \mathcal{N}(\psi y_{i1}, \sigma_w^2),$$

and all the ϵ_{it} and γ_i are assumed mutually independent conditional on ρ, σ, ψ, σ_w, and the initial conditions y_{i1}. We use the (improper) flat prior (with density proportional to a constant) for ρ and ψ, and independent scaled chi-square priors for the precisions:

$$\frac{1}{\sigma^2} \sim \frac{\chi^2(1)}{.01}, \qquad \frac{1}{\sigma_w^2} \sim \frac{\chi^2(1)}{.01}.$$

---

[2]Some households, where the head obtained no high school education or obtained an advanced degree, are therefore omitted from the remainder of the analysis.

Here $\chi^2(1)$ indicates a chi-square distribution with one degree of freedom.

We employ an independent copy of this model for each subsample—high school dropouts, high school graduates, and college graduates. This model is straightforward to simulate using Markov chain Monte Carlo (MCMC) methods. Posterior distributions for the three subsamples, obtained by simulation, are summarized in Table 1. For each of the models, posterior means and standard deviations were based on a single chain of 10,000 iterations, discarding the first 1000 for burn-in and keeping every 10th iteration, but we also used short multiple chains with different starting values to assess convergence. The rows indicated by "Normal" correspond to the parametric model. The autoregressive parameter ρ, which is one of the key determinants of the persistence of earnings innovations, varies dramatically by educational level; the posterior mean for ρ increases from about 0.34 for the high school dropouts to 0.51 for the college graduates; even accounting for uncertainty about ρ in the various models, there is strong evidence of systematic differences in persistence, although the overall effect is ambiguous because σ and σ_w also vary across the educational subsamples.

TABLE 1. Posterior Inference in the Random Effects Models

Sample/Model	ρ	σ	ψ	σ_w
High School Dropouts				
Normal	0.34	0.32	0.27	0.19
	(0.06)	(0.01)	(0.11)	(0.04)
NP Errors	0.33	–	0.27	0.22
	(0.04)	–	(0.11)	(0.03)
High School Graduates				
Normal	0.42	0.29	0.32	0.15
	(0.04)	(0.01)	(0.05)	(0.02)
NP Errors	0.53	–	0.26	0.13
	(0.05)	–	(0.05)	(0.02)
College Graduates				
Normal	0.51	0.24	0.23	0.13
	(0.04)	(0.01)	(0.04)	(0.02)
NP Errors	0.75	–	0.10	0.07
	(0.03)	–	(0.03)	(0.01)

Posterior means (standard deviations) obtained by MCMC simulation. "Normal" refers to the parametric random effects model; "NP Errors" refers to the model with a nonparametric error term.

20.4 A Semiparametric Model

We wish to extend the parametric random effects model by allowing the model disturbances ϵ_{it} to have a general form. To do so we adopt the model of Ferguson (1983), Lo (1984), and Escobar and West (1995), in which the ϵ_{it} are viewed as arising from a mixture of normals representation, where the mixing distribution is given a Dirichlet process prior.

An alternative generalization of the parametric model, examined in Hirano (1997), would allow the distribution of the individual-specific intercept term γ_i to have a mixture of normals representation. This model is a special case of the models with nonparametric random effects developed in Müller and Rosner (1997, 1998) and Ibrahim and Kleinman (1998). However, we have found that generalizing the distribution of the error term ϵ_{it} rather than the distribution of γ_i, is more important in this application, especially in generating predictive distributions, so we will focus on this case in the remainder of this chapter.

Our model can be written as:

$$y_{it} = \gamma_i + \rho y_{i,t-1} + \epsilon_{it},$$

$$\gamma_i \sim \mathcal{N}(\psi y_{i1}, \sigma_w^2),$$

$$\epsilon_{it} \sim \mathcal{N}(\mu_{it}, \sigma_{it}^2),$$

$$(\mu_{it}, \sigma_{it}^2) \sim G,$$

$$G \sim \mathcal{D}(G_0, \alpha).$$

This model falls within the framework of Escobar and West's chapter in this volume (Escobar and West, 1998). Their $\zeta(i)$ corresponds to (γ_i, ρ) in our model, and their θ_i corresponds to $(\mu_{it}, \sigma_{it}^2)$. The specification of y_{it} in the model considered here has an explicit dependence on $y_{i,t-1}$, so that the likelihood has a sequential decomposition, rather than involving conditionally independent terms.

We will use the same prior distribution as in the parametric model for ρ, ψ, and σ_w. It remains to specify the prior Dirichlet base measure G_0 and either fix α at some value, or place a distribution on α. We will use a $Gamma(a_1, a_2)$ prior for α, and use a conjugate prior for G_0: under G_0, the $(\mu_{it}, \sigma^2_{it})$ are distributed as

$$\frac{1}{\sigma^2_{it}} \sim \frac{\chi^2(s)}{sQ}, \qquad \mu_{it}|\sigma^2_{it} \sim \mathcal{N}(m, b \cdot \sigma^2_{it}).$$

Hence the hyperparameters for the base measure (λ in Escobar and West's (1998) notation) are (s, Q, m, b). We set $s = 3$, $Q = .01$, $m = 0$, $b = 4/Q$, $a_1 = 2$, and $a_2 = .5$. Q can be regarded as a prior guess for σ^2_{it}, with the low value for s making the prior quite diffuse. Similarly, b is set quite high to make the prior for μ_{it} relatively diffuse.

One way to see what the prior distribution implies is to consider the implications of the prior for the distribution of ϵ_{it}. The Dirichlet process model is equivalent (by the alternative construction of the Dirichlet process in Sethuraman and Tiwari (1982) and Sethuraman (1994)) to the following setup: regard the ϵ_{it} as i.i.d. with density function q, and write

$$q(\cdot) = \sum_{j=1}^{\infty} p_j \phi(\cdot|\mu_j, \sigma^2_j),$$

where $\sum_{j=1}^{\infty} p_j = 1$, and $\phi(\cdot|\mu, \sigma^2)$ is the density function of a normal random variable with mean μ and variance σ^2. Place a prior distribution on q as follows. For each pair μ_j, σ^2_j, use an independent copy of G_0 as its prior distribution. Place a prior distribution on the mixture weights $p_1, p_2, \ldots$ by a stick-breaking argument. Let $r_1, r_2, \ldots$ be a sequence of i.i.d. random variables with a Beta$(1, \alpha)$ distribution. Form $p_1 = r_1$, and for $j = 2, \ldots$ set

$$p_j = r_j \prod_{l=1}^{j-1} (1 - r_l).$$

The Sethuraman-Tiwari construction also suggests a natural way to generate draws for q under its prior, by truncating the countable normal mixture at some large number of components. So it is straightforward to generate draws for q under the prior distribution and examine whether or not they are appropriately "smooth." (We are emphasizing the relationship between the mixture weights and the smoothness of the implied density for ϵ_{it}, but clearly the choice of G_0, especially s and Q, also matters for the amount of smoothing that our procedure will carry out.)

Figure 1 displays draws for q under our prior specification for G_0. For each of four values for α (0.1, 1, 5, and 20), we have generated five draws for q. Setting α to 20 results in densities that are implausibly jittery, given our prior belief that the distribution of the ϵ_{it} should be fairly smooth. At the same time, we prefer not to be too dogmatic about the amount of smoothness, so we will place a prior on α rather than fixing it at some value.

We have decided to use a $Gamma(2, .5)$ prior for α, which has mean 4 and variance 8; this choice for the prior is intended to downweight unreasonable forms for the density of ϵ_{it} without ruling out potentially interesting densities. MCMC methods developed by Escobar (1994), Escobar and West (1995), MacEachern (1994), Bush and MacEachern (1996), and West, Müller, and Escobar (1994) can be used to simulate posterior distributions for the model parameters. Computational methods for models with Dirichlet process priors are reviewed in the chapters by Escobar and West (1998), and MacEachern (1998), in this volume. In Table 1, rows denoted by "NP Errors" correspond to this semiparametric model. Because there are $n \times (T-1)$ pairs of latent variables $(\mu_{it}, \sigma^2_{it})$, the computations are very time-consuming. For both the high school graduates and the college graduates subsamples, the semiparametric estimates for the autoregressive coefficient ρ are larger than the parametric estimates, but this is accompanied by a smaller value for ψ, the coefficient relating the individual effect γ_i to the initial condition y_{i1}. So the implications for predictive distributions are not entirely clear from the table alone.

FIGURE 1. Draws from the Prior for q

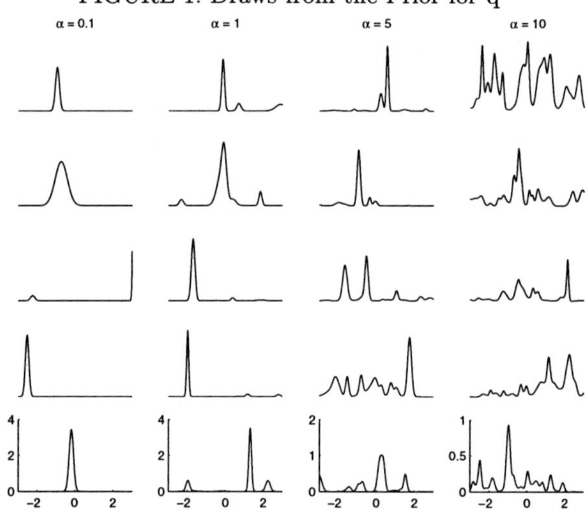

20.5 Predictive Distributions

Table 1 provides an incomplete view of the implications of the models. It would be possible to generate and plot draws for the density of ϵ_{it} from its posterior distribution, but it seems more fruitful to go back to our initial motivation. For applications to savings decisions, and for forecasting transitions in and out of poverty, we want to provide distributions for future earnings conditional on an earnings history. This means extending the models to future, hypothetically observable periods $t = T+1, \ldots, T+H$, and examining the implications of our models for the joint distribution of $y_{i,T+1}, \ldots, y_{T+H}$. Let π denote the model parameters, augmented by the latent γ_i, and, in the semiparametric model, the latent $(\mu_{it}, \sigma_{it}^2)$, and use Y to denote the data on individuals $i = 1, \ldots, n$ up to time T. Conditional on π, the model in equation 1 specifies a conditional distribution for future earnings given any earnings history, $p(y_{i,T+1}, \ldots, y_{T+H}|\pi, y_{i1}, \ldots, y_{iT})$, which has a multivariate normal form. Since π is not known, we need to integrate over the posterior distribution for unknown parameters:

$$p(y_{i,T+1}, \ldots, y_{i,T+H}|Y) = \int p(y_{i,T+1}, \ldots, y_{T+H}|\pi, y_{i1}, \ldots, y_{iT}) dP(\pi|Y).$$

These calculations are described in greater detail in Hirano (1997).

We will choose some interesting earnings histories, and examine their corresponding predictive distributions. First, consider two high school dropouts ($i = 14, i = 16$) whose average earnings over the sample period are quite different, but who have similar earnings in the final period. Predictive inference for these individuals is summarized in Figures 2 and 3. In each figure, the top left panel displays the earnings histories, while the bottom left panel displays the predictive density for $y_{i,T+1}$, the earnings residual one period ahead. The right panels display .1, .25, .5, .75, and .9 quantiles of the predictive distributions at horizons of one year to ten years, that is, for $t = T+1, \ldots, T+10$. In the model with normal disturbances, the predictive quantiles over the ten year horizon show some mean reversion, which is more noticeable for the individual with rising income over his earnings history. However, the predictive distributions do not fan out very much over the horizon. In comparison, the semiparametric model estimates the density of the disturbances to be extremely skewed, with a long left tail; this can be seen in the predictive densities for $y_{i,T+1}$. It seems that in order to capture the heavy tails, the parametric normal model has had to set the estimated variance rather high. So here the quantiles are much more compressed in the semiparametric model, even at relatively long horizons.

Next, we look at two high school graduates ($i = 49, i = 51$), chosen to have similar earnings histories, except in the last period. Figures 4 and 5 display predictive distributions for these individuals. Generally, the predictive quantiles at long horizons are similar, but at short horizons the quantiles in the semiparametric model are more compressed. The predictive densities for period $T+1$ are noticeably different, with the semiparametric model displaying tails that decay more slowly. Unlike with the high school dropouts subsample, here the densities seem to be more nearly symmetric.

Finally, we examine two college graduates ($i = 12, i = 90$) who have similar earnings histories except in the final period, when one individual

FIGURE 2. Predictive Inference, Parametric Model, Two High School Dropouts ($i = 14, 16$)

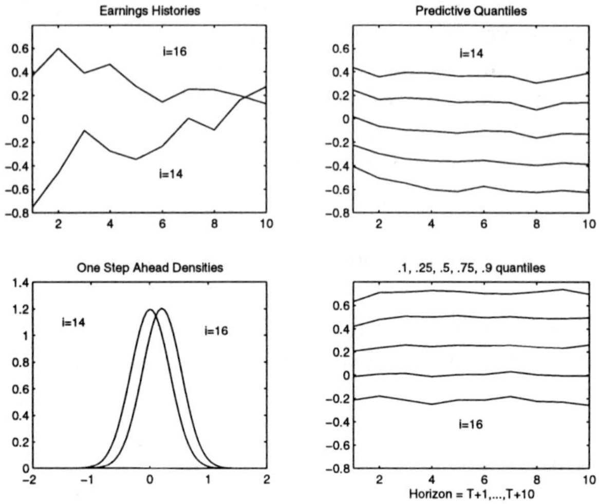

FIGURE 3. Predictive Inference, Semiparametric Model, Two High School Dropouts ($i = 14, 16$)

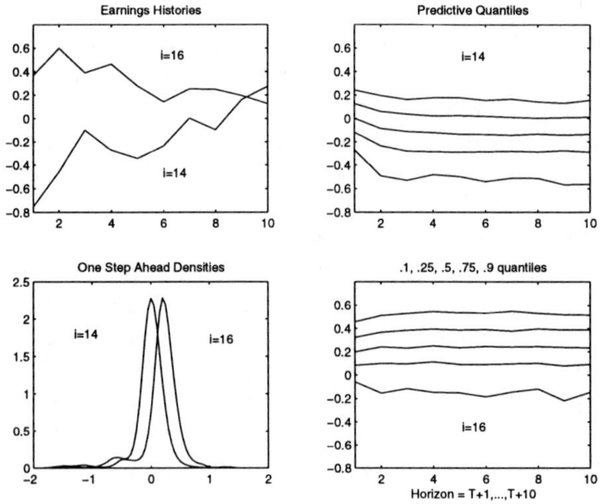

FIGURE 4. Predictive Inference, Parametric Model, Two High School Graduates ($i = 49, 51$)

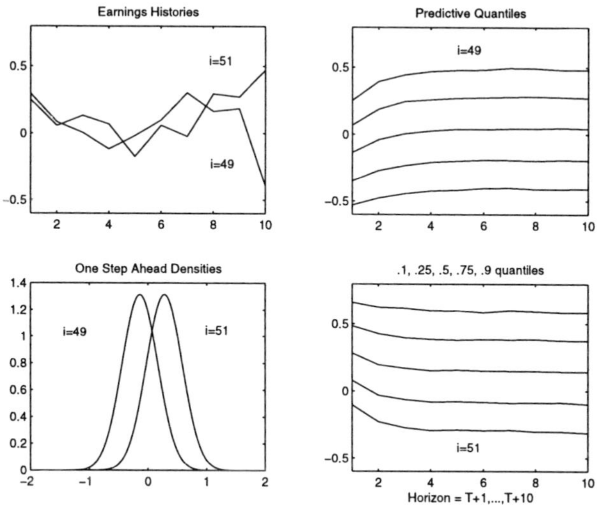

FIGURE 5. Predictive Inference, Semiparametric Model, Two High School Graduates ($i = 49, 51$)

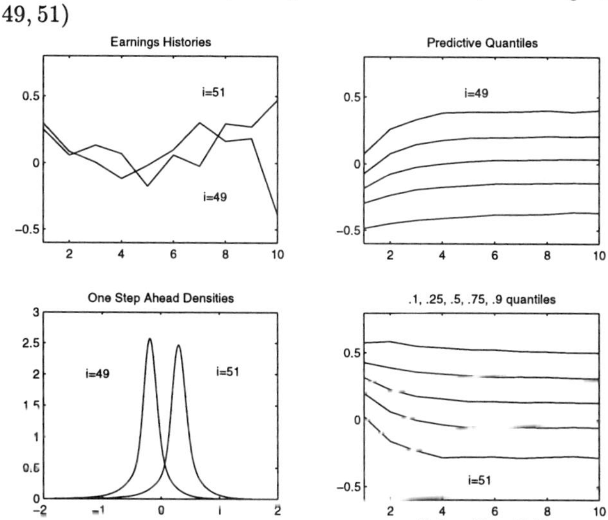

suffers an abrupt decline in earnings. Figures 6 and 7 display predictive inference for these two individuals. Figure 6 is based on the parametric random effects model, while Figure 7 is based on the semiparametric model.

FIGURE 6. Predictive Inference, Parametric Model, Two College Graduates $(i = 12, 90)$

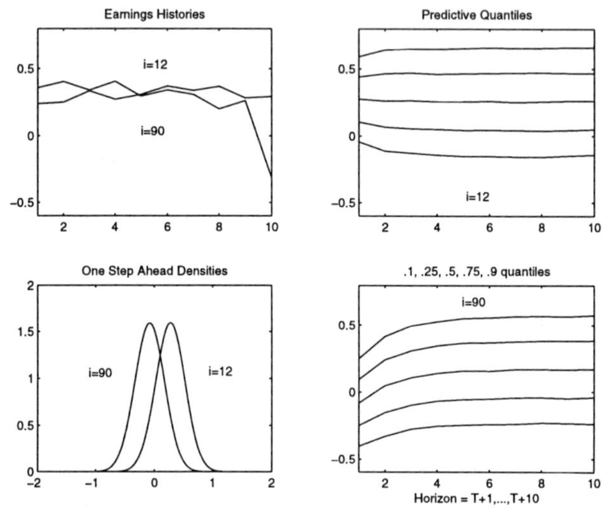

Again, the predictive distributions under the semiparametric model are strikingly different from the predictive distributions generated by the parametric model. Examining the one-period ahead predictive densities, in the semiparametric model more of the mass seems to be concentrated within about .25 of their modes, but as with the high school graduates, here the tail probabilities decay more slowly than in the parametric model. This is consistent with other studies (such as MaCurdy (1982) and Chamberlain and Hirano (1997)) that have found evidence of nonnormality and heavy tails in data on individual earnings. Also, there is weak evidence of skewness in that the density to the left of the mode appears to fall off more slowly than on the right. The predictive quantiles under the semiparametric model start out more concentrated near y_{iT}, but fan out rather quickly so that by ten years the quantiles are similar to those produced by the parametric model, if somewhat wider. But unlike in the parametric model, where the predictive quantiles at the four year horizon are very similar to the corresponding quantiles at the ten year horizon, in the semiparametric model the predictive quantiles change more gradually over the ten years.

FIGURE 7. Predictive Inference, Semiparametric Model, Two College Graduates ($i = 12, 90$)

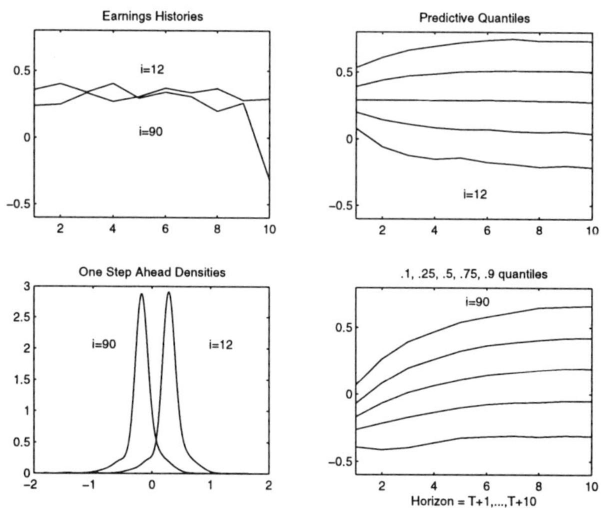

20.6 Conclusion

Focusing on predictive distributions, rather than parameter estimates, brings us closer to answering the questions that originally motivated the analysis. Relaxing conventional parametric assumptions can have a substantial impact on predictive distributions for earnings. Under the semiparametric model, these distributions are strikingly heavy-tailed, and seem to display greater persistence than in the model with normal errors. More work remains to be done to examine the implications of these models for policy issues and economic decision problems, and to extend these models to more flexible specifications.

References

Abowd, J. M. and Card, D. (1989), On the Covariance Structure of Earnings and Hours Changes, *Econometrica*, 57, 411–445.

Bush, C. A. and MacEachern, S. N., (1996), A Semi-parametric Bayesian Model for Randomized Block Designs, *Biometrika*, 83, 275–285.

Chamberlain, G. (1984), Panel Data, in *Handbook of Econometrics II*, Griliches, Zvi and Intriligator, M. (eds.), North–Holland, Amsterdam.

Chamberlain, G. and Hirano, K. (1997), Predictive Distributions Based on Longitudinal Earnings Data, draft.

Deaton, A. (1991), Saving and Liquidity Constraints, *Econometrica*, 59, 1221–1248.

Escobar, M. D. (1994), Estimating Normal Means With a Dirichlet Process Prior, *Journal of the American Statistical Association*, 89, 268–277.

Escobar, M. D. and West, M. (1995), Bayesian Density Estimation and Inference Using Mixtures, *Journal of the American Statistical Association*, 90, 577–588.

Escobar, M. D. and West, M., (1998), Computing Bayesian Nonparametric Hierarchical Models, in *Practical Nonparametric and Semiparametric Bayesian Statistics* (eds. Dey, D. and Müller, P. and Sinha, D.), New York: Springer-Verlag.

Ferguson, T. S. (1973), A Bayesian Analysis of Some Nonparametric Problems, *The Annals of Statistics*, 1, 209–230.

Ferguson, T. S. (1974), Prior Distributions on Spaces of Probability Measures, *The Annals of Statistics*, 2, 615–629.

Ferguson, T. S. (1983), Bayesian Density Estimation By Mixtures of Normal Distributions, in *Recent Advances in Statistics: Papers in Honor of Herman Chernoff on His Sixtieth Birthday*, pp. 287– 302.

Geweke, J. and Keane, M. (1997), An Empirical Analysis of the Male Income Dynamics in the PSID: 1968–1989, Research Department Staff Report, 233, Federal Reserve Bank of Minneapolis.

Hill, M. S. (1992), *The Panel Study of Income Dynamics: A User's Guide*, Sage Publications, Newbury Park, CA.

Hirano, K. (1997), Semiparametric Bayesian Models for Dynamic Earnings Data, Working paper.

Holtz–Eakin, D. and Newey, W. and Rosen, H. (1988), Estimating Vector Autoregressions with Panel Data, *Econometrica*, 56, 1371–1396.

Horowitz, J. and Markatou, M., (1996), Semiparametric Estimation of Regression Models for Panel Data, *Review of Economic Studies*, 63, 145–168.

Ibrahim, J. G. and Kleinman, K.P. (1998), Semi-parametric Bayesian Methods for Random Effects Models, in *Practical Nonparametric and Semiparametric Bayesian Statistics* (eds. Dey, D. and Müller, P. and Sinha, D.), New York: Springer-Verlag.

Lillard, L. A. and Willis, R. J. (1978), Dynamic Aspects of Earning Mobility, *Econometrica*, 46, 985–1012.

Lo, A. Y. (1984), On a Class of Bayesian Nonparametric Estimates: I. Density Estimates, *The Annals of Statistics*, 12, 351–357.

MacEachern, S. N. (1994), Estimating Normal Means With a Conjugate Style Dirichlet Process Prior, *Communications in Statistics, Part B–Simulation and Computation*, 23, 727–741.

MacEachern, S. N. (1998), Computational Methods for Mixture of Dirichlet Process Models, in *Practical Nonparametric and Semiparametric Bayesian Statistics* (eds. Dey, D. and Müller, P. and Sinha, D.), New York: Springer-Verlag.

MaCurdy, T. (1982), The Use of Time Series Processes to Model the Error Structure of Earnings in a Longitudinal Data Analysis, *Journal of Econometrics*, 18, 83–114.

Moffitt, R. A. and Gottschalk, P. (1995), Trends in the Autocovariance Structure of Earnings in the U.S.:1969–1987, Technical Report, Brown University.

Müller, P. and Rosner, G. (1997), A Bayesian Population Model with Hierarchical Mixture Priors Applied to Blood Count Data, *Journal of the American Statistical Association*, 92, 1279-1292.

Müller, P. and Rosner, G. (1998), Semi-parametric PK/PD Models, in *Practical Nonparametric and Semiparametric Bayesian Statistics* (eds. Dey, D. and Müller, P. and Sinha, D.), Springer-Verlag.

Sethuraman, J. (1994), A Constructive Definition of Dirichlet Priors, *Statistica Sinica*, 4, 639–650.

Sethuraman, J. and Tiwari, R. C. (1982), Convergence of Dirichlet Measures and the Interpretation of Their Parameter, in *Statistical Decision Theory and Related Topics III, in two volumes*, 2, pp. 305–315.

West, M. and Müller, P. and Escobar, M. D., (1994), Hierarchical Priors and Mixture Models, With Application in Regression and Density Estimation, in *Aspects of Uncertainty. A Tribute to D. V. Lindley*, pp. 363– 386.

Lecture Notes in Statistics

For information about Volumes 1 to 58 please contact Springer-Verlag

Vol. 59: S. Gupta, R. Mukerjee, A Calculus for Factorial Arrangements. vi, 126 pages, 1989.

Vol. 60: L. Gyorfi, W. Härdle, P. Sarda, Ph. Vieu, Nonparametric Curve Estimation from Time Series. viii, 153 pages, 1989.

Vol. 61: J. Breckling, The Analysis of Directional Time Series: Applications to Wind Speed and Direction. viii, 238 pages, 1989.

Vol. 62: J.C. Akkerboom, Testing Problems with Linear or Angular Inequality Constraints. xii, 291 pages, 1990.

Vol. 63: J. Pfanzagl, Estimation in Semiparametric Models: Some Recent Developments. iii, 112 pages, 1990.

Vol. 64: S. Gabler, Minimax Solutions in Sampling from Finite Populations. v, 132 pages, 1990.

Vol. 65: A. Janssen, D.M. Mason, Non-Standard Rank Tests. vi, 252 pages, 1990.

Vol 66: T. Wright, Exact Confidence Bounds when Sampling from Small Finite Universes. xvi, 431 pages, 1991.

Vol. 67: M.A. Tanner, Tools for Statistical Inference: Observed Data and Data Augmentation Methods. vi, 110 pages, 1991.

Vol. 68: M. Taniguchi, Higher Order Asymptotic Theory for Time Series Analysis. viii, 160 pages, 1991.

Vol. 69: N.J.D. Nagelkerke, Maximum Likelihood Estimation of Functional Relationships. V, 110 pages, 1992.

Vol. 70: K. Iida, Studies on the Optimal Search Plan. viii, 130 pages, 1992.

Vol. 71: E.M.R.A. Engel, A Road to Randomness in Physical Systems. ix, 155 pages, 1992.

Vol. 72: J.K. Lindsey, The Analysis of Stochastic Processes using GLIM. vi, 294 pages, 1992.

Vol. 73: B.C. Arnold, E. Castillo, J.-M. Sarabia, Conditionally Specified Distributions. xiii, 151 pages, 1992.

Vol. 74: P. Barone, A. Frigessi, M. Piccioni, Stochastic Models, Statistical Methods, and Algorithms in Image Analysis. vi, 258 pages, 1992.

Vol. 75: P.K. Goel, N.S. Iyengar (Eds.), Bayesian Analysis in Statistics and Econometrics. xi, 410 pages, 1992.

Vol. 76: L. Bondesson, Generalized Gamma Convolutions and Related Classes of Distributions and Densities. viii, 173 pages, 1992.

Vol. 77: E. Mammen, When Does Bootstrap Work? Asymptotic Results and Simulations. vi, 196 pages, 1992.

Vol. 78: L. Fahrmeir, B. Francis, R. Gilchrist, G. Tutz (Eds.), Advances in GLIM and Statistical Modelling: Proceedings of the GLIM92 Conference and the 7th International Workshop on Statistical Modelling, Munich, 13-17 July 1992. ix, 225 pages, 1992.

Vol. 79: N. Schmitz, Optimal Sequentially Planned Decision Procedures. xii, 209 pages, 1992.

Vol. 80: M. Fligner, J. Verducci (Eds.), Probability Models and Statistical Analyses for Ranking Data. xxii, 306 pages, 1992.

Vol. 81: P. Spirtes, C. Glymour, R. Scheines, Causation, Prediction, and Search. xxiii, 526 pages, 1993.

Vol. 82: A. Korostelev and A. Tsybakov, Minimax Theory of Image Reconstruction. xii, 268 pages, 1993.

Vol. 83: C. Gatsonis, J. Hodges, R. Kass, N. Singpurwalla (Editors), Case Studies in Bayesian Statistics. xii, 437 pages, 1993.

Vol. 84: S. Yamada, Pivotal Measures in Statistical Experiments and Sufficiency. vii, 129 pages, 1994.

Vol. 85: P. Doukhan, Mixing: Properties and Examples. xi, 142 pages, 1994.

Vol. 86: W. Vach, Logistic Regression with Missing Values in the Covariates. xi, 139 pages, 1994.

Vol. 87: J. Müller, Lectures on Random Voronoi Tessellations.vii, 134 pages, 1994.

Vol. 88: J. E. Kolassa, Series Approximation Methods in Statistics. Second Edition, ix, 183 pages, 1997.

Vol. 89: P. Cheeseman, R.W. Oldford (Editors), Selecting Models From Data: AI and Statistics IV. xii, 487 pages, 1994.

Vol. 90: A. Csenki, Dependability for Systems with a Partitioned State Space: Markov and Semi-Markov Theory and Computational Implementation. x, 241 pages, 1994.

Vol. 91: J.D. Malley, Statistical Applications of Jordan Algebras. viii, 101 pages, 1994.

Vol. 92: M. Eerola, Probabilistic Causality in Longitudinal Studies. vii, 133 pages, 1994.

Vol. 93: Bernard Van Cutsem (Editor), Classification and Dissimilarity Analysis. xiv, 238 pages, 1994.

Vol. 94: Jane F. Gentleman and G.A. Whitmore (Editors), Case Studies in Data Analysis. viii, 262 pages, 1994.

Vol. 95: Shelemyahu Zacks, Stochastic Visibility in Random Fields. x, 175 pages, 1994.

Vol. 96: Ibrahim Rahimov, Random Sums and Branching Stochastic Processes. viii, 195 pages, 1995.

Vol. 97: R. Szekli, Stochastic Ordering and Dependence in Applied Probability. viii, 194 pages, 1995.

Vol. 98: Philippe Barbe and Patrice Bertail, The Weighted Bootstrap. viii, 230 pages, 1995.

Vol. 99: C.C. Heyde (Editor), Branching Processes: Proceedings of the First World Congress. viii, 185 pages, 1995.

Vol. 100: Wlodzimierz Bryc, The Normal Distribution: Characterizations with Applications. viii, 139 pages, 1995.

Vol. 101: H.H. Andersen, M.Højbjerre, D. Sørensen, P.S.Eriksen, Linear and Graphical Models: for the Multivariate Complex Normal Distribution. x, 184 pages, 1995.

Vol. 102: A.M. Mathai, Serge B. Provost, Takesi Hayakawa, Bilinear Forms and Zonal Polynomials. x, 378 pages, 1995.

Vol. 103: Anestis Antoniadis and Georges Oppenheim (Editors), Wavelets and Statistics. vi, 411 pages, 1995.

Vol. 104: Gilg U.H. Seeber, Brian J. Francis, Reinhold Hatzinger, Gabriele Steckel-Berger (Editors), Statistical Modelling: 10th International Workshop, Innsbruck, July 10-14th 1995. x, 327 pages, 1995.

Vol. 105: Constantine Gatsonis, James S. Hodges, Robert E. Kass, Nozer D. Singpurwalla(Editors), Case Studies in Bayesian Statistics, Volume II. x, 354 pages, 1995.

Vol. 106: Harald Niederreiter, Peter Jau-Shyong Shiue (Editors), Monte Carlo and Quasi-Monte Carlo Methods in Scientific Computing. xiv, 372 pages, 1995.

Vol. 107: Masafumi Akahira, Kei Takeuchi, Non-Regular Statistical Estimation. vii, 183 pages, 1995.

Vol. 108: Wesley L. Schaible (Editor), Indirect Estimators in U.S. Federal Programs. viii, 195 pages, 1995.

Vol. 109: Helmut Rieder (Editor), Robust Statistics, Data Analysis, and Computer Intensive Methods. xiv, 427 pages, 1996.

Vol. 110: D. Bosq, Nonparametric Statistics for Stochastic Processes. xii, 169 pages, 1996.

Vol. 111: Leon Willenborg, Ton de Waal, Statistical Disclosure Control in Practice. xiv, 152 pages, 1996.

Vol. 112: Doug Fischer, Hans-J. Lenz (Editors), Learning from Data. xii, 450 pages, 1996.

Vol. 113: Rainer Schwabe, Optimum Designs for Multi-Factor Models. viii, 124 pages, 1996.

Vol. 114: C.C. Heyde, Yu. V. Prohorov, R. Pyke, and S. T. Rachev (Editors), Athens Conference on Applied Probability and Time Series Analysis Volume I: Applied Probability In Honor of J.M. Gani. viii, 424 pages, 1996.

Vol. 115: P.M. Robinson, M. Rosenblatt (Editors), Athens Conference on Applied Probability and Time Series Analysis Volume II: Time Series Analysis In Memory of E.J. Hannan. viii, 448 pages, 1996.

Vol. 116: Genshiro Kitagawa and Will Gersch, Smoothness Priors Analysis of Time Series. x, 261 pages, 1996.

Vol. 117: Paul Glasserman, Karl Sigman, David D. Yao (Editors), Stochastic Networks. xii, 298, 1996.

Vol. 118: Radford M. Neal, Bayesian Learning for Neural Networks. xv, 183, 1996.

Vol. 119: Masanao Aoki, Arthur M. Havenner, Applications of Computer Aided Time Series Modeling. ix, 329 pages, 1997.

Vol. 120: Maia Berkane, Latent Variable Modeling and Applications to Causality. vi, 288 pages, 1997.

Vol. 121: Constantine Gatsonis, James S. Hodges, Robert E. Kass, Robert McCulloch, Peter Rossi, Nozer D. Singpurwalla (Editors), Case Studies in Bayesian Statistics, Volume III. xvi, 487 pages, 1997.

Vol. 122: Timothy G. Gregoire, David R. Brillinger, Peter J. Diggle, Estelle Russek-Cohen, William G. Warren, Russell D. Wolfinger (Editors), Modeling Longitudinal and Spatially Correlated Data. x, 402 pages, 1997.

Vol. 123: D. Y. Lin and T. R. Fleming (Editors), Proceedings of the First Seattle Symposium in Biostatistics: Survival Analysis. xiii, 308 pages, 1997.

Vol. 124: Christine H. Müller, Robust Planning and Analysis of Experiments. x, 234 pages, 1997.

Vol. 125: Valerii V. Fedorov and Peter Hackl, Model-oriented Design of Experiments. viii, 117 pages, 1997.

Vol. 126: Geert Verbeke and Geert Molenberghs, Linear Mixed Models in Practice: A SAS-Oriented Approach. xiii, 306 pages, 1997.

Vol. 127: Harald Niederreiter, Peter Hellekalek, Gerhard Larcher, and Peter Zinterhof (Editors), Monte Carlo and Quasi-Monte Carlo Methods 1996, xii, 448 pp., 1997.

Vol. 128: L. Accardi and C.C. Heyde (Editors), Probability Towards 2000, x, 356 pp., 1998.

Vol. 129: Wolfgang Härdle, Gerard Kerkyacharian, Dominique Picard, and Alexander Tsybakov, Wavelets, Approximation, and Statistical Applications, xvi, 265 pp., 1998.

Vol. 130: Bo-Cheng Wei, Exponential Family Nonlinear Models, ix, 240 pp., 1998.

Vol. 131: Joel L. Horowitz, Semiparametric Methods in Econometrics, ix, 204 pp., 1998.

Vol. 132: Douglas Nychka, Walter W. Piegorsch, and Lawrence H. Cox (Editors), Case Studies in Environmental Statistics, viii, 200 pp., 1998.

Vol. 133: Dipak Dey, Peter Müller, and Debajyoti Sinha (Editors), Practical Nonparametric and Semiparametric Bayesian Statistics, xv, 408 pp., 1998.